# Production and Logistics

W0050965

Managing Editor

*Horst Tempelmeier,* University of Cologne, FRG

Editors

*Wolfgang Domschke,* Technical University of Darmstadt, FRG
*Andreas Drexl,* University of Kiel, FRG
*Bernhard Fleischmann,* University of Augsburg, FRG
*Hans-Otto Günther,* Technical University of Berlin, FRG
*Hartmut Stadtler,* Technical University of Darmstadt, FRG

# Erwin Pesch

# Learning in Automated Manufacturing

## A Local Search Approach

With 30 Figures and 30 Tables

**Physica-Verlag**

A Springer-Verlag Company

Professor Dr. Erwin Pesch
Institut für Gesellschafts-
und Wirtschaftswissenschaften
Universität Bonn
Adenauerallee 24-42
D-53113 Bonn, FRG

ISBN 978-3-7908-0792-9

Die Deutsche Bibliothek – CIP-Einheitsaufnahme
Pesch, Erwin:
Learning in automated manufacturing : a local search approach / Erwin Pesch. –
Heidelberg : Physica-Verl., 1994
(Production and logistics)
ISBN 978-3-7908-0792-9       ISBN 978-3-642-49366-9 (eBook)
DOI 10.1007/978-3-642-49366-9

The use of registered names, trademarks, etc. in this publication does not imply, even in the absence
of a specific statement, that such names are exempt from the relevant protective laws and regula-
tions and therefore free for general use.

88/2202-543210 - Printed on acid-free paper

# TABLE OF CONTENTS

# PROLOGUE

Machine learning has become a popular field of research during the last years. Several new areas have developed and the number of publications increased tremendously.[1]

"One of the most striking differences between how people and computers work is that humans, while performing any kind of activity, usually simultaneously expend efforts to improve the way they perform it. This is to say that human performance of any task is inseparable intertwined with a learning process, while current computers are typically only executors of procedures supplied to them. They may execute very efficiently, but they do no self-improve with experience." (Kodratoff / Michalski (1990)).

Machine learning became a popular field in Artificial Intelligence based on the idea to build computer programs capable to improve or to reason on existing knowledge such as examples, facts, rules, frames, constraints, etc. in order to guide a search process into a particular direction determined by a desired output (e.g. a high–quality solution of an optimization problem). Knowledge based systems (e.g. expert systems) are the outcome of this new way of thinking about problem solving by computers, system design and software implementation.

To date, one can identify four major *machine learning* paradigms: (i) inductive learning (e.g. acquiring concepts from examples), (ii) analytical learning (e.g. explanation based as well as case based learning and learning by analogy, (iii) local search based learning (in particular learning by genetic algorithms and classifier systems), (iv) connectionist learning (i.e. learning from neural networks). Although these paradigms emerged from different scientific roots they share a common goal of building machines (programs) that can learn in significant ways for a wide variety of task domains. Hereby learning can be defined as the ability to perform new tasks that could not be performed before, or to perform old tasks better in a certain sense, as a result of changes produced by the learning process (Weiss / Kulikowski (1990)).

There has been an increasing interest in experimental comparisons of various methods, and in theoretical analysis of learning algorithms. Applying new techniques to the same data sets of the same problems provides a necessary

---

[1]    See the bibliographies Eom / Lee (1990), Jaumard et al. (1988), Kedar–Cabelli (1983), Stefanski et al. (1990).

condition to understand the merits of different methods.

The central purpose of this book is to acquaint the reader especially with the cases of local search based learning as well as to introduce some methods arising in the field of analytical learning, both with respect to their use in automated manufacturing.

Production or manufacturing in our context basically means the transformation of an input (machine power, labor, resources, etc.) to some output (e.g. some final products for selling). There is a common characterization of production systems with respect to four different requirements: (i) repetition of tasks, (ii) layout, (iii) product structure, and (iv) order requirements. Considering the repetition of production tasks we restrict our attention to manufacturing of single (or small numbers of) products (e.g. special purpose machines, ships, etc.) in contrast to mass and serial production. The appropriate layout in single product manufacturing is the job shop, a collection of different production units all of which are supposed to process the particular products. Assembly lines are the opposite type of layout. The product structure is mainly converging from a larger input to a smaller output and production is usually initiated by order.[2] We restrict our attention to a particular type of job shops, the standard job shop scheduling problem (which also includes the case of flow shop scheduling), as well as to one–machine scheduling with sequence dependent setup times. Additionally some design and planning issues in flexible manufacturing systems are considered. The reason to focus on standard problems of scheduling is twofold; the proposed methods are comparable to previous methods. from literature, and, more importantly, standard problems provide a deeper insight into the behaviour of new self–adapting solution methods. That is, the standard problems serve as a bottom up approach in order to be able to construct complex intelligent (knowledge based) scheduling systems. General purpose search methods (e.g. for hard combinatorial optimization problems), which in particular include methods from local search such as simulated annealing, tabu search, and genetic algorithms, are the basic ingredients of such intelligent scheduling systems, enriched by a number of local decision rules in order to introduce problem specific knowledge. The local search algorithms serve as a general search strategy – a metaheuristic – in order to perfectly guide the decision rules to achieve a desired goal. The idea of a general purpose search strategy combines the advantage of a rapid implementation and an easily modifiable prototype together with the separation of problem specific

---

[2]    see Schneeweiß (1987), Zäpfel (1982)

(procedural) knowledge and its use in order to extend the knowledge base.
Chapter I provides a brief insight into the *basic concepts of local search*.
Conventional iterative improvement methods start with an initial feasible
solution (or configuration), repeatedly consider changes in the current
configuration, and accept only those that improve the objective function.
Iterative improvement methods characteristically converge to a local optimum. A
change of a feasible solution or a configuration usually is restricted to slight
modifications or perturbations of the configuration, each of which defines a
neighbour of the current configuration. The type of modifications defines the
neighbourhood structure and the new configurations are the neighbours of the
current solution. All local search methods are based on the definition of a
particular neighbourhood structure and on the particular solution or
configuration representation. Both, neighbourhood structure and representation
introduce problem specific knowledge into the search process. The neighbourhood
structure defining modifications are simple minded (and almost no time
consuming) heuristic rules.

In anticipation of locating a more global optimum, the *simulated annealing*
method probabilistically accepts configurations (from the neighbourhood) which
temporarily deteriorate the quality of the system. The acceptance probability is
a function of the change in the objective function and the temperature
parameter. As the parameter is slowly reduced, fewer non–improving moves are
accepted. Thus a coarse global search evolves into a fine local search for
optimality, and the probabilistic jumps provide an escape from non–global
optima.

*Tabu search* is a general proposal to guide deterministically the search process
out of local optima respectively into new regions of the search space, making use
of different strategies of diversification and regional intensification of the search.
Its basic concept is the use of a tabu list restricting the number of feasible
modifications to those which lead to more promising neighbour solutions, not
necessarily improving ones. The tabu list avoids cycling in a certain area of the
search space via the specification of attributes of forbidden (tabu) modifications.

*Genetic algorithms* work by maintaining a population of candidate solutions to
the given problem. Each solution is represented by an artificial string
(chromosome) encoded over some final alphabet. Development of the
representation is a non–trivial effort, because key features of the problem must
be captured in such a way that desirable solution characteristics are propagated,
and undesirable characteristics are suppressed. Hence, motivated by principles of

evolution a genetic algorithm aims at producing high–quality solutions by letting
a set of randomly generated configurations undergo a sequence of unary and
binary transformations (reproduction, crossover, mutation). The fitness of a
configuration determines the likelihood of a configuration to survive,
configurations representing worse solutions tend to die off.

In Chapter II after an extensive survey of the literature on the traveling
salesman problem (which is basic to a huge number of production scheduling
problems), we briefly review previous attempts to generate near–optimal
solutions of the traveling salesman problem by applying genetic algorithms.
Following the lines of Johnson (1990a) we discuss some possibilities for speeding
up classical local search algorithms by casting them into a genetic frame. In an
experimental study two such approaches, viz. genetic local search with 2–opt
neighbourhoods and Lin–Kernighan neighbourhoods, respectively, are compared
with the corresponding classical multi–start local search algorithms, as well as
with simulated annealing and threshold accepting, using 2–opt neighbourhoods.
As to be expected a genetic organization of local search algorithms can
considerably improve upon performance though the genetic components alone can
hardly counterbalance a poor choice of the neighbourhoods. We report how a
genetic strategy can guide a truncated local search in an excellent way in order
to be competitive even if severe time constraints are imposed.

The last section of Chapter II describes a fairly general tabu search approach
with a dynamically increasing tabu list, for the traveling salesman problem. The
method is known under the name ejection chain and generalizes and improves an
early idea of Lin and Kernighan (1973) and Glover (1969).

Machine scheduling problems arise in diverse areas such as flexible
manufacturing, production planning, computer design, logistics, communication
etc. A common feature of many of these problems is that no efficient solution
algorithms are known that solve each instance to optimality in a time bounded
polynomially in the size of the problem.[3] Approximation algorithms can help to
overcome these difficulties at the cost of a quality loss of the solutions. Local
decision rules are mostly the backbone of such approximation algorithms, i.e. the
latter try to find the best sequence of local decisions by deterministic[4] or
probabilistic[5] learning. The outcome of such an approach heavily depends on
the underlying meta–strategy. In Chapter III we provide a probabilistic learning

---

[3]     cf. Garey and Johnson (1979)

[4]     cf. Adams et al. (1988)

[5]     cf. Glover (1986)

strategy based on principles of evolution. We shall indicate how the method can be extended to consider local decision rules of any number and complexity in order that the resulting program will do better than the deterministic learning approach.

A class of approximation algorithms is described for solving the minimum makespan problem of job shop scheduling. A common basis of these algorithms is the underlying genetic algorithm that serves as a meta–strategy to guide an optimal design of local decision rule sequences. We consider sequences of dispatching rules for job assignment as well as sequences of one machine solutions in the sense of the shifting bottleneck procedure of Adams et al. (1988). Computational experiments show that our algorithm can find shorter makespans than the shifting bottleneck heuristic or a simulated annealing approach given the same amount of running time.

The minimum makespan problem of job shop scheduling recently attracks more attention from an artificial intelligence point of view. General learning strategies are combined with special purpose heuristics which introduce problem specific knowledge. Reasoning on a knowledge base of constraints (*constraint propagation*) is considered in Sections 2 and 3 of Chapter III. Constraint based reasoning heavily depends on the strength of the underlying constraints (e.g. the number of variables) as well as the power of the logic programming language that accepts a series of logic statements and constraints and then is capable to generate a feasible solution to the underlying constraint satisfaction problem. Informally, a constraint satisfaction problem is posed as follows: Given a set of variables and a set of constraints, each specifying a relation on a particular subset of the variables, find the relation on the set of all variables which satisfies all the given constraints. The required solution relation is a subset of the cartesian product of the variable domains. Traditionally backtrack search is used to solve constraint satisfaction problems [6]. In order to overcome the inefficiency of a simple backtrack search consistency checks among variable value assignments were incorporated introducing new knowledge by constraint based reasoning to reduce the search space and discover failures earlier. Most common are node– and arc–consistency checks the only ones which are also implicitly introduced in recent constraint based logic programming languages. Local consistency checks on the overall problem of job shop scheduling cannot increase the knowledge base (set of explicit constraints) substantially, however, constraint propagation on subproblem based knowledge bases can increase the size of the

---

[6]    see Pearl (1984)

knowledge base drastically, unfortunately on the cost of possibly excluding the optimal solution. In Section 3 of Chapter III a genetic algorithm serves as a metastrategy to guide an optimal design of schedule decompositions into subproblems. A subproblem based constraint propagation approach learns to find best bounds to tighten existing constraints and to fix new arc directions. The algorithm has solved the famous 10×10 problem formulated by Fisher and Thompson in 1963 which has defied solution for almost 25 years.

The last Chapter IV is dedicated to design and planning issues arising in flexible manufacturing. A widely spread method to create flexible manufacturing cells, i.e. grouping of machines and parts and assigning part families to machine cells, is by means of some clustering based on some measure of similarity between parts (or machines). We provide a new and generally applicable (to different objectivs and models) clustering method to address this problem, describe an extremely powerful, although simple, ejection chain heuristic as well as an exact method. Computational results show that both methods outperform all previous methods suggested in the literature. The heuristic leads to optimal solutions in all practical test problems known from literature. The heuristic is used to compute an initial lower bound as well as to guide branching in a branch and bound algorithm.

Section 2 of the last chapter addresses the problem of factory layout planning. For this particular problem we consider in more detail sophisticated data structures fundamental for the implementation of efficient local search procedures and ejection chains.

Section 3 of Chapter IV describes a special hierarchical clustering and location method which yields balanced workload on manufacturing cells, or suggests a best place for the location of a leitstand, control station, or tool magazine in a manufacturing environment. From the location point of view a number of customers in the considered network is located at fixed points. Each customer will purchase a commodity from the facility closer to his location more frequently than from a remote one. As a generalization of the Condorcet concept we define an optimal point as a location such that there exists no competitor with higher expected value. We show that the set of optimal points is finite. Suboptimal points where the maximal relative rejection by a rival point is minimal are determined in polynomial time. Hence the case of two competitors is polynomially solvable. Thus the generalization of this idea can lead to powerful local search methods.

## Acknowledgement

I would like to thank a number of people each of whom, in his own way, laid the foundations during the last years for a successful outcome of this book. First of all, without the support of Wolfgang Domschke and Andreas Drexl this work would actually not exist. I am very grateful for the opportunity to benefit from their great experiences while this work has been undergoing its preparation.

I very much appreciate the guidance of Wolfgang Domschke, his enthusiasm and interest and his sensitivity not only for scientific matters but for personal matters as well. Therefore I feel much indebted to him.

The support of Andreas Drexl became a major pillar of this research. I never made in vain a request for his time. His advice during lots of discussions and talks has been undoubtly invaluable and made me always wiser than I was before. I owe him many thanks.

Many thanks also to all people I collaborated with: To Antoon Kolen for introducing me into the topic of machine learning, to Fred Glover and Jacek Blazewicz for numerous fruitful discussions, in particular, while I was invited to the Business School of the University of Colorado at Boulder and to the Computing Science Department of the University of Technology at Poznan.

Many thanks also to Hans–Jürgen Bandelt and Yves Crama, they learned me some steps on the way of doing good research. I would also like to express my gratitude to Ulrich Dorndorf for a number of valuable discussions on various topics of this work.

Finally, I would like to thank all people who supported the implementation of the algorithms: U. Dorndorf, F.–J. Greger, A. Bauer, D. Applegate, B. Cook, B. Jurisch, J. Grefenstette, N. Ulder, E. Aarts, J. Leung, and M. Grötschel.

The preparation of the book was partially supported by the Deutsche Forschungsgemeinschaft.

# I.  LOCAL SEARCH AND EXTENSIONS


## 1.  Introduction - Local Search

Consider the minimization problem min $\{f(x) \mid x \in S\}$ where f is the objective function and S is the set of feasible solutions of the problem. One of the most intuitive solution approaches to this optimization problem is to start with a known feasible solution and slightly perturb it while decreasing the value of the objective function. In order to operationalize the concept of slight perturbation let us associate with every $x \in S$ a subset $N(x)$ of S, called neighbourhood of x. The solutions in $N(x)$, or neighbours of x, are viewed as perturbations of x. Now the idea of a local search algorithm is to start with some initial solution and move from neighbour to neighbour as long as possible while decreasing the objective value.[1] This local search approach can be seen as the basic principle underlying many classical optimization methods, like the gradient method for continuous nonlinear optimization or the simplex method for linear programming.[2] More importantly, it also best explains the dynamics of many classes of neural networks, like e.g. the sequential iterations of Hopfield nets. In this framework, the objective function corresponds to the energy (Lyapunov) function of the network, the feasible solutions are the different configurations, and two configurations are neighbours if they differ in the state of exactly one neuron (that is, the neuron is excited in one of the configurations and inhibited in the other).[3]

Some of the important issues that have to be dealt with when implementing a local search procedure are how to pick the initial solution, how to define neighbourhoods and how to select a neighbour of a given solution. In many cases of interest, finding an initial solution creates no difficulty. But obviously, the choice of this starting solution may greatly influence the quality of the final outcome. Therefore local search algorithms are usually run several times on the

---

[1]    For an excellent introduction see Papadimitriou / Steiglitz (1982). Tutorial introductions into local search can be found in Pirlot (1992) and Crama et al. (1993); parts of the latter contribute to this chapter.

[2]    Local search procedures in the continuous and nonlinear case are reviewed in Scales (1985).

[3]    See for instance Freeman / Skapura (1992), McCord Nelson / Illingworth (1992), or Brause (1991) in order to get into details.

same problem instance, using different (e.g. randomly generated) initial solutions. Whether or not the procedure will be able to significantly ameliorate a poor solution often depends on the size of the neighbourhoods. Small neighbourhoods (in the limit, empty ones) are easy to search, but offer little room for improvement. Large neighbourhoods (in the limit, encompassing all solutions) raise the odds of reaching an optimal solution, but may be very tedious to explore. The choice of neighbourhoods for a given problem is conditioned by this trade-off between quality of the solution and complexity of the algorithm, and is generally to be resolved by experimentation. Another crucial issue in the design of a local search algorithm is the selection of a neighbour which improves the value of the objective function. What neighbour should be picked? The best one (greedy strategy)? Or the first one found in the search of the neighbourhood and improving upon the current solution? Or still some other candidate? This question is rarely to be answered through theoretical considerations. In particular, the effect of the selection criterion on the quality of the final solution, or on the number of iterations of the procedure is often hard to predict (although, in some cases, the number of neighbours can rule out an exhaustive search of the neighbourhood, and hence, the selection of the best neighbour). Here again experimentation with various strategies is required in order to make a decision (see the vast literature on the selection of entering variables in the simplex method).

The attractiveness of local search procedures stems from their wide applicability and (usually) low empirical complexity.[4]  Indeed, local search can be used for highly intricate problems, for which analytical models would involve astronomical numbers of variables and constraints, or about which little theoretical knowledge is available. All that is needed here is a reasonable definition of neighbourhoods, and an efficient way of searching them. When these conditions are satisfied, local search can be implemented to quickly produce good solutions for large instances of the problem. Running the procedure many times, with various initial solutions, adds to its quality and flexibility. These features of local search explain that the approach has been applied to a wide diversity of

---

[4]    See Johnson et al. (1988), Krentel (1990), Evans (1987) and Yannakakis (1990) for more information on the theoretical complexity of local search; see Vaessens et al. (1992) for a local search template and a classification of local search algorithms. The reader who would like to go into detail concerning complexity issues and the efficiency of algorithms is referred to the excellent work of Garey and Johnson (1979), the elementary work of Bachem (1980), Papadimitriou and Steiglitz (1982), or Karp (1986), or to the fundamental paper of Karp (1975). We also recommend Johnson (1990b) for the advanced reader.

situations. This will be illustrated, in the next section, on combinatorial optimization problems arising in the area of scheduling.

Nevertheless, local search also knows its drawbacks. Most notably, the procedure stops as soon as it encounters a local optimum, i.e., a solution x such that $f(x) \leq f(y)$ for all y in N(x). In general, such a local optimum is not a global optimum. Even worse, there is usually no guarantee that the value of the objective function at an arbitrary local optimum comes close to the optimal value. This inherent shortcoming of local search can be palliated in some cases by the use of multiple starts. But, because NP–hard problems often possess many local optima, even this remedy may not be potent enough to yield satisfactory solutions. In view of this difficulty, several extensions of local search have been recently proposed, which offer the possibility to escape local optima by accepting occasional degradations of the objective function. This is the case for certain types of neural networks, e.g. of Boltzmann machines with probabilistic update rules. In Sections 3 and 4, we discuss two other successful approaches based on related ideas, namely simulated annealing and tabu search. Another interesting extension of local search works with a population of feasible solutions (instead of a single one) and tries to detect properties which distinguish good from bad solutions. These properties are then used to construct a new population which hopefully contains a better solution than the previous one. This technique, known under the name of genetic algorithm will be discussed in Section 5. But before this, we will first illustrate the concepts introduced above for a few well–known combinatorial optimization problems.

## 2. Infamous Scheduling Problems

Scheduling problems are part of the larger set of combinatorial optimization problems in which the set S of feasible solutions is finite. The problem is to find an element s in S of minimum objective function value, i.e. $f(s) = \min \{f(x) \mid x \in S\}$. Usually the number of elements in S (the cardinality of S) is extremely large so that complete enumeration is computationally impossible.

**Example 1. The Traveling Salesman Problem**

Consider n jobs which have to be processed on one machine which can handle only one job at a time. Let $p_j$ denote the processing time of job j, j = 1, ..., n. Furthermore, assume there is a switch–over (or setup) time $c_{ij}$ required between jobs i and j, i,j = 0,1, ..., n, where 0 corresponds to the rest state of the machine. The objective is to complete all jobs as soon as possible. An instance with n = 4 jobs is presented in Figure 1.

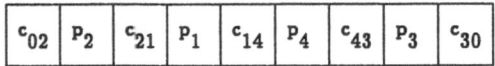

Figure 1.   A one machine schedule for 4 jobs.

Since the sum of the processing times $\sum_{j=1}^{n} p_j$ is always included in the total processing time, the latter is determined by the switch–over time. Therefore the problem can be viewed as the problem of finding a permutation $\pi : \{0,1, ..., n\} \rightarrow \{0,1, ..., n\}$ which minimizes

$$f(\pi) = \sum_{i=0}^{n-1} c_{\pi(i)\pi(i+1)} + c_{\pi(n)\pi(0)}$$

over the set S of all permutations. The latter combinatorial optimization problem is called the (symmetric) traveling salesman problem; when $c_{ij}$ ($= c_{ji}$) is viewed as the distance between two cities i and j the problem translates into finding the shortest tour which visits each city exactly once.

The tour corresponding to the schedule in Figure 1 can be represented by the edges [0,2], [2,1], [1,4], [4,3], and [3,0], as is illustrated in Figure 2.

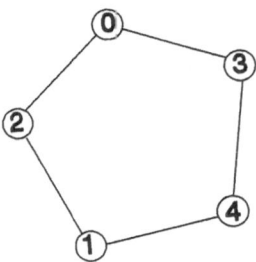

Figure 2.   The tour corresponding to the schedule of Figure 1.

Probably the best known neighbourhood structure for the symmetric traveling salesman problem is determined by the concept of r-exchange. Two tours are neighbours with respect to an r-exchange if they differ in exactly r edges. Figure 3 describes a 2-exchange where the edges [1,2] and [3,4] are replaced by the new edges [1,3] and [2,4]. The change in the length of the tour is easy to calculate as $c_{13} + c_{24} - c_{12} - c_{34}$. The deletion of three edges does not uniquely determine three new edges which have to be inserted in order to get a feasible tour. In Figure 4 four possible 3-exchanges result when the edges [1,2], [3,4], and [5,6] are replaced by three new ones. Newly introduced edges are either [1,3], [2,5], [4,6], or [1,4], [2,5], [3,6], or [1,4], [2,6], [3,5], or [1,5], [2,4], [3,6].

A Lin-Kernighan exchange is an r-exchange, where r is chosen such that exchanging r edges in the tour by r other edges results in an improvement, but it is not possible to add another edge of the tour to the r edges such that exchanging the r+1 edges results in an improvement.[5]

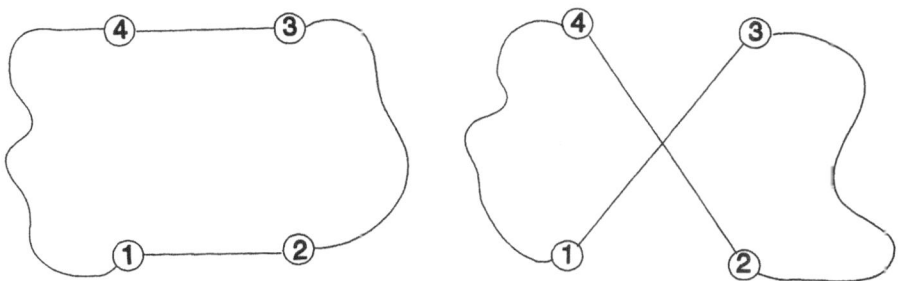

Figure 3.    A 2-exchange of edges [1,2], [3,4] by [1,3], [2,4].

5     cf. Lin / Kernighan (1973), Lawler et al. (1985), see Chapter II.

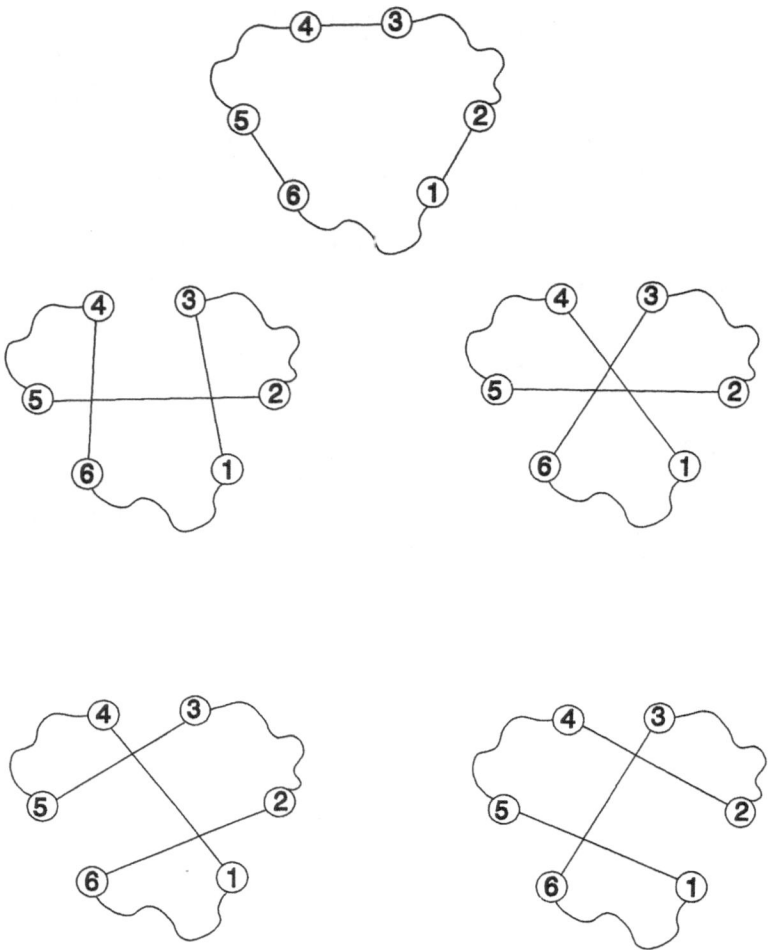

Figure 4.    All possible 3—exchanges for the edges $[1,2]$, $[3,4]$, and $[5,6]$.


## Example 2.  Job Shop Scheduling

A job shop consists of a set of different machines that perform operations on jobs. Each job has a specified processing order through the machines; that is, a job is an ordered list of operations, each of which is determined by the machine it requires and by its processing time. Operations cannot be interrupted (non—preemption), each machine can handle only one job at a time, and each job can be performed on only one machine at a time. The operation sequences on the machines are unknown and have to be determined so as to minimize the makespan, i. e. the time required to complete all jobs. An illuminating problem representation is the disjunctive graph model due to Roy and Sussmann (1964). Let $V = \{0,1, \ldots ,n\}$ denote the set of operations where 0 and n are considered as dummy operations "start" and "end", respectively. Let M denote the set of machines; A is the set of pairs of operations constrained by the precedence relations for each job. For each machine k, the set $E_k$ describes the set of all pairs of operations to be performed on machine k, i.e. operations which cannot overlap. In the disjunctive graph there is a vertex for each operation $i \in V$ and vertices 0 and n representing the start and the end, respectively, of a schedule. For every two consecutive operations of the same job there is a directed arc; the start vertex 0 is considered to be the first operation of every job and the end vertex n is considered to be the last operation of every job. For each pair of operations $\{i,j\} \in E_k$ that require the same machine there are two arcs (i,j) and (j,i) with opposite directions. Thus, single arcs between operations represent the precedence constraints on the operations and opposite directed arcs between two operations represent the fact that each machine can handle at most one operation at the same time. Each arc (i,j) is labeled by a positive weight $p_i$ corresponding to the processing time of operation i . All arcs from 0 have a label 0. Figure 5a illustrates the disjunctive graph for a problem instance with 3 machines M1, M2, M3 (machine numbers are indicated by the vertex labels) and 3 jobs J1, J2, J3. The machine sequences of job J1, J2, and J3 (see the rows of Figure 5a) are M1→M2→M3, M3→M2, and M2→M1→M3, respectively. The processing times are presented in Table 1.

| M1 | 3 | - | 3 |
|----|---|---|---|
| M2 | 2 | 4 | 6 |
| M3 | 3 | 3 | 2 |
|    | J1 | J2 | J3 |

Table 1. Processing times of a 3 job 3 machine instance.

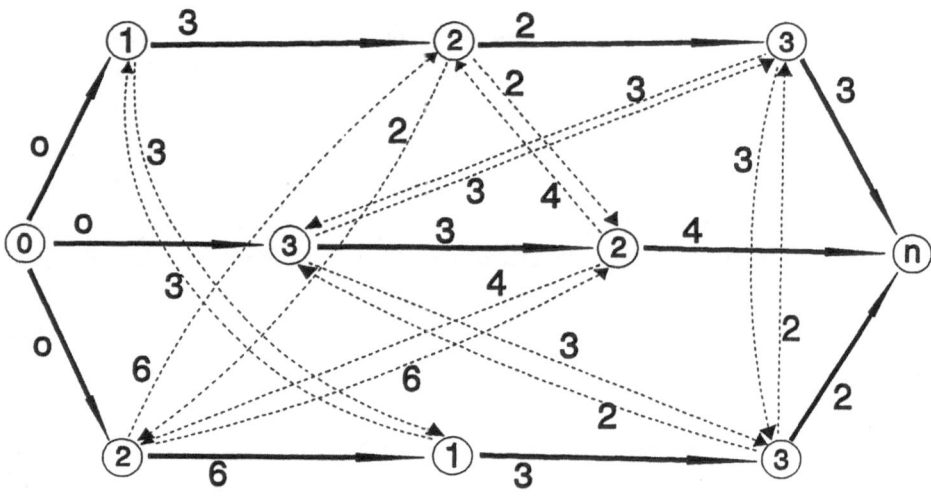

Figure 5a.    The disjunctive graph for the problem instance of Table 1.

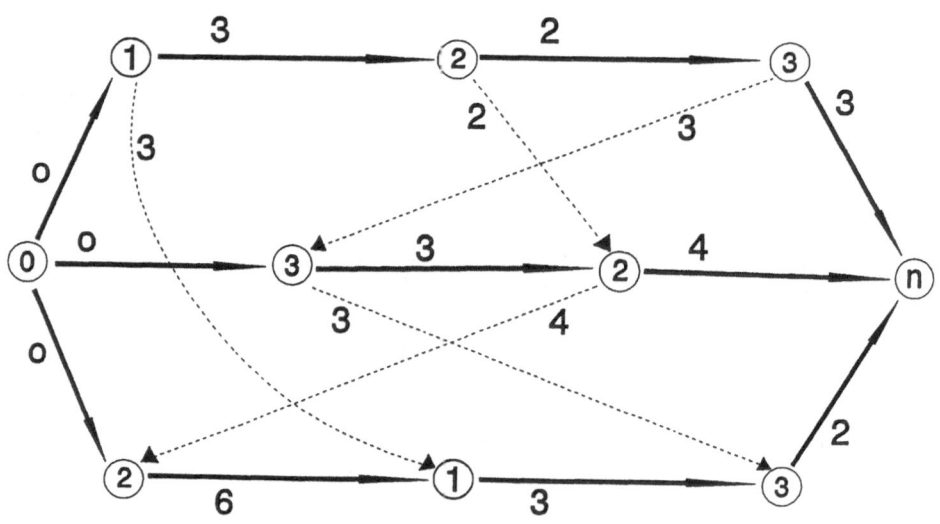

Figure 5b.    A solution for the problem instance of Table 1 and Figure 5a.

The job shop scheduling problem requires to find an order of the operations on each machine, i.e. to select one arc among all opposite directed arc pairs such that the resulting graph is acyclic (i. e. there are no precedence conflicts between operations) and the length of the maximum weight path between the start and end vertex is minimal. The length of a maximum weight (or longest) path determines the makespan.

In order to improve a current schedule, we have to modify the machine order of jobs (i.e. the sequence of operations) on longest paths. Therefore a neighbourhood structure can be defined by (i) reversing an edge (between operations on the same machine) on a longest path in the graph, and (ii) reversing an edge on a longest path in the graph such that this edge is incident (on a longest path) to an edge of the arc set A.[6]

For the problem instance of Figure 5a let us consider the schedule defined by the jobs processing sequence J1→J3 on machine M1, and J1→J2→J3 on machine M2 and M3. Hence all jobs are lying on a longest path of length 26 (see Figure 5b). Reversing the processing order of jobs J2 and J3 on machine M2 yields a reduced makespan of 16 for the new schedule.

### Example 3. Minimizing the Sum of Weighted Completion Times

Consider n jobs which have to be processed on one machine. Each job has a processing time $p_j$ and a weight $w_j$, $j = 1, ..., n$. All jobs are available at the start of the planning period, say at time 0. The completion time $C_j$ of job j, $j = 1, ..., n$, is defined as the time at which the processing of job j is finished. The objective is to find a sequence of the jobs which minimizes the weighted sum of completion times $\sum_{j=1}^{n} w_j \cdot C_j$.

Consider an instance involving 5 jobs, the processing times and weights of which are presented in Table 2. A schedule with job sequence 1,2,3,4,5 is represented in the Gantt-chart [7] of Figure 6. Its objective function value is $2 \cdot 8 + 3 \cdot 2 + 6 \cdot 9 + 10 \cdot 10 + + 13 \cdot 5 = 241$.

---

6    For details we refer to Matsuo et al. (1988), Aarts et al. (1991), and van Laarhoven et al. (1992). The importance of the neighbourhood structure for the final outcome of a search process is also pointed out by Brucker et al. (1993). They suggest a neighbourhood structure for certain scheduling problems consisting of only locally optimal solutions.

7    see Gantt (1919)

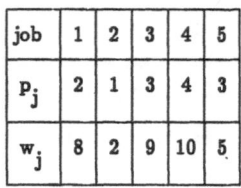

Table 2.   Processing times and weights of a 5 job problem instance.

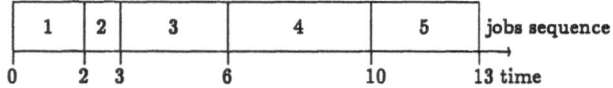

Figure 6.   A schedule for the problem instance of Table 2.

Two job sequences are defined to be neighbours if one can be obtained from the other by interchanging two consecutive jobs. If job i is an immediate predecessor of job j then interchanging i and j does not affect the completion time of the other jobs. Therefore the new schedule, where j precedes i, is an improvement if
$w_i \cdot (T + p_i) + w_j \cdot (T + p_i + p_j) >$
$w_j \cdot (T + p_j) + w_i \cdot (T + p_j + p_i)$, where T is the starting time of job i in the first schedule. Thus the interchange gives an improvement if $w_j/p_j > w_i/p_i$. For instance, interchanging jobs 2 and 3 in the schedule of Figure 6 leads to an improvement since $9/3 > 2/1$ (the schedule 1,3,2,4,5 has an objective function value of 238). As a matter of fact, an optimal solution can always be found by ordering the jobs in non–increasing order of $w_j/p_j$ [8]. Hence, in contrast with the preceding examples, the neighbourhood structure defined above guarantees that every local minimum is a global minimum, and that local search always leads to an optimal solution.

---

[8]    This is Smith's ratio rule (1956).

# 3. Simulated Annealing

In the development of suboptimal heuristics for combinatorial optimization, innovative progress has been made by Kirkpatrick, Gelatt and Vecchi (1983) [9] and, independently, by Cerny (1985) [10], who perceived an analogy between the behaviour of a physical substance in low energy states and the nature of the iterative improvement that can be made in a large and complex system that is in a nearly optimal configuration. During the physical annealing process temperature is reduced slowly in order to maintain system equilibrium with respect to temperature. States of low energy in the physical system are viewed as being analogous to the nearly optimum configuration (measured by the objective function) in a minimization problem. Hence simulated annealing – although originally devised by Metropolis et al. (1953) to simulate the annealing (or slow cooling) of solids, after they have been heated to their melting point – was proposed as a framework for the solution of intractable optimization problems [11]. In simulated annealing procedures, the sequence of solutions does not roll monotonically down towards a local optimum, as was the case with local search. Rather, the solutions trace an up–and–down random walk through the feasible set S, and this walk is loosely guided in a "favourable" direction. To be more specific, let us now describe the k–th iteration of a typical simulated annealing procedure, starting from a current solution $x \in S$. First, a neighbour of x, say $y \in N(x)$, is selected (usually, but not necessarily, at random). Then, based on the amplitude of $\Delta = f(x) - f(y)$, a transition from x to y (i.e. an update of x by y) is either accepted or rejected [12]. This decision is made nondeterministically: the transition is accepted with probability $p_k(\Delta)$, where $p_k$ is a probability distribution depending on the iteration count k. The intuitive justification for this rule is as follows. In order to avoid getting trapped early in

---

9      see also Vecchi and Kirkpatrick (1983), Jespen / Gelatt (1983), Kirkpatrick (1984), Kirkpatrick / Swendsen (1985)

10     see also Burkard / Rendl (1984)

11     An enormous number of publications has appeared handling completely diverse application areas and we refer only to the books of Johnson (1988), Otten / Van Ginneken (1989), the bibliography of Collins, Eglese and Golden (1988) which can serve as a reference manual, and the review paper of Rutenbar (1989). The interested reader is referred to Van Laarhoven / Aarts (1987) for a thorough treatment of the theory of simulated annealing and for a description of various applications. An introductory course is presented in Aarts / Van Laarhoven (1987, 1989), Eglese (1990), Osman (1991), Press et al. (1992), or Kuhn (1992).

12     See the modification of Stern (1992) in employing a penalty function.

a local optimum, transitions implying a deterioration of the objective function (i.e., with $\Delta < 0$) should be occasionally accepted, but the probability of acceptance should nevertheless increase with $\Delta$. Moreover, the probability distributions are chosen so that $p_{k+1}(\Delta) \leq p_k(\Delta)$. In this way, escaping local optima is relatively easy during the first iterations, and the procedure explores the set S freely. At first, almost any perturbation to a solution is accepted. But, as the iteration count increases, only improving transitions tend to be accepted, and the solution path is likely to terminate in a local optimum. The procedure stops if the value of the objective function remains constant in L (a termination parameter) consecutive iterations, or if the number of iterations becomes too large [13].

In most implementations, and by analogy with the original procedure of Metropolis et al. (1953), the probability distributions $p_k$ take the form:

$$p_k(\Delta) = \begin{cases} 1 & \text{if } \Delta \geq 0 \\[2em] e^{c_k \Delta} & \text{if } \Delta < 0, \end{cases}$$

where $c_{k+1} \geq c_k \geq 0$ for all k, and $c_k \rightarrow \infty$ when $k \rightarrow \infty$.

A popular choice for the parameter $c_k$ is to hold it constant for a fixed number (T) of consecutive iterations, and then to increase it by a constant factor:

$$c_{i \cdot T + t} = \alpha^i \cdot c_0 \quad \text{for } t = 1, 2, ..., T \text{ and } i = 0, 1, 2, ...$$

Here, $c_0$ is a small positive number, and $\alpha$ is slightly larger than 1. It is clear that the choice of the termination parameter and of the distributions $p_k$ (k = 1, 2, ...) (the so-called cooling schedule) strongly influences the performance of the procedure. If the cooling is too rapid (e.g. if T is small and $\alpha$ is large), then simulated annealing tends to behave like local search, and gets trapped in local optima of poor quality. If the cooling is too slow, then the running time becomes prohibitive. Greene and Supowit (1986) suggested a rejectionless method in order to save time in the later part of the run when the majority of the moves are rejected [14]. Under some reasonable assumptions on the cooling schedule, theoretical results can be established concerning convergence to a

---

[13]    cf. Otten / Van Ginneken (1988, 1990)

[14]    see also Fox (1993a,b,c)

global optimum or the complexity of the procedure [15,16]

In practice, determining appropriate values for the parameters is part of the fine tuning of the implementation, and still relies on experimentation.[17,18]    In particular, many researchers tested the performance of simulated annealing approaches to the traveling salesman problem (the seminal papers by Kirkpatrick et al. (1983) and Cerny (1985) already handled this problem) [19]. The neighbourhood structure used in these implementations is generally based on 2- or 3-exchanges [20]. This seems to lead to algorithms which are more effective than repeated applications of simple local search, but less than the Lin–Kernighan heuristic (1973). However, when Golden and Skiscim (1986) tested some simulated annealing algorithms against more traditional methods for the traveling salesman problem and p–median problems, the simulated annealing methods did not perform well by comparison. Tovey (1988) suggested to use problem specific knowledge to identify the promising subsets of the neighbourhood and to choose a neighbourhood correspondingly. Simulated annealing has also been applied to job shop scheduling, with neighbourhood structures of the type described in Section 2 [21]. The resulting algorithms perform again better than multiple–start local search or simple–minded heuristics. Given

[15]  See Mitra et al. (1986), Lundy / Mees (1986), Anily / Federgruen (1987a,b), Van Laarhoven and Aarts (1987), Faigle / Schrader (1988), Gidas (1988), Aarts and Korst (1989a), Otten / Van Ginneken (1990), Faigle / Kern (1991, 1992) further Vanderbilt / Louie (1984), Dekker / Aarts (1991), Bélisle et al. (1990), and Gelfand / Mitter (1993) who also cover the continuous case.

[16]  Necessary and sufficient conditions for the simulated annealing scheme to converge with probability one to a global optimum have been derived in Van Laarhoven / Aarts (1987). Several results have been derived giving sufficient conditions for asymptotic convergence to the set of global optima. Hajek (1988) showed that the rate of cooling required for asymptotic convergence depends on the depth of the local mimima, i.e. the topology that is imposed by the neighbourhood structure, see Cheh et al. (1991). Chiang / Chow (1988) give an inhomogeneous continuous Markov–chain version of simulated annealing, with a detailed analysis of the rate of convergence to the optimal set.

[17]  We refer to the extensive computational studies by Johnson et al. (1989, 1991) for the wealth of details on this topic.

[18]  Simulated annealing has been applied to several types of combinatorial optimization problems, with various degrees of success (see van Laarhoven and Aarts (1987), Aarts and Korst (1989a), and Johnson et al. (1989, 1991)). An application to general zero–one programming problems (which can also be considered as multi–constraint knapsack problems) can be found in Drexl (1988) and Helming / Jörnsten (1991).

[19]  See van Laarhoven and Aarts (1987) and Johnson (1990a) for an overview of the literature.

[20]  see Section 2 of this chapter

[21]  see Matsuo et al (1988), Aarts et al. (1991), and van Laarhoven et al. (1992)

sufficient (very high) running time, they can produce better solutions than the efficient bottleneck procedure due to Adams et al. (1988).

As a general rule, one may say that simulated annealing is a reliable procedure to use in situations where theoretical knowledge is scarce or appears difficult to apply algorithmically. Even for the solution of complex problems, simulated annealing is relatively easy to implement, and usually outperforms a local search procedure with multiple starts.

## 4. Tabu Search

Tabu search is a general framework for the solution of discrete optimization problems, which was originally proposed by Glover, and subsequently expanded in a series of papers (Glover (1969, 1972, 1977, 1986, 1989a,b,c, 1990a), Glover and McMillan (1986), etc.).[22] One of the central ideas in this proposal is to guide deterministically the local search process out of local optima (in contrast with the non–deterministic approach of simulated annealing). This can be done using different criteria, which ensure that the loss incurred in the value of the objective function in such an "escaping" step is not too important, or is somehow compensated for.

For instance, assume that several numerical criteria, say $f_1$, ..., $f_m$ are relevant to evaluate the quality of candidate solutions. A weighted combination of these criteria, $f = \Sigma\ w_i \cdot f_i$ , can then be used as objective function in a classical local search procedure to produce a local optimum x with respect to f. If the weights $w_i$ are now modified, another combination of the criteria, say f', is obtained, for which x is (in general) no longer a local optimum. Local search can then proceed with this new objective function, to generate alternative solutions to the problem.[23]

Another, more straightforward criterion for leaving local optima is to replace the improvement step in the local search procedure by a "least deteriorating" step. One version of this principle was proposed by Hansen (independently of Glover's work on tabu search), under the name <u>steepest descent mildest ascent</u> [24]. In its

---

[22]   See also the recent book on new tabu search applications and strategies edited by Glover et al. (1993) and the introductory courses by Glover (1990c), Laguna (1992), Glover / Laguna (1993), Glover et al. (1992, 1993), and De Werra / Hertz (1989).

[23]   This type of approach was used by Glover and McMillan (1986) in their solution of very large employee scheduling problems, see also Schaffer / Grefenstette (1985).

[24]   see Hansen and Jaumard (1990), as well as Glover (1989c)

simplest form, the resulting procedure replaces the current solution x by a solution y $\in$ N(x) which maximizes $\Delta$ = f(x) - f(y). If during L (a termination parameter) iterations no improvements are found, the procedure stops. Notice that $\Delta$ may be negative, thus resulting in a deterioration of the objective function. Now, the major defect of this simple tabu search procedure is readily apparent. If $\Delta$ is negative in some transition from x to y, then there will be a tendency in the next iteration of the procedure to reverse the transition, and go back to the local optimum x (since x improves on y). Such a reversal would cause the procedure to oscillate endlessly between x and y. To prevent this phenomenon (which is likely to occur in every version of tabu search), Glover and Hansen propose to maintain throughout the search a (dynamic) list of forbidden transitions, called tabu list (hence the name of the procedure). The purpose of this list is not to rule out cycling completely (this would in general result in heavy bookkeeping and loss of flexibility), but at least to make it improbable.

In the framework of the steepest descent mildest ascent procedure, we may for instance implement this idea by placing a solution x in a tabu list T after every transition away from x. In effect, this amounts to deleting x from S. But, for reasons of flexibility, a solution would only remain in the tabu list for a limited number of iterations, and then should be freed again.

Another possible implementation would be to create a tabu list T(y) for every solution y $\in$ S. After a transition from x to y, x would be placed in the list T(y), meaning that further transitions from y to x are forbidden (in effect, this amounts to deleting x from N(y)). Here again, x should be dropped from T(y) after a number of transitions.[25]

Tabu search encompasses many features beyond the possibility to avoid the trap of local optimality and the use of tabu lists. Even though we cannot discuss them all in the limited framework of this introduction, we would like to mention two of them, which provide interesting links with artificial intelligence and with genetic algorithms (to be discussed in the next section). In order to guide the search, Glover suggests to record some of the salient characteristics of the best solutions found in some phase of the procedure (e.g., fixed values of the variables in all, or in a majority of those solutions, recurring relations between the values of the variables, etc.). In a subsequent phase, tabu search can then be restricted

---

25    For still other possible definitions of tabu lists, see e.g. Glover (1986, 1989a,c, 1990a. 1991a,b, 1992a,b,c), Pesch and Glover (1993), Glover and Greenberg (1989), Hansen and Jaumard (1990), Herts and De Werra (1990), Dammeyer and Voß (1993), Dammeyer et al. (1991).

to the subset of feasible solutions presenting these characteristics. This enforces what Glover calls a "regional intensification" of the search in promising "regions" of the feasible set. An opposite idea may also be used to "diversify" the search. Namely, if all solutions discovered in an initial phase of the search procedure share some common features, this may indicate that other regions of the solution space have not been sufficiently explored. Identifying these unexplored regions may be helpful in providing new starting solutions for the search. Both ideas, of search intensification or diversification, require the capability of recognizing recurrent patterns within subsets of solutions. Techniques developed in the fields of pattern recognition or learning may clearly be relevant for this purpose (Glover (1986, 1989c, 1990a)).

Variants of tabu search have been successfully applied to a large diversity of optimization problems [26]: scheduling, clustering, generalized bin packing [27], graph coloring [28], maximum satisfiability [29], quadratic semi–assignment problem [30], etc. Applications to the traveling salesman problem are reported in Malek et al. (1989); various neighbourhood structures are presented by Glover (1991a, 1992a,c). Taillard (1989) has implemented a parallel tabu search approach for the job shop scheduling problem, while Widmer (1991) applied tabu search to a generalized version of the same problem. Taillard uses a neighbourhood structure defined by reversing an edge between operations on the same machine on the longest path in the disjunctive graph, see Section 2 of this chapter. Such a reversed edge becomes tabu for a certain number of iterations. Widmer, on the other hand, defines a neighbour of the current schedule by selecting a machine M and an operation J, and by shifting the position of J in the operations sequence of machine M. If J is shifted from position i to position k in the sequence, then the couple (J,i) is included in the tabu list, meaning that operation J may not return to position i in the operations sequence for machine M (until (J,i) is removed from the tabu list).

Like simulated annealing (or, maybe, more than it), tabu search has established itself as a successful general–purpose heuristic for combinatorial optimization problems. But the full potential of the method, as well as its theoretical properties, largely remain to be understood.

---

[26]     cf. Voß (1994)

[27]     see Glover (1986, 1990a), Glover and McMillan (1986)

[28]     Hertz and de Werra (1987)

[29]     Hansen and Jaumard (1990)

[30]     Domschke et al. (1992)

# 5.  Genetic Algorithms

Genetic algorithms (GAs) [31]   have been designed as general purpose search strategies and optimization methods; the GA catechism is laid down in Goldberg (1989a,b). The very name, though, might be misleading since the word "algorithm" alludes to a special method of solving a certain kind of problem, but what is really meant – from a users point of view – is a general strategy or metaheuristic calling on principles of evolution. Roughly speaking, a genetic algorithm aims at producing near–optimal solutions by letting a set of random solutions undergo a sequence of unary and binary transformations governed by a selection scheme biased towards high–quality solutions.

Genetic algorithms have been designed as general purpose search strategies and optimization methods working on populations of feasible solutions [32]. Contrary to simulated annealing or tabu search which are based on manipulating one feasible solution, working with populations permits to identify and explore properties which good solutions have in common (this is similar to the regional intensification idea mentioned in our discussion of tabu search). Solutions are encoded as strings consisting of elements chosen from a finite alphabet and a property is defined by fixing certain elements of the string. Operations on these strings often create infeasibilities for tightly constrained problems. So encoding schemes are created which still maintain the idea of a genetic algorithm as general problem solver but guaranty feasibility if the representation is mapped to a set of local decisions [33],[34]

---

[31]     Genetic algorithms date back to the early work of Rechenberg (1973), who had already experimented with a kind of GAs dubbed "erweiterte Evolutionsstrategie", Holland (1973, 1975), Schwefel (1977), De Jong (1975, 1980, 1990), Bethke (1980), and Brindle (1981), see also Davis (1987, 1991), Booker et al. (1989), Goldberg (1989a,b), and Liepins and Hilliard (1989), Bäck et al. (1990), Hoffmeister / Bäck (1991), Michalewicz (1992), Michalwicz et al. (1991), Rawlins (1991), Nix / Vose (1992), the state of the art of current research can be found in the proceedings of the international conferences on Genetic Algorithms as well as in the proceedings of the international conferences on Parallel Problem Solving from Nature (see also the bibliography of Goldberg / Thomas (1986)). An easy to read introduction and sample genetic algorithm for the solution of nonlinear systems of equations is described in Wayner (1991), further easy readings are provided by Mertens (1991), Bäck et al. (1991), Schwefel / Bäck (1992).

[32]     or feasible solutions of a nonlinear Lagrangian relaxation, the genetic algorithm of which can be shown to converge towards the optimal solution of the primal problem (Bean / Hadj–Alouane (1992))

[33]     see Chapter III

[34]     Such an idea is also used by Bean (1992) who assigned random keys to variables and the key value determines the variable value under certain rules of evaluation.

A genetic algorithm aims at producing near–optimal solutions while a set of solutions representing strings transforms, governed by a selection scheme favouring "survival of the fittest". Therefore the quality or <u>fitness</u> value of an individual in the population, i.e. a string, has to be defined. Usually it is the value of the objective function or some scaled or penalized version of it [35]. The transformations on the individuals of a population constitute the <u>recombination</u> steps of a genetic algorithm and are performed by three simple operators. The effect of the operators is that implicitly good properties [36] are identified and combined into a new population which hopefully has the property that the value of the best individual (representing the best solution in the population) and the average value of the individuals are better than in previous populations. The process is then repeated until some stopping criteria are met. It can be shown that the process converges to an optimal solution with probability one [37].

The three basic operators of a genetic algorithm when a new population is constructed are reproduction, crossover and mutation. Via <u>reproduction</u> a new temporary population is generated where each member is a replica of a member of the old population. A copy of an individual is produced with probability proportional to its fitness value, i.e. better strings probably get more copies. The intended effect of this operation is to improve the quality of the population as a whole. In particular reproduction aims at increasing the number of individuals in the population satisfying properties of high quality. Hereby, we measure the quality of a property as the average fitness of all individuals satisfying the property, i.e. the average fitness of all individuals which have string entries identical to the corresponding entries of a property. Of course we can only explore properties which are represented by individuals of the population. Even for a property which is represented we are not able to calculate its quality exactly because not necessarily all solutions satisfying it are present in the population. Therefore the quality of a property is estimated by the average value of the solutions in the population satisfying it. Since the number of different properties is exponential in the number of elements in the string it would be too time consuming to calculate the quality of every property explicitly. So the effect of reproduction is that implicitly properties of higher (lower) quality are

---

[35]    see Richardson et al. (1989)

[36]    called schemata, (Battle / Vose (1993), Vose (1991))

[37]    See Eiben et al. (1991), Rabinovich / Wigderson (1991), Davis / Principe (1991);
        Kingdon (1992) and Liepins / Vose (1990) for the description of GA–hard problems.
        The proof is very similar to the convergence proof of simulated annealing, see Aarts
        and Korst (1989a).

increased (decreased) in their number in the new temporary population.
However, no genuinely new solutions and hence no new information are created
in the process. The generation of such new strings is handled by the crossover
operator.

In order to apply the crossover operator the temporary population is randomly
partitioned into pairs. Next, for each pair, the crossover operator is applied with
a certain probability by choosing a position randomly in the string and
exchanging the tails (defined as the substring starting at the chosen position) of
the two strings [38]. The effect of the crossover is that certain properties of the
individuals are combined to new ones or other properties are destroyed [39].
Considering a pure genetic algorithm (lateron we are going to consider hybrids)
there is lots of evidence that for many applications the crossover operator is the
"key to the success of a genetic algorithm" (Mühlenbein (1989), De Jong (1980),
Holland (1975)).[40]   If we define the length of a property as the number of string
entries between the first and the last fixed element of the property, then the
probability for a property to survive the one–point crossover is proportional to
its length. Properties of small length, called building–blocks, then are combined
to form new solution representing strings. Consequently, the representation of a
solution must be such that building blocks have a meaningful interpretation with
respect to the original problem.

The mutation operator which makes random changes to single elements of the
string only plays a secondary role in genetic algorithms. Mutation serves to
maintain diversity in the population [41].

Besides unary and binary recombination operators, one may also introduce
operators of higher arities such as consensus operators, that fix variable values
common to most solutions represented in the current population. Selection of
individuals during the reproduction step (to undergo crossover) can be realized
in a number of ways: one could adopt the scenario of Goldberg (1989a) or use
deterministic ranking. Further it matters whether the newly recombined
offspring compete with the parent solutions or simply replace them; if selection
is done over the whole population or only with respect to a local population
around a certain individual [42]. Mühlenbein (1989) says that selection restricted to

---

[38]   This is the simplest version of a crossover, called one–point crossover.

[39]   De Jong / Spears (1992), Liepins / Vose (1992)

[40]   A type of unary crossover is suggested by Kelly et al. (1990).

[41]   See the previous section on tabu search.

[42]   Bäck / Hoffmeister (1992)

subpopulations of the complete population gives "much more variability than in a random mating population".

The traditional genetic algorithm, based on a binary string representation of solutions, is often unsuitable for combinatorial optimization problems because it is very difficult to represent a solution in such a way that substrings (building-blocks) have a meaningful interpretation. For instance consider the first attempt by Grefenstette et al. (1985) [43], to solve the traveling salesman problem by a tradional genetic algorithm using the ordinal representation.

The ordinal representation starts with an ordering of the cities on a list, say 1,2,...,n where 1 is assumed to be the starting and ending city of the tour. The first coordinate of the ordinal representation is 1 indicating that city 1 is first on the list. Then city 1 is deleted from the list. After the k-th city of the tour has been assigned the list contains the remaining n-k cities. The k+1-th coordinate of the ordinal representation is the position on the current list of the k+1-th city in the tour. So the tour (1,3,2,5,4) corresponds to the ordinal representation 12121 as shown in Figure 7.

$$
\begin{array}{ccccc}
1 & 2 & 2 & 4 & 4 \\
2 & 3 & 4 & 5 & \\
3 & 4 & 5 & & \\
4 & 5 & & & \\
5 & & & &
\end{array}
$$

|            |   |   |   |   |   |
|------------|---|---|---|---|---|
| City     : | 1 | 3 | 2 | 5 | 4 |
| Position : | 1 | 2 | 1 | 2 | 1 |

Figure 7.   Calculating ordinal representation.

The ordinal representation results in a string over the alphabet $\{1,2,..,n\}$ where the k-th element has a value less than or equal to n-k+1, k=1,..,n. Conversely any string with this property represents a tour (simply reverse the construction). This representation is motivated by the simplicity of the crossover (cut the two sequences and interchange the tails). The newly obtained solutions still represent tours. However the building blocks have no direct relationship to subsets of edges in the tour. So it is no surprise that this first attempt by Grefenstette et al. (1985) to solve the traveling salesman problem by a traditional genetic algorithm led to solutions as far as 25% above the optimum, even for small

---

[43]   see also Grefenstette (1987b)

problem sizes up to 50 cities. A more natural representation, e.g. a sequence of the edges of a tour, defines a property as fixing certain edges. The length of a property is the maximum number of edges between two fixed edges in the tour and a building block is a sequence of consecutive edges. Hence a crossover advances clustering of the cities such that cities closer to each other are probably more often adjacent in the tours [44]. However, choosing a more natural representation of solutions, for instance, a permutation of the cities for the traveling salesman problem or a list of operation sequences per machine for job shop scheduling, involves more intricate recombination operators, in particular domain dependant crossover operators, in order to get feasible offspring.[45] The construction of a crossover operator should also take into consideration that fitness values of offspring are not too far from those of their parents, and that offspring should be closely genetically related to their parents. Let us illustrate this discussion on some examples.

For the traveling salesman problem, the Grefenstette–crossover [46] constructs one new tour from two parent tours as follows:

(i) Randomly choose a city as the current city of the tour and label it "visited".

(ii) Consider all the edges incident to the current city in both parents and choose among these edges a shortest one leading to an unvisited city. If all edges lead to an already visited city, randomly choose an edge (which is not in one of the parents) to one of the unvisited cities. Say j is the unvisited endpoint of this edge. Label j "visited", and repeat (ii) with j as the new current city, until all cities have been visited.

The procedure can be repeated to generate two offspring from the two parents. Variations are possible, for instance in step (ii) we may select edges at random or with a probability inversely proportional to their length.

The Mühlenbein–Gorges–Schleuter–crossover [47] chooses a path in one of the parents and incorporates this path in the other parent while leaving as many as possible of the edges undisturbed. The length of the path is randomly chosen within the interval $[n/3, n/2]$; the first vertex of the path is also randomly

---

[44]    Reinelt (1992)

[45]    This tradeoff has been, for instance, noticed by Aarts et al. (1991) for the job shop scheduling problem and by Mühlenbein et al. (1987, 1988), Gorges–Schleuter (1989), or Kolen and Pesch (1991) for the traveling salesman problem.

[46]    Grefenstette (1987b)

[47]    Mühlenbein et al. (1988) and Gorges– Schleuter (1989)

chosen. We illustrate the Mühlenbein–Gorges– Schleuter–crossover by an
example. Assume that we wish to implant the path (1,2,3) from parent
(1,2,3,4,5,6,7,8) into parent (1,8,4,6,3,5,2,7), called the receiving parent. The first
step to perform is to create a new tour such that both endpoints of the path,
city 1 and city 3 in our case, are adjacent. Adjacency can be reached by either
of two 2–exchanges. In the first one, edges [1,8] and [3,5] are replaced by the
new edges [1,3] and [5,8], while in the other 2–exchange the edges [1,7] and
[3,6] are replaced by the new edges [1,3] and [6,7]. Thus two tours are
obtained where cities 1 and 3 are adjacent. In both of them all cities of the path
that has to be implanted are removed from their positions while the order of all
other cities remains untouched. In our case city 2 will be dropped from both
tours and an edge [5,7] is introduced in both. Finally the path is implanted
between the two endpoints, i.e. city 2 becomes adjacent to cities 1 and 3 in both
tours. Hence, we get two new tours (1,2,3,6,4,8,5,7) and (1,2,3,5,7,6,4,8), the
best of which is chosen as a result of the crossover. Similarly we get the second
offspring when we start choosing a path in the other parent.

The crossover operator used by Aarts et al (1991) in case of job shop scheduling
is also based on a natural solution representation. The idea is to implant a
subset of edges from one parent into the receiving parent. More specifically, an
arc (i,j) sequencing two jobs on the same machine in the first parent is
randomly chosen. If this arc occurs on a longest path in the receiving parent,
then it is reversed in the latter and the longest paths are recomputed. This
process is repeated k times where k is at most the number of operations in the
underlying job shop scheduling problem. A second offspring is obtained by
interchanging the roles of the parents.

Problems from combinatorial optimization are well within the scope of genetic
algorithms and early attempts closely followed the scheme of what Goldberg
(1989a) calls a simple genetic algorithm. Compared to standard heuristics, for
instance for the traveling salesman [48] or the job shop scheduling problem [49],
"genetic algorithms are not well suited for fine–tuning structures which are very
close to optimal solutions" (Grefenstette (1987b)). Therefore it is essential, if a
competitive genetic algorithm is desired, to compensate for this drawback by
incorporating (local search) improvement operators into the basic scheme [50].

---

[48]    cf. Lawler et al. (1985), Grötschel and Holland (1991)

[49]    cf. Baker (1974), French (1982), Adams et al. (1988)

[50]    see Mühlenbein et al. (1987, 1988), Mühlenbein (1989), Jog et al. (1989) and Suh and
         Van Gucht (1987)

Then the power of genetic algorithms derives largely from their implicit parallelism, i.e. the simulataneous allocation of search effort to many regions of the search space [51]. The resulting algorithm has then been called genetic local search heuristic or genetic enumeration; for the traveling salesman we refer to the papers of Ulder et al. (1991), Johnson (1990a), and Kolen and Pesch (1991); for the job shop scheduling problem we refer to Dorndorf and Pesch (1992a, 1993c). Local search improvement algorithms that were used for the traveling salesman problem and applied to some or all of the individuals in the population are 2-opt, i.e. repeated 2-exchanges, tabu search, and the algorithm of Lin and Kernighan (1973), i.e. repeated Lin-Kernighan exchanges. Each individual of the population is then replaced by a locally improved one or an individual representing a locally optimal solution, i.e. an improvement procedure is applied to each individual either partially (to a certain number of iterations) or completely. [52]

Liepins et al. (1987) noticed "A criticism of conventional genetic algorithms as optimizers is that they fail to incorporate problem structure in their formulation. Their only tie to the specific function being optimized is through the reward structure. The incorporation of the greedy algorithm allows problem specific information to be used in the crossover operator".

In any case the improvement step as well as the crossover operator heavily depend on the representation of the solution. Usually a simple representation requires more sophisticated recombination operators and vice versa. To overcome these difficulties Dorndorf and Pesch (1992a) proposed a completely different encoding scheme for the job shop scheduling problem. In this scheme, each individual of the population is a string of n − 1 entries $(p_1, p_2, ..., p_{n-1})$ where n − 1 is the number of operations in the underlying problem instance. The entry $p_i$ represents a rule from a set of priority rules [53]; this rule is then used to determine the i-th operation to be processed. Such a solution representation enables to use the simplest type of crossover as well as to incorporate problem specific knowledge, i.e. as Davis (1985) claimed "to examine the workings of a good deterministic program in that domain"; the resulting algorithm is competitive with special purpose heuristics. Putting things in a more general

---

[51]     Grefenstette / Baker (1989), Whitley (1992)

[52]     Some type of improvement heuristic may also be incorporated into the crossover operator (see Kolen and Pesch (1991)).

[53]     see Panwalkar and Iskander (1977)

framework, a genetic meta–strategy controls a sequence of local decisions [54]  in order to find best combinations.

We would like to conclude this section with some hints on available genetic algorithm software packages. The systems GENESIS and GENITOR developed by Grefenstette (1984, 1987a) and Whitley (1989) or Whitley / Kauth (1988), respectively, both, are public domain. They provide a reasonable framework and test environment in order to rapidly implement a genetic algorithm general problem solver. OOGA is a commercial package developed by L. Davis (1991), and finally SPLICER is a NASA tool developed by S.E. Bayer in 1991.

---

[54]    such as priority rules or even more complicated ones, see Chapter III

# II. THE TRAVELING SALESMAN PROBLEM

## 1. Introduction and Survey

The standard traveling salesman problem is to find a shortest tour from a home city to visit a given set of n cities exactly once and then return to the home city, provided that any pair of cities is connected and at a certain distance which is supposed to be symmetrical. Furthermore, it is supposed that the triangle inequality is satisfied which is automatically the case for problems in the plane. The problem arises in diverse areas such as one machine scheduling, flow shop scheduling (with no wait in process) [55], vehicle routing [56], computer wiring [57], some cutting problems [58], hole punching in metallic sheet manufacturing [59], printed circuit board production [60], flexible manufacturing [61], crystallography [62], in high energy physics, e.g. the collision of accelerated particles at a very high energy [63], etc. [64]. The problem is NP–complete (Garey / Johnson (1979)) also in the Euclidean case (Papadimitriou (1977)). Hundreds of articles have been written on that problem all of which we are not going to discuss here. An excellent and comprehensive survey of the important articles until about 1985 provides the book by Lawler et al. (1985).[65] The traveling salesman turned out to be just the right problem for polyhedral

---

[55] Pensini et al. (1991), Piehler (1960), Reddi / Ramamoorthy (1972)

[56] Laporte (1992b), Savelsbergh (1992)

[57] Lenstra / Rinnooy Kan (1975)

[58] Garfinkel (1977)

[59] Reinelt (1992)

[60] Grötschel et al. (1991), Chan / Mercier (1989)

[61] Tang / Denardo (1988a,b)

[62] Bland / Shallcross (1989)

[63] Beyer (1992)

[64] see Gensch (1978)

[65] A more introductory survey can be found in the book of Domschke (1989) and the recent paper by Laporte (1992a); a survey on early approaches to the traveling salesman problem is given in the book of Müller–Merbach (1970) and the articles of Bellmore / Nemhauser (1968) and Burkard (1979). In our survey we very briefly focus attention on recent (exact) methods based on polyhedral combinatorics, some historical papers, and certain local search heuristics, cf. Pesch (1993c).

methods which appeared to be superior to any previous exact solution method.
Dantzig et al. (1954) formulated the problem as a zero–one linear programming
model involving $O(n^2)$ variables and $O(2^n)$ linear constraints.[66] Various
researchers have proposed different formulations in order to reduce the
exponential number of constraints. For instance, Miller et al. (1960) and
Desrochers / Laporte (1991), Fox et al. (1980), and Claus (1984) suggest
formulations under a polynomial number of constraints however at the expense
of increasing the number of variables. Padberg / Sung (1991) address the issue if
the more compact formulations provide better characterizations of the traveling
salesman polytope than the standard formulation. They investigate the question
whether the traveling salesman problem can be solved more effectively with
linear programming based solution methods such as branch and bound,
Lagrangian relaxation or branch and cut under the more compact formulations.[67]
Only a few *special classes* of graphs are known for which the traveling salesman
problem is *polynomially solvable*, e.g. series–parallel graphs [68] as a generalization
of the graphs considered in Ratliff / Rosenthal (1983), TSP–perfect graphs [69],
some planar graphs [70], and in the special case of an Euclidean traveling salesman
problem cases where the points lie on a fixed number of nearly parallel lines in
the plane. For the last case Rote (1988, 1992) suggested a polynomial dynamic
programming approach.

There cannot be an efficient (polynomial time) approximation algorithm for the
traveling salesman problem in case of general distances that achieves some
bounded approximation ratio, unless P = NP (Sahni / Gonzalez (1976). For the
symmetric traveling salesman problem with distances satisfying the triangle
inequality, however, it is known that approximation algorithms can achieve a
ratio not worse than 3/2 of the optimum value. This bound is guarantied by the
Christofides (1976) heuristic. Improving over this performance is in general an
open problem, but under certain assumptions polynomial approximation
algorithms can be found that perform better. Papadimitriou / Yannakakis (1993)
considered such a special case where all distances are 1 or 2, a generalization of

---

[66]  See also Yannakakis (1991) who proved that a symmetric linear program for the
      traveling salesman problem — the vertices of the complete graph are treated the same
      way — requires exponential size.

[67]  See also the problem classification of Langevin et al. (1990).

[68]  Cornuéjols / Naddef / Pulleyblank (1985)

[69]  Fonlupt / Naddef (1992)

[70]  Chiba / Nishizeki (1989)

the Hamiltonian cycle problem and therefore still NP–complete.

A number of authors have suggested *branch and bound* methods. They are based on the elimination of subtours and lower bounds are obtained by the solution of the associated assignment problem as a relaxation [71]. Branch and bound algorithms applying 1–tree relaxation lower bounds [72] are described in a number of papers [73]. The 1–tree relaxation introduced by Held and Karp is much more effective than the assignment relaxation. Shapiro (1991) solved a traveling salesman problem as a sequence of shortest route problems using Lagrange relaxation.[74] New lower bounding and reduction procedures are described by Smith et al. (1990). They improved a 2–matching relaxation lower bound through a restricted Lagrangian 2–matching approach. Their outcomes are compared with lower bounds obtained from Lagrangian 1–tree relaxation.

The successful optimization of traveling salesman tours lies in the use of polyhedral theory [75], the roots of which date back to the seminal work by Dantzig et al. (1954, 1959).[76]   The description of the symmetric traveling salesman problem by linear inequalities has received a lot of attention and finding valid inequalities and facet–inducing inequalities for this polytope remains an interesting challenge.

In recent years a tremendous number of publications has dealt with *polyhedral descriptions* of the traveling salesman polytope, see the excellent survey by Grötschel / Padberg (1985). The idea is to define families of valid inequalities which define facets of the polytope, i.e. inequalities that are, loosely speaking, as tight as possible. By a theorem of Weyl (1935) we know that there exists a finite linear system of inequalities and facet defining for the symmetric traveling salesman polytope. However, it is very unlikely that such a system can be completely described and that it can be given by classes of inequalities on

---

[71]   Little et al. (1963), Bellmore / Mallone (1971), Garfinkel (1973), Smith et al. (1977), Carpaneto / Toth (1980), Balas / Christofides (1981), and Miller / Pekny (1991)

[72]   Kruskal (1956)

[73]   See Christofides (1970), Held / Karp (1970, 1971), Hansen / Krarup (1974), Smith / Thompson (1977), Volgenant / Jonker (1982, 1983), Jonker / Volgenant (1984), Gavish / Srikanth (1986), and Carpaneto et al. (1989); see also Volgenant (1990) for one of the best solution methods.

[74]   see also Houck et al. (1980)

[75]   Nemhauser / Wolsey (1988), Schrijver (1986)

[76]   see also Grötschel / Padberg (1977, 1979a,b), Grötschel (1977, 1980), and Padberg / Hong (1980)

problems which are only NP–descriptive [77]. Although there is nowadays only a partial description of the linear system known this incomplete characterization of the polytope can efficiently be used to solve large problem instances to optimality by polyhedral cutting plane algorithms. Grötschel / Holland (1991), Padberg / Rinaldi (1987, 1991) and Cook (1993) [78] report of the exact solution of 1000, 2392, and more than 4000 city instances, respectively. The integer linear programming formulation of the 2392 city instance contains 2,859,636 binary variables and an equally huge amount of constraints. Its solution required about 27 hours on a Cyber–205 computer. On an IBM 3090/600 with a better linear programming code and an improved algorithm the time is brought down to 2.6 hours. However, the 532–city instance took about twice that time on the same computer, indicating that the mere size of the problem is not the determining factor for running time.

According to the structure of the inequalities of a polyhedral description they are named as comb inequalities [79], which are generalizations of the 2–matching inequalities of Edmonds (1965), clique tree inequalities [80], path inequalities [81], path trees [82], and hyperstar inequalities [83] all of which are generalized in the binested inequalities of Naddef (1992) [84]. This is the largest family of valid inequalities known so far. Inequalities that do not belong to the above families are the ladder inequalities [85] and a few other inequalities which have only been shown to be facet defining for the 8– or 9–dimensional polytope [86].

The first nontrivial class of valid inequalities was dicovered by Chvàtal (1973) and later generalized by Grötschel / Padberg (1979a,b, 1985), the so–called comb inequalities. Again a generalization was discovered by Grötschel / Pulleyblank (1986), the clique tree inequalities. Comb inequalities have been generalized by path inequalities (Cornuéjols et al. (1985), Fleischmann (1988)). Clique trees have been generalized by path trees (Naddef / Rinaldi (1991)), by

---

[77]   Karp / Papadimitriou (1980)

[78]   private communication

[79]   Chvàtal (1973)

[80]   Grötschel / Pulleyblank (1986)

[81]   Cornuéjols et al. (1985)

[82]   Naddef / Rinaldi (1991)

[83]   Fleischmann (1988)

[84]   see also Naddef / Rinaldi (1992, 1993)

[85]   Boyd / Cunningham (1991)

[86]   Christof et al. (1991), Queyranne / Wang (1993)

hyperstar inequalities (Fleischmann (1988)), and by the bipartition inequalities (Boyd / Cunningham (1991)). In Grötschel / Padberg (1979a,b) the subtour elimination inequalities [87] are proven to be facet defining.

In order to run all inequalities in a cutting plane algorithm, respectively a *branch and cut* algorithm [88], for solving the traveling salesman problem exactly, the separation problem has to be solved. The separation problem for a class of linear inequalities which are valid for a polyhedron in $\mathbb{R}^n$ can be stated as follows: Given a point x in $\mathbb{R}^n$, find, if it exists, an inequality from the above class, which is violated by x. Solving efficiently the separation problem for a class of linear inequalities valid for the polyhedron is a crucial step in a polyhedral cutting plane algorithm [89]. Nothing is known on the separation problem for a large set of valid inequalities. Only for the comb inequalities both exact and heuristic algorithms have been studied [90]. The separation problem for the subtour elimination constraints can be solved in polynomial time [91].

The first solution method with respect to a tour minimization over the subtour polytope was the branch and bound technique based on 1–trees used by Held and Karp (1970, 1971). They use a Lagrangian relaxation of the traveling salesman problem to obtain a lower bound for the optimal solution. Solving this Lagrangian relaxation is equivalent to minimizing over the subtour polytope [92]. Christofides (1979) showed empirically that the Lagrangian bounds were approximately 99% of the optimal value for the traveling salesman problem on randomly generated problems.

Padberg / Rinaldi (1987) report on the exact solution of a 532–city traveling salesman problem involving the optimization over more 140,000 binary variables. It had been a new record for the traveling salesman problem after a previous exact solution of the 318–city involving more than 50,000 binary variables [93].

Because of its intractability for larger problem instances fast heuristics with a good empricial performance are indispensible [94]. There are nearly no approximation algorithms with a worst case bound independent on the problem

---

[87] Domschke (1989)

[88] Rayward–Smith / Clare (1986)

[89] Padberg / Rinaldi (1991), Fleischmann (1985)

[90] Lawler et al. (1985), Padberg / Rinaldi (1990)

[91] Padberg / Hong (1980)

[92] Boyd / Pulleyblank (1990)

[93] see Crowder / Padberg (1980)

[94] Stadler / Schnabl (1992)

size. So, most of the heuristics rely on the performance of a number of benchmark problems in order to show if they are competitive. Reinelt (1991) describes a traveling salesman problem library which provides the researchers with a broad set of test problems with various properties including bounds and references on computational runs. Roughly speaking heuristics for the traveling salesman problem can be classified as constructing heuristics which are building a tour from chains or which are increasing a tour by successivly insertion of new cities and improvement heuristics which are modifying an existing complete tour [95],[96]  Most of these heuristics are very special cases of an *ejection chain algorithm* [97], the origin of which is already indicated by Glover (1969, 1972) as well as Lin / Kernighan (1973). Criteria for increasing or inserting are for instance "nearest neighbour" to the latest inserted city, or the city yielding the least increase of the subtour, farthest or random insertion or addition of a city etc.[98] The improvement heuristics tentatively modify the current tour locally via a predescribed neighbourhood defining the feasible and usually simple modifications [99]. Most promising modifications (or moves to neighbour solutions) are accepted either deterministically or probabilistically. The r–opt algorithm [100] of which 2–opt is a special case [101]  are deterministic exchange algorithms. A neighbour is obtained while replacing r edges of the current tour by r new ones. Lin and Kernighan (1973) dynamically adapted r to a locally best value [102]. The Lin–Kernighan algorithm and its cousin the ejection chain procedures by Glover (1991a, 1992a,c) substantially outperform all r–opt based approaches, both in speed and in quality of the solution found. They are sophisticated successors of the 3–opt algorithm; they find better tours, and are orders of magnitude faster. But there exists no polynomial bound on the number of Lin–Kernighan iterations, a fact which in particular justifies limitation of the number of Lin–Kernighan iterations in Johnson's (1990a) implementation or in the genetic

---

[95]     see also Müller–Merbach (1974, 1981)

[96]     There are composite heuristics (Perttunen (1992).

[97]     as described in Glover (1991a, 1992a,c)

[98]     A comprehensive study is provided by Rosenkrantz et al. (1977) and Bentley (1992), further Webb (1971), Or (1976), and Norback / Love (1977).

[99]     Gendreau et al. (1992)

[100]     see Chapter I, Lin (1965), Müller–Merbach (1961), Christofides / Eilon (1972), Müller–Merbach (1970), and for implementation concerns see Johnson (1990a) or Margot (1992)

[101]     Croes (1958)

[102]     For the asymmetric case see Kanellakis / Papdimitriou (1980).

hybrid algorithm described in Kolen / Pesch (1991). In other words, there exist starting tours for which the Lin–Kernighan algorithm is required to take an exponential number of steps. It is said that the Lin–Kernighan neighbourhood structure is complete for the class PLS of all polynomial–time local search neighbourhood structures as defined in Johnson et al. (1988), Papadimitriou (1992), or Schäffer / Yannakakis (1990). For a PLS–complete neighbourhood structure local optima can be found in polynomial time only if they can be found in polynomial time for all neighbourhood structures in PLS. If also 2–opt and 3–opt neighbourhood structures are in PLS is currently unknown, but there is an r > 3 such that r–opt is PLS–complete. Golden et al. (1980) and Johnson (1990a) report on comprehensive computational comparisons of different deterministic local search improvement heuristics. Papadimitriou and Steiglitz (1977, 1978) conclude that local search algorithms are not always as effective as they seem to be on random or typical test problems. They constructed instances for which local search heuristics are ineffective motivated by the following observation. If an instance has a very large number of local optima (with respect to some neighbourhood structure), and a unique global optimum that is much better, then this is a difficult instance with respect to local search heuristics operating on the underlying neighbourhood structure. They constructed instances of the traveling salesman problem consisting of $n = 8 \cdot k$ vertices. These instances have a unique global optimum and the next best tour is arbitrarily bad, there exist $2^{k-1} \cdot (k-1)!$ of them, and they differ from the optimal tour in exactly $3k$ edges. This holds for instances without obeying the triangle inequality but if the triangle inequality holds then local optima can reach arbitrarily close the value of an optimal tour. Anyway, a form of limited r–opt where r is rather small cannot always overcome local optima of such difficult traveling salesman problems, unless r is chosen dynamically also allowing substantial deteriorations. That is the case for ejection chain methods, tabu search, or some probabilistic search procedures.

A number of authors realized that r–opt heuristics are more effective if the set of possible exchange or insert edges is limited to a preferred candidate list, for instance, to the k shortest edges incident to each vertex, or a certain number of minimal 1–trees of an instance [103]. It is based on the observation that a large number of the edges are superfluous and all edges of an optimal tour already belong to such a preferred candidate list. This is close to the idea of tabu search. Glover (1991a) mentioned that the quality of the solutions heavily depends on

---

[103]    cf. Pesch / Glover (1993) and Stewart (1987)

the amplification factor, i.e. the number of edges from a preferred candidate list compared to the number of all edges involved into a move. In order to get fast solutions for large problems clustering and partitioning methods, e.g. Delaunay triangulations and nearest neighbour partitioning, are employed to reduce the problem size and solve hierachically embedded traveling salesman problems [104].

Johnson (1990) reports on runs on 10,000 to 50,000 city problems and claims that Lin–Kernighan finds tours that are within 2% of optimality. For 10,000 cities the running time is little more than an hour on a VAX 8550 and for 100 cities, it is only three seconds. Simulated annealing (as proposed by Bounds (1987)) offers Lin–Kernighan little competition, neither does a neural net approach of Hopfield and Tank (1985) [105], a genetic approach of Brady (1985), or an approach based on the analogy with rubber bands [106]. As to the latter, an elastic rubber band is allowed to expand to touch all cities. The variables are the coordinates on the band, which vary with a gradient descent prescription on a cleverly chosen energy function. It can even not guaranty convergence to feasible local optima [107]. The biggest instance Hopfield and Tank (or Aarts and Korst (1989b)) considered was 30 cities, and in all their many neural net simulations on this instance, Hopfield and Tank (1985) never came within 17% of the optimal tour length. Durbin and Willshaw (1987) found an optimal solution to the 30–city instance. For 100 cities, however, they apparently performed no better than 3–opt, worse than Lin–Kernighan.

Simulated annealing has been applied to the traveling salesman problem by a number of authors [108]. It has its origin in the famous paper of Kirkpatrick et al. (1983) who ran problem instances of up to 400 cities. They showed that annealings appeared to provide reasonably good results for traveling salesman problems up to hundreds of cities. Catthoor et al. (1988) observed that their simulated annealing routine is better suited for clustered traveling salesman problems, which exhibit an inherent hierarchical partitioning, than for non–clustered ones. They conclude "that the presence of some form of inherent 'hierarchical' partitioning, which is not necessarily visible, represents an

[104]    Bentley (1992), Reinelt (1992), Bartholdi / Platzman (1988)

[105]    see also Wilson / Pawley (1988)

[106]    Durbin / Willshaw (1987), Durbin et al. (1989), Burke / Damany (1992)

[107]    Simmen (1991)

[108]    among others by Bonomi / Lutton (1984), Randelman / Grest (1986), Rossier et al. (1986), Golden / Skiscim (1986), Aarts et al. (1988), Malek et al. (1989), Nahar et al. (1989), Johnson (1990a), and Cheh et al. (1991)

extremely important property for 'suited' annealing algorithms" (Catthoor et al. (1988)).

The algorithm of Lin–Kernighan as the most famous representative of tabu search is the first application of a tabu search type idea to the traveling salesman problem.[109] In recent years the number of publications on genetic algorithms increased dramatically and it comes not as a surprise that the traveling salesman problem became one of the most favourite victims in order to test all thinkable kinds of parameter settings for genetic algorithms.[110] A number of people studied different parallel implementations of local search approaches such as simulated annealing, tabu search, elastic nets, genetic algorithms, as well as hybrids of those.[111]

## 2. Effective Genetic Local Search

Problems from combinatorial optimization are well within the scope of genetic algorithms, so it was inevitable that the traveling salesman eventually became a victim of GA activities. Early attempts closely followed the scheme of what Goldberg (1989a) called a simple genetic algorithm and were actually rather discouraging when compared with standard traveling salesman problem heuristics; for instance the experiments of Grefenstette et al. (1985) led to solutions as far as 25% from the optimum, in case of a 50–city traveling salesman problem.

The conclusion, however, that "genetic algorithms are not well suited for tuning structures which are very close to optimal solutions" (Grefenstette (1987b)) is a bit precipitate. As Suh and Van Gucht (1987) emphasize "it is ... essential if a competitive genetic algorithm is desired, to incorporate ... local improvement operators into the recombination step of a genetic algorithm". A resulting

---

[109]   Meanwhile many papers followed (see the survey of Laporte (1992a)), e.g. Malek et al. (1989), Glover (1991a, 1992a,b,c), Pesch / Glover (1993).

[110]   Some of the results are cited in succeeding sections and the following three papers are exceptional among the vast number of articles, Mathias / Whitley (1992), Whitley et al. (1991), Norman / Moscato (1990).

[111]   We refer for details to the papers of Chamberlain et al. (1988), Roussel–Ragot / Dreyfuss (1990), Peterson (1990), Malek et al. (1989), Allwright / Carpenter (1989), Mühlenbein / Kindermann (1989), Mühlenbein (1989, 1991, 1992), Hoffmeister (1991), Verhoeven et al. (1992), Colorni et al. (1992), Fox (1993a,b), Voß (1993), and Kindervater / Lenstra (1989) concerning complexity issues of parallel traveling salesman heuristics.

algorithm has then been called heuristic genetic algorithm, which in a way is a
pleonasm since every genetic algorithm incorporates – at least implicitly –
heuristic information about the problem. Since Lin and Kernighan (1973) the
prevalent local improvement operator is a 2–exchange. Jog et al. (1989) further
improve their genetic algorithm by incorporating Or–exchanges [112].

Equally essential is the careful selection of the binary recombination operator,
the crossover that entails heuristic information [113]. For the traveling salesman
problem, Mühlenbein et al. (1988) propose a binary recombination operator that
transplants a subpath of the first tour into the appropriately modified second
tour.

In this section [114] we address the question as to what extent concepts from
population genetics can improve the performance of classical local search
algorithms. For this we concentrate on a numerical study for the traveling
salesman problem in which the performance of genetic algorithms is compared
with that of more classical search algorithms such as multi–start local search,
simulated annealing and threshold accepting. The remainder of the section is
organized as follows. First we give a template of a general genetic local search
algorithm and show how it can be tailored to the traveling salesman problem.
Next we describe the setup of our numerical study and present the results that
were obtained. The section is concluded with a discussion of the potentials of
genetic local search algorithms.

It is desirable to put the previous approaches to the traveling salesman problem
using genetic algorithms into appropriate perspective. Every successful strategy
to produce near–optimal solutions necessarily relies upon some efficient iterative
heuristic, typically a local search technique. Well–known local search algorithms
for the traveling salesman problem are the 2–opt algorithm because of its
efficiency [115], the Lin–Kernighan algorithm because of its effectiveness [116], and
special variants of r–opt algorithms such as the Or–opt algorithm. All these
algorithms differ with respect to their neighbourhood structures. Any such
structure specifies a set of neighbouring solutions that are in some sense close to
that solution. The associated local improvement operator replaces a current

---

[112]    Or (1976)

[113]    see again Suh / Van Gucht (1987)

[114]    which is based on Ulder et al. (1991)

[115]    that means the running time of the algorithm

[116]    with respect to the quality of the results

solution by a neighbouring solution of better value if possible. Then local search – starting from some initial solution – proceeds by applying this operator until a local optimum is reached .

In practice, multi–start local search is used rather than a single run, i.e. the local search algorithm is repeated several times, retaining the best local optimum found. It is plausible that independent multiple runs of a local search algorithm generally will not constitute an effective procedure since, losely speaking, every individual solution has to find its own way to near–optimal regions. Cooperation and competition between individual solutions should certainly contribute to the overall performance of an algorithm. Several authors have therefore devised a collective organization of local search algorithms, drawing ideas from population genetics [117]. These approaches can be schematized as is shown in Table 1 [118]. This schema is just a template, requires further refinements in order to design a successful algorithm. We will now briefly mention a number of options in each step.

| | | |
|---|---|---|
| 1. | **Initialise.** | Construct an initial population of solutions. |
| 2. | **Improve.** | Use a local search algorithm to replace each solution in the current population by a better solution, e.g., a local optimum. |
| 3. | **Recombine.** | Extend the current population by adding solutions obtained by recombining two or more solutions in the current population. |
| 4. | **Improve.** | Use a local search algorithm to replace each offspring solution in the current population by a better solution, e.g., a local optimum. |
| 5. | **Select.** | Reduce the extended population to its original size according to prescribed selection rules. |
| 6. | **Evolve.** | Repeat steps 3 to 5 until some stopping criterion is met. |

Table 1.   Genetic local search.

As to the intialization, one would often generate random populations. At least in the case of the traveling salesman problem, there is a wealth of tour construction heuristics that could be used to make up an initial population of medium quality [119].

---

[117]   see e.g. Ackley (1987), Suh / Van Gucht (1987), Mühlenbein et al. (1987, 1988), Mühlenbein / Kindermann (1989), Mühlenbein (1989), Gorges–Schleuter (1989), Jog et al. (1989), Liepins et al. (1990), etc.

[118]   see Ulder et al. (1991)

[119]   see Lawler et al. (1985) or Johnson (1990a)

The local search algorithm of choice in the improvement step should simple be the best one available that meets given time capacity constraints. For the traveling salesman problem this is – beyond doubt – the heuristic due to Lin and Kernighan (1973). In case that severe time restrictions are imposed one can still use a truncated version of the local search algorithm such that it goes through only a small number of iterations.

Besides the carefully designed binary recombination operators one may also introduce operators of higher arities such as consensus operators, that fix edges common to most traveling salesman tours of the current population [120].

Selection of an individual is based on the fitness (possibly with respect to the overall fitness of the whole population). In order to emphasize an intensification strategy the value of the objective function matters; in contrast a diversification strategy might prefer those individuals which have as less edges in common with the other individuals of the population as possible. This is close to the idea of deviding the population into clusters of individuals as homogeneous as possible [121] and representing each cluster in the new population according to its size. A promising modification of recombination and selection on base of clusters involves the design of a population structure that defines proximity between positions of individuals, resulting in overlapping cliques (with respect to the traveling salesman problem defined on a complete graph, i.e. without loss we assume there is a direct connection between any two cities), called demes. Then recombination and selection is restricted to take place only among the individuals from each deme [122].

## 2.1  Numerical Results

We have tested two basic versions of genetic local search algorithms for the traveling salesman problem. Both algorithms depart from random populations of solutions, the population sizes being variable and dependent on the problem instances. The first one uses the 2–opt neighbourhood structure for the local search in the improvement step, so that the standard 2–opt heuristic is performed on each individual tour. The second one uses the more complicated Lin–Kernighan neighbourhood structure, thus yielding a pair of improvement operators, viz. the dynamical r–exchange and the additional 4–exchange as

---

[120]    cf. the reduction procedure of Lin and Kernighan (1973)

[121]    see Section IV.1

[122]    see Gorges–Schleuter (1989)

described in the original paper of Lin and Kernighan (1973). We adopted the
implementation due to Lageweg (CWI Amsterdam), disregarding the optimal
reduction part. In both algorithms crossover is done by taking two tours at
random in the current population and implanting a randomly chosen subpath of
one of the tours – containing at most one third of all cities – into the other one,
in essentially the same way as was proposed by Mühlenbein et al. (1988) and
Gorges–Schleuter (1989) and described in Chapter I.

Selection is executed by simply collecting the best (shortes) tours of the
extended population (the old population extended by the newly created
offspring). The algorithm stops when either all tours in the current population
have the same length or the length of the shortest tours did not improve within
five successive generations. We compared the performance of the above two
algorithms with that of the corresponding multi–start local search algorithms, as
well as with simulated annealing (SA) and its deterministic variant threshold
accepting (TA) due to Dueck and Scheuer (1990). Both SA and TA use the
2–opt neighbourhood structure. For Lin–Kernighan and TA the original
FORTRAN code was translated to PASCAL, (in a straightforward manner), so
that all six programs were in PASCAL. Moreover, care was taken to have
identical data structure and subroutines wherever possible. Our experimental
study is based on a comparison of the statistical averages of the tour lengths of
the final solutions obtained by applying the six algorithms five times each to
eight well–known instances of the traveling salesman problem, ranging from 48
up to 666 cities. For each instance, the algorithms are all allowed an almost
equal amount of running time. So we focus on effectiveness rather than
efficiency. The reference points are given by SA according to the cooling
schedule of Aarts / Van Laarhoven (1985), with the parameter value $\alpha = 1$. In
order to have the stopping criterion for the two genetic local search algorithms
fulfilled just within the time bound provided by each run of SA, we adjusted the
free parameter, the population size, accordingly. Indeed, the larger the
populations become, the more diversity we get and thus longer running times.
Table 2 gives the average deviations from the known optimal solutions. The
genetic versions Gen2–Opt and GenLK of 2–opt and Lin–Kernighan, respectively,
perform clearly better than their multiple–run companions. Moreover, GenLK is
superior to the other algorithms. In contrast to the 2–Opt and the LK variant,
the outcomes for SA and TA do not change considerably with the problem sizes;
the average deviations from an optimum are 2.4% for SA and 2.0% for TA over
all instances.

| Instance | t | SA | TA | Mult2-Opt | MultLK | Gen2-Opt | GenLK |
|----------|------|------|------|-----------|--------|----------|-------|
| GRO48 | 6 | 1.89 | 1.65 | 1.35 | 0 | 0.19 | 0 |
| TOM57 | 10 | 1.94 | 2.88 | 1.34 | 0 | 0.50 | 0 |
| EUR100 | 60 | 2.59 | 3.41 | 3.23 | 0 | 1.15 | 0 |
| GRO120 | 86 | 2.94 | 2.01 | 4.57 | 0.08 | 1.42 | 0.05 |
| LIN318 | 1600 | 2.37 | 1.27 | 6.35 | 0.37 | 2.02 | 0.13 |
| GRO442 | 4100 | 2.60 | 1.31 | 9.29 | 0.27 | 3.02 | 0.19 |
| GRO532 | 8600 | 2.77 | 1.79 | 8.34 | 0.37 | 2.99 | 0.17 |
| GRO666 | 17000 | 2.19 | 1.70 | 8.67 | 1.18 | 3.45 | 0.36 |

Legend to the table:

| | |
|---|---|
| $\bar{t}$ | Average running time in seconds on a VAX 8650 under VMS 5.1 |
| SA | Simulated annealing with 2-opt neighbourhoods |
| TA | Threshold accepting with 2-Opt neighbourhoods |
| Mult2-Opt | Multi-start local search with 2-Opt neighbourhoods |
| MultLK | Multi-start local search with Lin-Kernighan neighbourhoods |
| Gen2-Opt | Genetic local search with 2-Opt neighbourhood |
| GenLK | Genetic local search with Lin-Kernighan neighbourhoods |

| | |
|---|---|
| GRO48 | Instance with 48 cities due to Götschel |
| TOM57 | Instance with 57 cities due to Karg & Thompson |
| EUR100 | Instance with 100 cities due to Aarts & Van Laarhoven |
| GRO120 | Instance with 120 cities due to Grötschel |
| LIN318 | Instance with 318 cities due to Lin & Kernighan |
| GRO442 | Instance with 442 cities due to Grötschel |
| GRO532 | Instance with 532 cities due to Grötschel |
| GRO666 | Instance with 666 cities due to Grötschel |

Table 2.    Performance comparison of six local–search–based algorithms: average relative deviation from the optimal tour length in % for eight well–known instances of the traveling salesman problem.

Now, let us have a closer look at the numbers of iterations (trials or runs) that were needed to arrive at the solutions from Table 2. It is interesting to compare Mult2-Opt and Gen2-Opt – and the two LK versions – in this respect. See Table 3:   Gen2-Opt allows 3.2 to 9.2 more single runs of 2-Opt than Mult2-Opt – the corresponding numbers for the LK versions are 1.2 and 2.7 respectively – although the genetic variants still have to spend additional time on recombination and selection. It thus pays off when intermediate solutions are of higher quality. Note that the population sizes forced by the experimental design are quite small for GenLK the sizes range from 8 to 10 while the range is 14 to 56 in the case of Gen2-Opt.

| Instance | SA | TA | Mult2-Opt | MultLK | Gen2-Opt | GenLK |
|----------|-----|-----|-----------|--------|----------|-------|
| GRO48  | 89,112      | 180,000     | 40  | 19 | 140   | 24  |
| TOM57  | 145,236     | 300,000     | 41  | 14 | 140   | 32  |
| EUR100 | 668,250     | 1,800,000   | 67  | 31 | 216   | 40  |
| GRO120 | 1,135,260   | 2,400,000   | 64  | 30 | 216   | 48  |
| LIN318 | 15,221,706  | 45,000,000  | 99  | 37 | 390   | 100 |
| GRO442 | 34,988,499  | 120,000,000 | 108 | 67 | 720   | 100 |
| GRO532 | 61,442,010  | 246,000,000 | 116 | 77 | 954   | 120 |
| GRO666 | 112,494,060 | 450,000,000 | 122 | 45 | 1,120 | 100 |

Table 3:  Total numbers of trials (SA, TA), numbers of single runs (Mult2–Opt, MultLK), and population sizes times generation numbers (Gen2–Opt, GenLK), respectively, for the data in Table 2.

Similar experiments have been carried out for other time conditions, for example using $\alpha = 0.1$ in the cooling schedule of the simulated annealing algorithm. The obtained results show a similar behaviour as those obtained for $\alpha = 1$. Furthermore, we have run experiments where heavy time constraints were imposed. Under such circumstances truncations of 2–Opt or Lin–Kernighan were needed in the improvement step in order to achieve satisfactory results, see the next section.

## 2.2   Discussion

To investigate the potentials of genetic local search in combinatorial optimization, we applied two pertinent algorithms to the traveling salesman problem. The traveling salesman problem offers a great challenge since there exists a plethora of approximative algorithms for this problem that serve as good comparative standard. Among these, the algorithms based on exchange heuristics are the widest used ones, with the Lin–Kernighan heuristic as the uncontested champion; cf. Johnson (1990a). We decided to implement genetic local search – incorporating the 2-opt and Lin–Kernighan heuristics – in a straightforward way: only a single parameter, i.e. the population size, has to be chosen by the user in advance. It is therefore not necessary to first calibrate a whole bunch of parameters in order to get reasonably good solutions. Our experimental study indicates that genetic local search is consistently superior to the notorious multi–start local search. With larger problem sizes it becomes apparent that this simplistic strategy tends to strand at local optima of only moderate quality, so that the SA and TA algorithms eventually beat Gen2–Opt in this experiment.

Certainly, both Gen2–Opt and GenLK can be improved further by integrating
more subroutines inferred from population genetics [123].

Still GenLK seems to outperform the "champions" ASPARAGOS [124] and TA on
the traveling salesman problem. This is not really surprising since the much
poorer 2–opt heuristic – or a truncation thereof – is embedded in the latter two
algorithms. The additional use for Or–opt, or Or–exchanges, does not seem to
provide strinkingly better results either [125]. We have refrained from including
simulated annealing and threshold accepting endowed with dynamic r–exchange
instead in our experimental study since this would have required extra testing of
appropriate cooling schedules and threshold sequences. Anyway, genetic local
search should not be viewed as being opposed to SA or TA because elements of
these strategies can be implemented in genetic local search at the improvement
or selection step. A genetic organisation of some basic local search algorithms
would constitute only one out of many implementation devices that are
necessary in order to cope with very large problem sizes [126]. In the case of the
traveling salesman problem with thousands of cities, some hierarchical
structuring of the solution strategy seems to be unavoidable [127]. Then genetic
local search techniques may enter into any level of a hierarchically organized
strategy to further improve intermediate solutions.

# 3.  Bounded Genetic Local Search

In this section [128]  we will present some computational results comparing genetic
algorithms with local search algorithms for the traveling salesman problem given
a time bound on the running time of the algorithm. Local search as well as
genetic operators will be incorporated and we investigate whether the local
search phase should be carried on to optimality or whether it should be
truncated.

---

[123]    see Mühlenbein (1989)

[124]    Gorges–Schleuter (1989)

[125]    cf. Mühlenbein et al. (1988) and Jog et al. (1989)

[126]    See concluding remarks and announced research in Johnson (1990a).

[127]    For a pertinent approach commencing by a geometrical clustering of the cities see
        Reinelt (1992).

[128]    The results of this section can be found in Kolen / Pesch (1991).

## 3.1 Implementation Details and Numerical Results

Following the line of Section 2 we tested two basic versions of genetically guided local search algorithms for the traveling salesman problem, using either 2–exchanges or Lin–Kernighan exchanges. Either algorithm starts with a randomly generated population of tours. Then either algorithm performs $i^3$ 2–exchanges or Lin–Kernighan exchanges, respectively, applied to each solution of the i–th population. Only exchanges which improve the tour were counted and we always chose the best (most improving) 2–exchange as well as the ("locally best") Lin–Kernighan exchange. The Lin–Kernighan procedure is that one described in Lawler et al. (1985) without the special 4–exchange as used in the previous section and without the various refinements discussed in Lin and Kernighan (1973). We depart from a randomly chosen city which initiates one Lin–Kernighan exchange. A selection of a tour to undergo crossover is done according to its (unscaled) fitness, i.e. the probability of selecting a tour is inversly proportional to its length. Hence shorter tours are more often preferred. To let the 10% best tours of the old population survive crossover and to be included in the new population did not yield a substantial improvement in our experiments. Nor did the runs where each tour of the old population was chosen exactly twice for crossover. The crossover rate [129] is always kept to one. The Grefenstette–crossover [130] is applied with a probability which is linearly decreased in the number of generations from one to almost zero (exactly 0.01). The reason is that the Grefenstette–crossover is only useful to reduce the average length of the tours in a population until a certain average value is reached. The Mühlenbein–Gorges–Schleuter–crossover [131] is used with a probability increasing from zero to one. The mutation rate [132] is kept zero. The population is ranged from 8 to 20.[133] The following list describes the parameters as we chose them for all problem instances.

---

[129]  The crossover rate time 100 is the probability for a pair of individuals of the old population to undergo the crossover operator.

[130]  see Chapter I

[131]  see Chapter I

[132]  Population size times the mutation rate gives the number of offspring that are undergoing the mutation operator.

[133]  see also Grefenstette (1986)

- population sizes : 8, 14, and 20
- number of generations : 40, 35, and 30 (corresponding to the population sizes)
- crossover rate : 1.0
- crossover rate of the Grefenstette–crossover in the first generation : 1.0
- reduction of the crossover rate of the Grefenstette–crossover from the
  i–th to the (i+1)–th generation : 5/(number of generations)
- minimum crossover rate of the Grefenstette–crossover : 0.01
- crossover rate of the Mühlenbein–Gorges–Schleuter–crossover :
  1 – crossover rate for Grefenstette–crossover
- mutation rate : 0.0
- fitness evaluation : tour length without scaling
- number of runs : 5 runs on GRO120 and GRO442, and 3 runs on LIN318, GRO532, and
  GRO666
- number of improving (2– or Lin–Kernighan) exchanges in the i–th generation : $i^3$

We compared our results with those obtained when we applied the local search heuristic to each tour of the population up to suboptimum. Implementation was done in PASCAL. The runs were performed on a VAX 8650 under VMS. We were mainly interested in a comparison of performance when approaching the optimum. The stopping criteria are to reach the maximum number of generations or when the genetic algorithm with a complete local search on each tour of each population yields results exceeding the best result found in the "truncated version". We considered the following 5 problem instances:

GRO120    : instance with 120 cities due to Grötschel

LIN318    : instance with 318 cities due to Lin & Kernighan

GRO442    : instance with 442 cities due to Grötschel

GRO532    : instance with 532 cities due to Grötschel

GRO666    : instance with 666 cities due to Grötschel

To GRO120, LIN318, GRO442 and GRO532 we applied the (see Table 4)

(1) genetic algorithm with complete 2–opt (Complete 2–Opt)

(2) genetic algorithm with truncated 2–opt (Truncated 2–Opt)

(3) genetic algorithm with complete Lin–Kernighan (Complete LK)

(4) genetic algorithm with truncated Lin–Kernighan (Truncated LK)

For the GRO666 instance we renounced the 2–opt version.

The results are presented in Table 4. For each instance we report the average

time $\bar{t}$ (in seconds) needed over the five (or three) runs to come as close as $\overline{\%}$
percent to the optimum where $\overline{\%}$ is the average of the values found in the runs.
Among all the population sizes we only give the results for that population size
that gave the best performance. The size of the population is included in
brackets beside the name of the problem instance. While smaller problems yield
better results for larger population sizes, the best results for the large problems
are obtained with the smallest population size of 8. The reason might be that
the standard deviation of the fitness values in small populations of small
problem instances tends to become zero while this is not the case for the larger
problem instances. Probably some scaling of fitness values could slow down
convergence with small population sizes. In Table 4 the behavior of the runs for
the more important fitness changes is presented. Table 5 presents the 120 cities
problem in more detail. In both tables of each instance the first row of the
"complete" columns is the result when the first population is made locally
optimal. In all instances a truncated Lin–Kernighan version in a genetic
algorithm is worse than the complete version at the beginning, and worse at the
end, though it performs better in between. This tends also to be true for larger
problem instances and the 2–opt versions of the genetic local search algorithm.
Compared to the 2–opt version the Lin–Kernighan version did not behave well in
case of LIN318 and needed surprisingly a huge amount of time. It probably
points up the danger of conclusions from only a small number of problem
instances.

Under severe time constraints simulated annealing [134], threshold accepting [135]
or multi–start local searches generally did not perform better than the truncated
Lin–Kernighan algorithm incorporated into a genetic algorithm as can be seen by
comparing the results given in Table 4 with those presented in Section 2.

If tight time bounds are imposed Table 4 and 5 show that the tours in each
population should not be made locally optimal with respect to 2–opt or
Lin–Kernighan. To get good results it is sufficient to perform a limited number
of exchanges. They need less time than a complete local search algorithm applied
to each tour of the population. A genetic algorithm with a truncated 2–opt may
even be superior in some cases to one that incorporates Lin–Kernighan
exchanges. However, truncation only saves time for tour lengths up to about
1/4% (GRO210), 1% (GRO442), 3/2% (GRO532), 5/2% (GRO666) above the

---

[134]   Aarts and van Laarhoven (1985)

[135]   Moscato / Fontanari (1989), Dueck and Scheuer (1990)

optimum. Further progress can only be made if each solution is locally optimal with respect to the best available local search method, i.e. Lin–Kernighan (possibly in its more refined version) in the case of the traveling salesman problem.

We tried other experiments where we changed a few of the above mentioned parameters. Using a simple scaling where the scaled fitness of each tour equals the length of the tour minus the length of the optimal tour led to premature convergence on small problem sizes for our population sizes. A mutation rate of 0.001, where the mutation is simply the exchange of the positions of two randomly chosen cities in the tour, did not lead to some improvements. A lot of runs, in particular on the problems GRO120 and GRO442, with different crossover rates between 1.0 and 0.6 and different reductions for the Grefenstette–crossover (ranging from no reduction to only use of the Mühlenbein–Gorges– Schleuter–crossover) led to our choice of parameters. A genetic algorithm, without a local search heuristic, that uses only the Grefenstette–crossover leads to populations with average tour lengths of about 20% above the optimum (depending on the problem size). The tours become more and more similar. The Mühlenbein–Gorges–Schleuter–crossover does not decrease the average tour lengths; it maintains diversification of the tours in the population. Other increasing functions for the number of Lin–Kernighan exchanges like $[i^2/a]$, where a ranges from 0.1 to 10, usually gave better results for the GRO120 problem (see Table 5 for $i^2$). For the smallest problem instance, GRO120, runs on a population size of at least 50 and about 100 generations using quadratically increased Lin–Kernighan interchanges $(i^2/2)$ found the optimum in more than 95% of our runs.

| | Truncated LK | | Complete LK | | Truncated 2-Opt | | Complete 2-Opt | |
|---|---|---|---|---|---|---|---|---|
| | t | % | t | % | t | % | t | % |
| GRO120 (8) | 0.8 | 5.68 | 7.8 | 0.77 | | | | |
| | 1.5 | 3.73 | 9.1 | 0.68 | | | | |
| | 2.4 | 1.60 | 11.1 | 0.53 | | | | |
| | 5.0 | 1.09 | 13.1 | 0.39 | | | | |
| | 6.5 | 0.39 | 16.8 | 0.26 | | | | |
| | 7.7 | 0.24 | 24.4 | 0.19 | | | | |
| GRO120 (14) | | | | | 11.1 | 3.79 | 10.2 | 4.71 |
| | | | | | 13.7 | 3.14 | 12.5 | 2.41 |
| | | | | | 15.7 | 1.78 | 17.1 | 1.88 |
| | | | | | 19.1 | 1.49 | 19.3 | 1.54 |
| LIN318 (20) | 348.8 | 5.41 | 282.8 | 5.82 | 298.1 | 7.66 | 288.3 | 6.22 |
| | 415.0 | 4.50 | 1001.8 | 4.89 | 356.6 | 6.79 | 338.4 | 4.93 |
| | 838.5 | 4.17 | 1677.8 | 4.26 | 424.9 | 5.31 | 382.3 | 4.29 |
| | 1028.1 | 3.97 | 1822.7 | 3.92 | 626.7 | 3.93 | 508.1 | 3.83 |
| GRO442 (14) | 320.9 | 2.07 | 401.7 | 1.60 | 550.6 | 4.93 | 517.7 | 8.38 |
| | 726.1 | 1.95 | 730.6 | 1.26 | 663.5 | 4.00 | 638.0 | 4.14 |
| | 752.5 | 1.07 | 1192.9 | 1.07 | 718.2 | 2.74 | 969.1 | 3.81 |
| GRO532 (8) | 2124.9 | 2.40 | 854.8 | 2.32 | 1915.8 | 8.75 | 1954.2 | 7.63 |
| | 3127.7 | 2.19 | 5389.2 | 1.77 | 2136.7 | 7.15 | 2365.4 | 5.70 |
| | 4212.7 | 1.61 | 6199.6 | 1.57 | 3557.2 | 5.32 | 4121.0 | 5.07 |
| GRO666 (8) | 3089.0 | 2.95 | 4112.4 | 2.72 | | | | |
| | 4821.8 | 2.40 | 6104.8 | 2.39 | | | | |
| | 5822.1 | 2.31 | 7370.0 | 2.29 | | | | |

Table 4.   Results of a genetic local search algorithm under tight time bounds.

| Truncated LK | | Complete LK | |
|---|---|---|---|
| t | % | t | % |
| 2.2 | 5.67 | 17.9 | 0.52 |
| 4.1 | 3.30 | 23.2 | 0.28 |
| 7.0 | 2.21 | 25.4 | 0.24 |
| 10.4 | 1.31 | 31.5 | 0.18 |
| 14.4 | 0.34 | 34.2 | 0.11 |
| 19.0 | 0.27 | 47.7 | 0.05 |
| 26.3 | 0.14 | | |
| 29.9 | 0.12 | | |
| 33.6 | 0.08 | | |
| 45.3 | 0.06 | | |
| 55.5 | 0.02 | | |

Table 5.    GRO120 (20), $i^2$ Lin–Kernighan interchanges in population i.

## 3.2 Conclusions

The success of the genetic based algorithm for the traveling salesman problem is based on the following factors: A natural representation of a tour, a reasonable local search procedure (preferable the best local search procedure available for the problem) and a crossover which has the property that it changes the structure of a local optimal tour at the cost of a small increase in the length of the tour. The change of the structure is such that applying the local search procedure results in an improvement.

If tight time constraints are imposed a complete (in particular time consuming) local search algorithm has to be truncated. For smaller problems (less than 60 cities) the truncated versions were not superior to others. We easily reached the optimum. The truncated version can be guided by a metaheuristic as a genetic algorithm. In this way with a built–in heuristic genetic algorithms compete with simulated annealing. Genetic local search seems to perform better for larger problem instances, however for thousands of cities some clustering of the cities seems to be unavoidable and truncations of local search heuristics might lead to drastic time reductions.[136]

---

[136]    For more information on genetic organization of local search algorithms as well as comparison to tabu search see Johnson (1990).

# 4.  Variable Depth Search Based Learning

## 4.1  Ejection Chains

All neighbourhood structures that have been employed in local search procedures for the traveling salesman problem can be considered to be r–exchanges where r edges in the current tour are replaced by r new ones. This neighbourhood structure is a generalization of the 2–opt neighbourhood which is based on 2–exchanges. Even more, any r–exchange can be obtained by a sequence of (not necessarily improving) 2–exchanges. The 2–opt neighbourhood is connected, i.e. for any two solutions (including the optimal one) x and y there is a sequence of moves (2–exchanges) connecting x to y.

A 2–exchange is a minimal move with respect to the traveling salesman problem, i.e. there is no other neighbourhood structure affecting less than 2 edges such that its induced moves are feasible (or, equally, a move leads to a feasible solution). Actually the 2–exchange is a compound move of two 1–exchanges both of which solely are not feasible. The first one introduces an edge in the current tour and ejects another one leading to an infeasible solution (only the number of edges is maintained). The second one removes the infeasibility without affecting the two edges involved in the first 1–exchange. Generalizing this idea, there is a sequence or ejection chain of 1–exchanges each of which is incomplete, each subsequence leads to an infeasible solution, and a final 1–exchange produces a new feasible solution. The whole chain of 1–exchanges may be considered as a special type of r–exchange such that a sequence of moves (e.g. 1–exchanges) is compressed into a single compound move. The component moves (1–exchanges) carried forward from one level to the next are supposed to lead to a reference structure (a representation of an almost feasible solution) that is close to a feasible solution. In our case a final 1–exchange transforms the infeasible tour representing reference structure into a feasible tour. Actually after every subsequence of component moves such a final infeasibility transformation is performed temporarily. Hence we get a sequence of tours of which only the best one is memorized and finally replaces the current tour. Thus the neighbourhood becomes embedded level by level, in successivly larger neighbourhoods that represent more complex moves. The moves at each level cannot be obtained by a collection of independent and non–intersecting moves of previous levels. It is a

variable depth search [137] consisting of a simple neighbourhood structure at each
depth level which is composed to complex and powerful moves. The method can
be seen as a special case of a more general approach introduced by Glover
(1991a, 1992a,c). The basic idea is similar to the one used in tabu search, the
main difference being that the list of forbidden (tabu) moves grows dynamically
during a variable depth search iteration and is reset at the beginning of the next
iteration. The algorithm is outlined in Figure 1 with respect to 1–exchange
component moves; in our case f(x) is the objective function value (tour length)
of tour x; it might also be a scaled version of it or some other criterion which is
met by an optimal tour. With x(d) we denote the tour reference structure after
performing d component moves (1–exchanges); thus x = x(0).

*begin*
  Start with an initial solution x*.
  x := x*;
  Let s be any city in x.
  k* := s;
  *repeat*

  T1 := ∅;  T2 := ∅;  {T1, T2 are tabu lists}
  d := 0;  {d is the current search depth}

  *while* there are non–tabu edges in x(d) *do*

  i := k*;
  d:= d + 1;

  Find the best component move, i.e. the edge pair $[i,j^*]$, $[j^*,k^*]$ for
  which the gain $g(i,j^*,k^*) = \max \{c_{ij} - c_{jk} \mid [i,j]$ is not an edge in
  x(d–1) and $[j,k]$ is an edge in x(d–1); $[i,j]$ is not in T1; $[j,k]$ is not in
  T2}; {Note that $g(i,j^*,k^*)$ can be negative.}

  Perform this move, i.e. introduce edge $(i,j^*)$ and remove edge $[j^*,k^*]$ thus
  obtaining x(d) as a new reference structure at search depth d;

  T1 := T1 $\cup \{[i,j^*]\}$;      T2 := T2 $\cup \{[j^*,k^*]\}$;

  Let s' be a neighbour of s in x such that the component move which
  ejects the edge $[s',s]$ and inserts the edge $[k^*,s']$ yields a feasible tour
  x*(d).

  Let d* denote the search depth at which the best solution x*(d*) with
  $f(x^*(d^*)) = \min \{f(x^*(d)) \mid 0 < d \leq n\}$ has been found;

  *if* d* > 0  *then begin* x* := x*(d*); x := x* *end*

  *until* d* = 0;
*end*

Figure 1.   An ejection chain procedure based on 1–exchange component moves.

---

[137]   cf. Papadimitriou / Steiglitz (1982)

The above procedure describes in its inner *repeat ... until* loop one iteration of an ejection chain search. The *while ... do* describes one component move. Starting with an initially best solution x*(0), the procedure looks ahead for a certain number of component moves and then sets the new currently best solution (or solution reference) x*(d) for the next ejection chain iteration to the best solution found in the look–ahead phase at depth d*. The iterations are repeated as long as an improvement is possible. The maximum look–ahead depth is reached if all edges in the current solution x are set tabu. The step leading from a solution x to a new solution consists of a varying number d* of component moves in the neighbourhood, hence the name variable depth search where a complex compound move results from a sequence of compressed simpler moves each of which introduces and ejects an edge. The algorithm can escape local optima because moves with negative gain are possible. Continuously growing tabu lists avoid cycling of the search procedure. As an extension of the algorithm, the whole *repeat ... until* part could easily be embedded in yet another control loop (not shown here) leading to a multi–level (parallel) search algorithm, see Glover (1992a). Obviously, the algorithm of Lin and Kernighan (1973) is a special instance of the above procedure, where the neighbour city s' of the starting city s is always the same. This is not necessarily the case for the procedure of Figure 1. Both neighbours of s in x may be considered as finally "visited" city s'. Figure 2 illustrates the procedure for two 1–exchanges. The labels of the cities describe the execution. Starting with city 1 there is an edge added to city 2 which ejects the existing edge [2,3]. After a feasibility and improvement check from 3 to a neighbour of 1 a new edge connecting 3 to 4 is introduced (both cases are indicated, either 4 lies on the path part of the stem–and–cycle reference structure, or 4 lies on the cycle part). Edge [4,5] is ejected and a final 1–exchange connects 5 to one of the two neighbours of s in x. Generalizations with respect to the tabu lists' management are straightforward. Consider, for instance, that each edge introduced into the current reference structure is allowed to be removed again. Thus it is not immediately set tabu. Only, if this edge is introduced a second time it is set tabu and may not be removed once more. If we allow only one edge of the tabu list T1 to be removed exactly once then we can get the Lin / Kernighan's exceptional 4–exchange by the procedure of Figure 1. This is illustrated in Figure 3 where an edge [3,4] inserted in an earlier step is removed again (see labeling 8, 9 in Figure 3). Finally the compound move leading to an infeasibility is to replace edges [2,3], [4,5], and [6,7] by the new edges [1,2], [5,6], and [7,4/8]. A final component move guarantees a feasible outcome. Thus the procedure of Figure 1 can easily be extended to include also the limited 4–opt neighbourhood which is separately treated in Lin / Kernighan (1973).

The procedure of Figure 1 is not at all limited to 1–exchange component moves or a 2–opt based neighbourhood structure. It can easily be generalized to a large number of different neighbourhood definitions the component moves of which may include or eject edges, vertices (of the tour representing graph), or even both, alternating.

For instance, consider the case of a vertex exchange move based neighbourhood which is defined as an exchange of the positions of two vertices in the tour. Note, this leads to a limited case of a 4–opt. A component move consists of replacing a vertex, j say, by a vertex i such that the neighbours j' and j" of j become new neighbours of i. The reference structure consists of a path connecting the former neighbours i' and i" of i via the edges [j',i] and [i,j"], and the isolated vertex j. A check on improvement and feasibility makes the vertex j adjacent to i' and i". The chain continuous vertex j ejects a vertex, say k, from its current tour position, i.e. [k',j] and [j,k"] become new edges within the modified reference structure and vertex k is the newly isolated vertex. The procedure continuous like that until in the last step of an ejection chain iteration the reference structure is transformed into a feasible tour. That means, that the finally isolated vertex becomes a new neighbour of vertices i' and i". Figure 4 illustrates an ejection chain based on vertex exchange moves. Eight new edges are included into the new solution, namly [1',2], [2,1"], [2',3], [3,2"], [3',4], [4,3"], [4',1], and [1,4"]. Eight edges of the old tour are deleted, namely [1',1], [1,1"], [2',2], [2,2"], [3',3], [3,3"], [4',4], and [4,4"].

There are a number of possible modifications of ejection chains even with respect to one particular neighbourhood structure. The interested reader is referred to Glover (1991a, 1992a,c). One may also think of modifications with respect to the component moves. Each component move might be considered as one step in a simple local search procedure which is not necessarily based on feasible solutions but on reference structures closely related to feasible solutions. A component move is performed greedily avoiding tabu moves. A probabilistic acceptance (as in simulated annealing) is also thinkable. Different tabu lists' management, such as including chains of an arbitrary length or some asperation criteria (i.e. a tabu move is allowed to be performed if it leads to a new suboptimal reference structure (or solution)), present ejection chains as a much more general search strategy which may contain in each iteration a limited simulated annealing or tabu search procedure. Furthermore, ejection chains can easily be included into a genetic framework serving as improvement procedure.

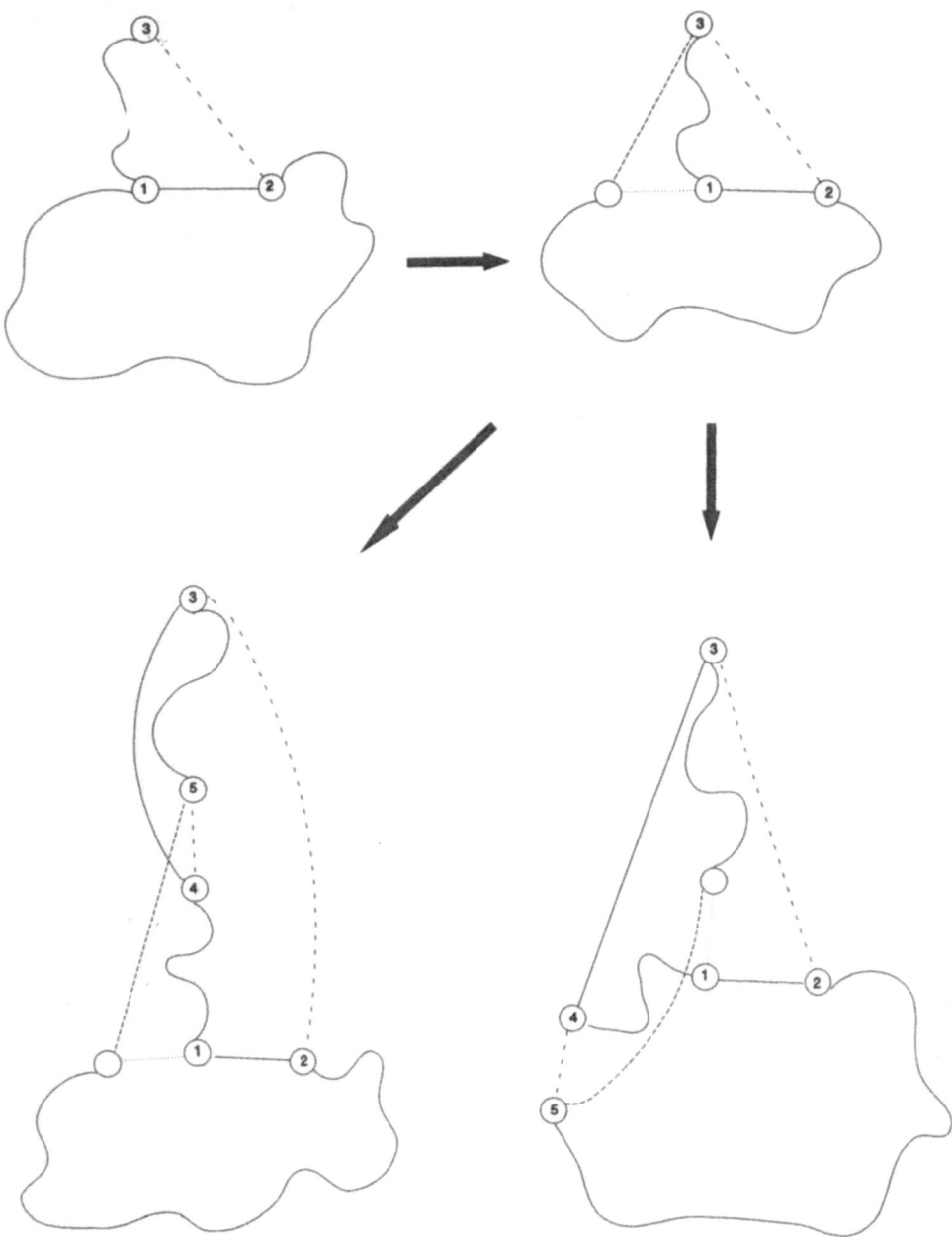

Figure 2.          An ejection chain based on 1—exchanges and a
                   stem—and—cycle reference structure.

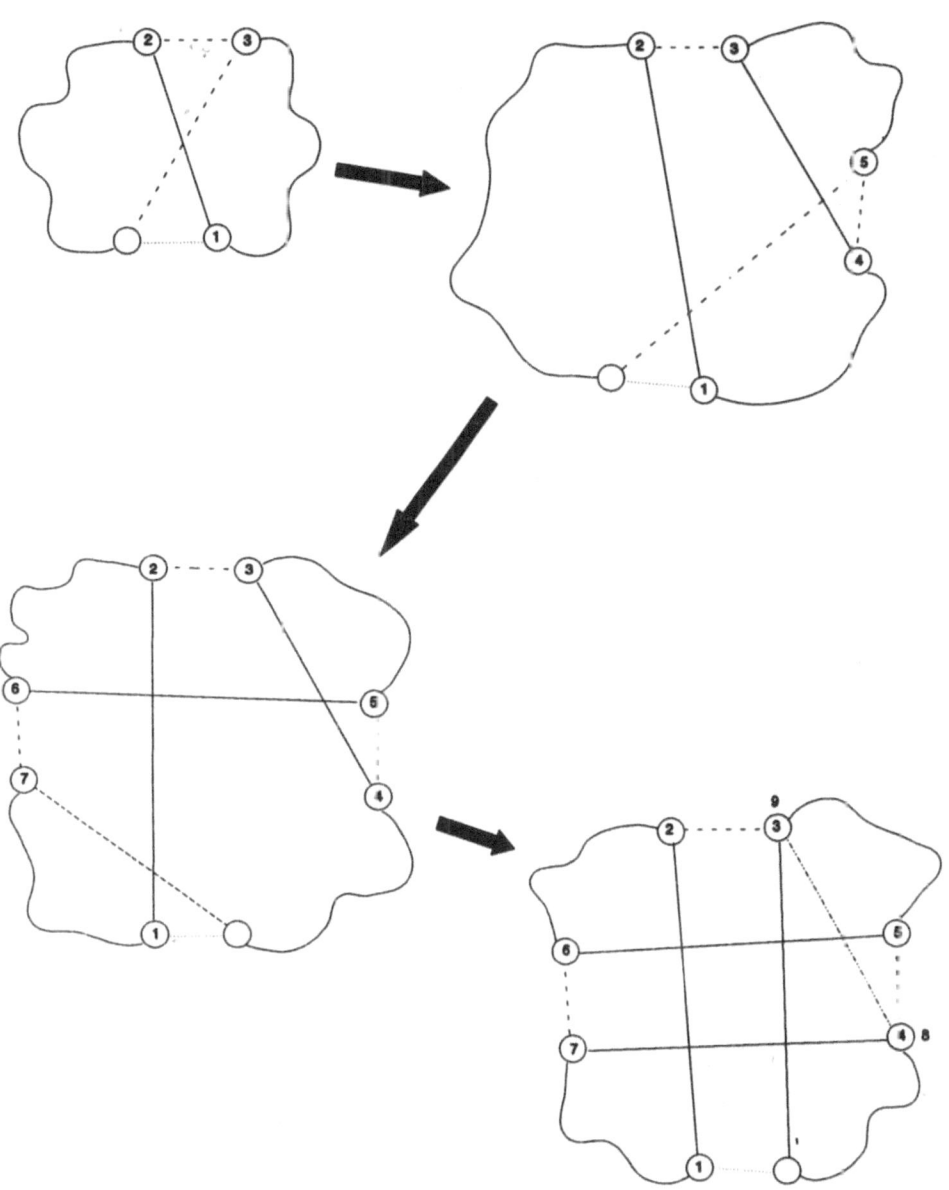

Figure 3.   A limited 4–opt case.

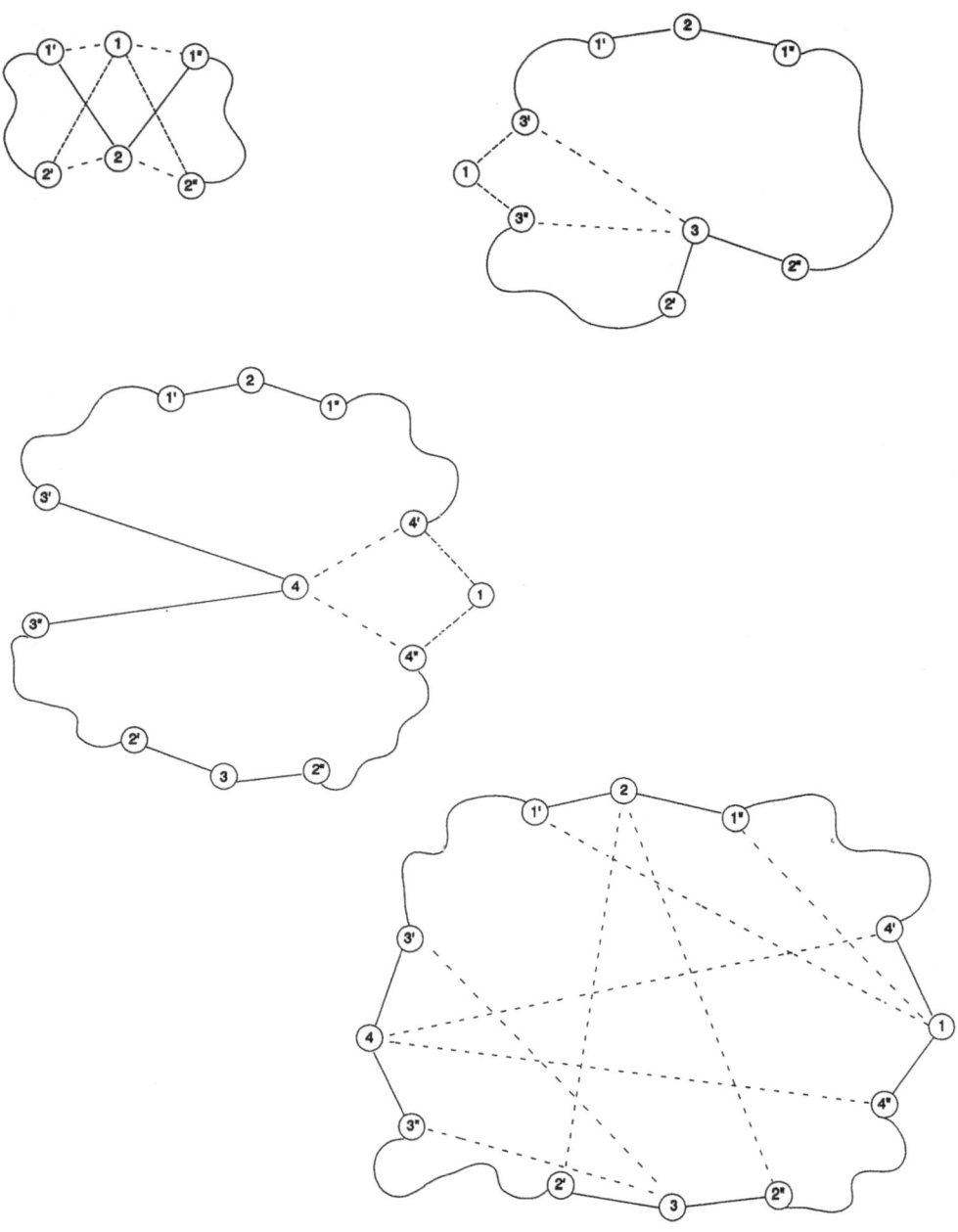

Figure 4.    An ejection chain based on vertex exchange moves.

## 4.2   Computational Results

The variable depth search or ejection chain procedures described are implemented in PASCAL and have been tested on a VAX 8650 under VMS. Hence the results can easily be compared to the genetic algorithms as well as to the simulated annealing results from the previous sections. As test beds we used the same instances as in the previous section, ranging from 48 to 666 cities, see Table 6. Each algorithm has been run five times on each problem instance, respecting a running time limit as provided in Section 2 of this chapter for the particular instances. Table 6 presents the results as average percentages over the average number of search processes or iterations (indicated in brackets) that could be performed over the five runs. The columns Mult2–Opt and MultLK are the results of Table 2 from this chapter which have been obtained for a 2–opt local search and the Lin–Kernighan procedure. Column MultLK2 contains results of a modified Lin–Kernighan procedure. It is based on the observation that optimal or near–optimal solutions often can be constructed for traveling salesman problems by limiting consideration to a small number of shortest edges incident to each vertex (for instance 5 to 20 depending on the problem size). Hence moves are performed on a neighbourhood that is restricted in a certain sense, i.e. the set of candidate moves is reduced to the preferred attribute candidate list. Only those move candidates are considered that meet a preferred attribute, e.g. including some of the shortest edges. Such moves can be classified by their amplification factor which is the number of edges added by a move devided by the number of edges that belong to a "k–shortest" category [138]. The Lin–Kernighan modification as described in column MultLK2 of Table 6 is as follows.

(1)   Randomly divide  the vertex set in subsets (not necessarily non–overlapping) of 50 to 100 vertices (respectively cities). Consider $n/10$ to $n/5$ subsets where n is number of the cities in the underlying problem instance. (Size and number of the subsets is chosen randomly.)

(2)   Consider each subset of cities as a traveling salesman problem which is solved using the algorithm of Lin and Kernighan (1973).

(3)   The edges of all tours obtained in (2) define a preferred attribute candidate list. Solve the complete traveling salesman problem on these preferred edges using the algorithm from Lin / Kernighan, i.e. in each

---

[138]   Any edge which belongs to the k shortest edges incident to a city belongs to this category.

step of a Lin–Kernighan iteration an edge is allowed to be introduced
into the currently modified tour only if it belongs to the set of
preferred edges. In the final step of a Lin–Kernighan iteration (the step
that achieves a feasible tour) an arbitrary edge may be chosen to be
introduced into the newly generated tour.

Table 6 shows that MultLK2 did not create a subproblem for the smallest
problem instance consisting of 48 cities. In all remaining cases the number of
runs of MultLK2 starting from a randomly generated initial solution is, in
particular for the larger instances, substantially higher than only for MultLK.
This is an effect of the restricted number of tests for the component moves that
have to be performed during the search process, because the number of edges is
enormously reduced. The quality of the outcome has increased which is not only
an effect of inheriting edges from the suboptimal solutions of the subproblems.
We assume that a main reason for the improved outcome is the diversification
effect of the preferred edge candidate list. The restricted number of edges
available for the local (infeasible) moves within a Lin–Kernighan iteration forces
to accept temporarily much stronger deterioration steps. The overall
achievement in a Lin–Kernighan iteration will compensate for those
deteriorations. The last column MultEC describes the results obtained with an
edge based ejection chain procedure as it was described above (hence a
generalization of Lin and Kernighan's ejection chain). At most one edge was
allowed to be removed from the tabu list T1 after it had been inserted earlier.
We can see that the number of runs reduces to almost half the number of
MultLK or MultLK2 runs. This might be an effect of the increased search space
resulting from the increased number of possible local exchange steps. The tour
lengths are the best obtained from all possible methods. Even for the biggest
instance with 666 cities the worst tour length is less than half percent from the
optimal length. The two methods of ejection chains were the best we tried.
Different methods based on vertex ejection chains (see above) or subpaths (of
length at 3) ejection chains, or ejection chains on reference structures pretty far
from representing a feasible tour performed much worse. For instance, vertex
based ejection chains, even restricted to some preferred attribute candidate list,
could not reach results better than 10% above the optimal tour length. A major
reason for that behaviour might be that edge ejection chains in the sense of
Figure 1 consist of minimal steps, that means, all local 1–exchanges are minimal
modifications of the reference structure. This is not the case for vertex based
ejection chains. Hence edge based ejection chains are more suitable for fine

tuning.

| Instance | t | Mult2-Opt | | MultLK | | MultLK2 | | MultEC | |
|----------|-----|-----------|-------|--------|------|---------|------|--------|------|
| GRO48 | 6 | 1.35 | (40) | 0 | (19) | 0 | (19) | 0 | (16) |
| TOM57 | 10 | 1.34 | (41) | 0 | (14) | 0 | (17) | 0 | (14) |
| EUR100 | 60 | 3.23 | (67) | 0 | (31) | 0 | (33) | 0 | (24) |
| GRO120 | 86 | 4.57 | (64) | 0.08 | (30) | 0.02 | (36) | 0.05 | (27) |
| LIN318 | 1600 | 6.35 | (99) | 0.37 | (37) | 0.21 | (43) | 0.11 | (29) |
| GRO442 | 4100 | 9.29 | (108) | 0.27 | (67) | 0.09 | (84) | 0.04 | (48) |
| GRO532 | 8600 | 8.34 | (116) | 0.37 | (77) | 0.28 | (85) | 0.10 | (44) |
| GRO666 | 17000 | 8.67 | (122) | 1.18 | (45) | 0.82 | (81) | 0.41 | (21) |

Table 6.        Performance of the ejection chain approaches, including the number The legend of this table is presented in Table 2 of this chapter.

## 4.3   Conclusions

We have presented an ejection chain algorithm for the traveling salesman problem that substantially performs better than any other type of local search procedure. Ejection chains contain a large potential of different types of local search learning structures. They can be thought of either local improvement procedures in order to accelerate, for instance, genetic algorithms, or search strategies guiding the local exchange steps in each iteration. Each iteration can be implemented as a tabu search or simulated annealing iteration. The real power of ejection chains results from their general applicability. We can think of chains based on edges, vertices, subpaths, alternations of edges and vertices, etc. which depend on the solution representing reference structure. Considering reference structures of infeasible solutions provides a much greater flexibility for the definition of moves with respect to their neighbourhood. They are in particular useful for the development of powerful compound moves which can exploit a search space effectively of a magnitude an order larger than their neighbourhood search space. The real potential of ejection chains is largely unknown and remains to be examined from a theoretical and application point of view.

# III. JOB SHOP SCHEDULING


## 1. Introduction – Conventional and New Solution Techniques

A job shop consists of a set of different machines (like lathes, milling machines, drills etc.) that perform operations on jobs. Each job has a specified processing order through the machines, i.e. a job is composed of an ordered list of operations each of which is determined by the machine required and the processing time on it. There are several constraints on jobs and machines: (i) There are *no precedence constraints* among operations of different jobs; (ii) operations cannot be interrupted (*non−preemption*) and each machine can handle only one job at a time.

Consequently, each job can be performed on only one machine at a time. The operation sequences on the machines are unknown and have to be determined in order to *minimize the makespan*, i.e. the time required to complete all jobs. It is well known that the problem is NP−hard [139] and belongs to the most intractable problems considered [140].

A huge amount of literature on machine scheduling including scheduling of job shops, has been published within the last 30 years, among others the book by Conway et al. (1967), who wrote the first book on scheduling theory [141]. A variety of scheduling rules for certain types of job shops have been considered in an enormous number of publications, see the excellent surveys, also covering complexity issues and mathematical programming formulations [142]. Among the different problem formulations [143] we have adopted that one presented by

---

[139]    see Lenstra / Rinnooy Kan (1979)

[140]    Lawler et al. (1989)

[141]    Ashour (1972), Baker (1974), Ecker (1977), Brucker (1981), the dissertations Rinnooy Kan (1976), Lenstra (1977), Van de Velde (1991), and the edited books by Muth / Thompson (1963) and Coffman (1976). A broader view on production and operation scheduling is provided in the books by Domschke et al. (1993), Blazewicz et al. (1993), Kistner / Steven (1990), Hax / Candea (1984), Johnson / Montgomery (1974)

[142]    by Blazewicz et al. (1991, 1993), Lawler et al. (1989, 1982), Rodammer / White (1988), Blazewicz (1987), Lageweg et al. (1982), Johnson (1983), Lawler (1982), Graham et al. (1978, 1979), Salvador (1978), Day / Hottenstein (1970), Bakshi / Arora (1969), Elmaghraby (1968), Moore / Wilson (1967), and Mellor (1966)

[143]    Those by Bowman (1959), Wagner (1959), and the mixed integer formulation of Manne (1960) are the first ones published; see also Fisher (1973), Fisher et al. (1983), Blazewicz et al. (1991), Brah et al. (1991) as well as extensions in Stafford (1988).

Adams et al. (1988). Let $V = \{0,1, \ldots ,n\}$ denote the set of operations where 0 and n are considered as dummy operations "start" (the first operation of all jobs) and "end" (the last operation of all jobs), respectively. Let M denote the set of machines; A is the set of pairs of operations constrained by the precedence relations for each job. For each machine k, the set $E_k$ describes the set of all pairs of operations to be performed on machine k, i. e. operations which cannot overlap (cf. (ii)). For each operation i, its processing time $p_i$ is fixed, and the earliest possible process start of i is $t_i$ , a variable that has to be determined during the optimization. Hence, the job shop scheduling problem can be modelled as

$$\min t_n$$
$$\text{s. t.}$$

$$\begin{array}{lll}
(1) & t_j - t_i \geq p_i & (i,j) \in A, \\
(2) & t_j - t_i \geq p_i \text{ or } t_i - t_j \geq p_j & \{i,j\} \in E_k , k \in M, \\
(3) & t_i \geq 0 & i \in V .
\end{array}$$

Restrictions (1) ensure that the processing sequence of operations in each job corresponds to the predetermined order. Constraints (2) demand that there is only one job on each machine at a time, and (3) requires an earliest starting time for the jobs. Any feasible solution to the constraints (1), (2), and (3) is called a schedule.

An illuminating problem representation is the disjunctive graph model due to Roy and Sussmann (1964). It has mostly replaced the solution representation by Gantt charts as described in Gantt (1919), Clark (1922), and Porter (1968). Figure 5 from Chapter I illustrates the disjunctive graph for a problem with 3 jobs and 8 operations on 3 machines. The job shop scheduling problem is to find an order of the operations on each machine, i.e. to select one arc among all opposite directed arc pairs such that the resulting graph is acyclic (i.e. there are no precedence conflicts between operations) and that the length of the maximum weight path between the start and end vertex is minimal. The length of a maximum weight (or longest) path determines the makespan.[144]

The minimum makespan problem of job shop scheduling is a classical combinatorial optimization problem that has received considerable attention in the literature. It belongs to the most intractable problems considered. Only a

---

[144] Extensions of the disjunctive graph representation including additional job shop constraints are discussed in White / Rogers (1990), see also Bruns / Appelrath (1991).

few particular instances are efficiently solvable:  (i) *scheduling two jobs* by the graphical method as described in Brucker (1988) and first introduced by Akers (1956); in general this idea can be used to compute good lower bounds sometimes superior to the one–machine bounds of Carlier (1982) [145];  (ii) the *two machine flow shop* case, i.e. the machine sequences of all jobs are the same (Johnson (1954)) [146] [147];  (iii) the *two machine job shop* problem where each job consists of at most two operations (Jackson (1956));  (iv) the two machine job shop case with unit processing times (Hefetz / Adiri (1982)). Slight modifications turn out to be difficult. The two machine job shop problem where each job consists of at most three operations, the three machine job shop problem where each job consists of at most two operations are NP–hard [148].  The job shop problems with two and three machines and operation  processing times equal to 1 or 2, and equal to 1, respectively, are NP–hard even in the case of preemption [149].  Permutation flow shop problems, i.e. a schedule with identical job processing order on all machines, are in some cases polynomially solvable [150], or admit to develop good lower bounds [151]  or dominance rules [152]  applied in early enumeration approaches [153].  Although it is common practice to focus attention on permutation schedules that can be costly with respect to makespan minimization. Potts et al. (1991) showed that there are instances for which the objective value of the optimal permutation schedule is much worse (in a factor of the number of machines) than that of the true optimal flow shop schedule. The current champions for solving permutation flow shops are the fast insertion method of Nawaz et al. (1983), cf. Turner / Booth (1987), the extension of Palmer's heuristic (1965) by Hundal / Rajgopal (1988), the effective simulated

---

[145]    see Brucker / Jurisch (1993)

[146]    see Gonzalez / Sahni (1978)

[147]    The idea of Johnson can be extended to a special case of three machine flow shop scheduling (Burns / Rooker (1978) and Jackson (1956)), however in general the three machine case is NP–complete (Röck (1984) and Sotskov (1991)) even in the case of preemption, i.e. if operations need not be continuously processed on a machine (Gonzalez / Sahni (1978)). Flow shop scheduling on two machines with operations release dates is NP–hard (Garey / Johnson / Sethi (1976)) also in the preemption case (Cho / Sahni (1981)).

[148]    see Lenstra et al. (1977) and Gonzalez / Sahni (1978)

[149]    see Lenstra / Rinnooy Kan (1979)

[150]    Rinnooy Kan (1976), Monma / Rinnooy Kan (1983)

[151]    Lageweg et al. (1978), Ignall / Schrage (1965)

[152]    McMahon (1969), Szwarc (1971, 1973)

[153]    Potts (1980a), Gupta /Reddi (1978), and Szwarc (1978)

annealing algorithm of Osman / Potts (1989), and the heuristical minimization of gaps between successive operations by Ho / Chang (1991). The latter consider a neighbour of a permutation schedule as a schedule that can be obtained by moving a single job to another position.

Widmer / Hertz (1989) and Taillard (1990) solved the flow shop sequencing problem using tabu search. Neigbours are defined mainly as in the traveling salesman case: exchanging the order of processing for two operations. Computational results are compared to a number of previously published procedures. They did not make use of features such as different strategies of intensification and diversification of the search with respect to the dynamics of the tabu list in order to reinforce attributes of attractive solutions and on driving the search into new regions, as demonstrated by Hübscher / Glover (1992) for parallel machine scheduling.

The history of the job shop scheduling problem, starting more than 30 years ago, is also the history of a well known benchmark problem consisting of 10 jobs and 10 machines and introduced by Fisher and Thompson in 1963. This particular instance of a 10 job 10 machine problem opposed its solution for 25 years leading to a competition among researchers for the most powerful solution procedure. Since then branch and bound procedures have received substantial attention from numerous researchers.[154]        A long time the algorithm of McMahon / Florian (1975) was the best exact solution method. Instead of using the worse bounds of Charlton / Death (1970) they combined the bounds for the one machine scheduling problem with operation release dates and the objective function to minimize maximum lateness with the enumeration of active schedules [155]   among which are also optimal ones. An alternative approach whereby at each stage one disjunctive arc of some crucial pair is selected leads

---

[154]   Early work was performed by Brooks and White (1965), followed by Greenberg (1968), who's method was based on Manne's integer programming formulation, Balas (1969), Charlton / Death (1970), Florian et al. (1971), Ashour et al. (1974), Ashour / Hiremath (1973), and Fisher (1973) who obtained lower bounds by the use of Lagrange multipliers.

[155]   see Giffer / Thompson (1960)

to a computationally inferior method [156],[157]

During the last ten years substantial algorithmic improvements were achieved and accurately reflected by the stepwise optimum approach for the notorious 10–job 10–machine problem. Lageweg et al. (1982) applied computationally costly surrogate duality relaxations, weighting and aggregating into a single constraint, either machine capacity constraints or job operation precedence constraints. A first attempt to obtain bounds by polyhedral techniques is from Balas (1985). The neighbourhood structure used in some recent local search algorithms is also mainly employed as branching structure in the exact method of Barker / McMahon (1985). They rearrange operations on a longest path if the operations use the same machine. Lawler et al. (1989) report that, with respect to the famous 10×10 problem "Lageweg (1984) found a schedule of 930, without proving optimality; he also computed a number of multi–machine lower bounds, ranging from a three–machine bound of 874 to a six–machine bound of 907". So he was the first who found an optimal solution. Optimality of a schedule of length 930 was first proven by Carlier and Pinson (1989). Their algorithm is based on bounds obtained for the one machine problems with precedence constraints, release dates and allowed preemption. This problem is polynomially solvable. Additionally, they used several simple but effective inference rules on operation subsets.

Currently the job shop champions among the exact methods are besides the excellent branch and cut implementation of Applegate and Cook (1991), the branch and bound algorithms of Carlier / Pinson (1990), Brucker / Jurisch / Sievers (1991, 1992), and Brucker / Jurisch / Krämer (1992), see also Pinson (1988, 1990). The power of their methods basically results from some inference rules which describe simple cuts, and a branching scheme such that operations

---

[156]   Lageweg et al. (1977)

[157]   A survey on earlier approaches in order to schedule flow shops exactly can be found in Elmaghraby / Elshafei (1976), Baker (1975), and Kawaguchi / Kyan (1988). Dudek et al. (1992) review flow shop sequencing research since 1954.
Considerable effort has been invested in the empirical testing of various priority rules, see Gere (1966), and the survey papers of Day / Hottenstein (1970), Panwalker / Iskander (1977), Haupt (1989). Noteworthy flow shop heuristics, paralleling the work on optimal procedures for the makespan criterion are those of Campbell et al. (1970) and Dannenbring (1977). Both used Johnson's algorithm, the former to solve a series of two machine approximations to obtain a complete schedule, which the latter locally improved by switching adjacent jobs in the sequence. For a survey on problem solutions on which Johnson's algorithm and underlying ideas have had an influence see Proust (1992). Giffler et al. (1963) applied a probabilistic priority rule or random sampling for operation selection in order to build a feasible schedule.

which belong to a block (a sequence of operations on a machine) on the longest path are moved to the block ends, hence improving an idea of Grabowski et al. (1986).

Nowadays, tailored approximation methods viewed as an opportunistic problem solving process can yield optimal or near–optimal solutions even for problem instances up to now considered as difficult [158]. Hereby opportunistic problem solving or opportunistic reasoning characterizes a problem solving process where local decisions, such as which operations, jobs, or machines should be considered next, are concentrated on the most promising aspects of the problem, e.g. job contention on a particular machine. Hence subproblems often defining bottlenecks are extracted and separately solved and serve as a basis from which the search process can expand. Breaking down the whole problem into smaller pieces takes place until, eventually, sufficiently small subproblems are created for which effective exact or heuristic procedures are available. However the way in which a problem is decomposed affects the quality of the solution reached. Not only the type of decomposition such as machine / resource (Adams et al. (1988)), job / order (Pesch (1993a)), or event based (Sadeh (1991)) has a dramatic influence onto the outcome but also the number of subproblems and the order of their consideration. In fact, an opportunistic view suggests that the initial decomposition be reviewed in the course of problem solving to see if changes are necessary. The shifting bottleneck heuristic from Adams et al. (1988) and its improving modifications from Balas et al. (1992) and Dauzere–Peres / Lasserre (1993) are typical representatives of opportunistic reasoning. It is resource based as there are sequences of one machine schedules successively solved and their solutions introduced into the overall schedule.

In recent years local search based scheduling became very popular [159]. These algorithms are all based on a certain neighbourhood structure how to obtain a new solution from existing ones. The first efforts to implement powerful general problem solvers such as simulated annealing [160], parallel tabu search [161], and genetic algorithms [162]     finally culminated   in the excellent tabu search

---

[158]   cf. Adams / Balas / Zawack (1988), Ow / Smith (1988), Sadeh (1991), Balas et al. (1992), and Dauzere–Peres / Lasserre (1993)

[159]   For a survey see Barnes et al. (1992) as well as Glass et al. (1992).

[160]   Van Laarhoven / Aarts / Lenstra (1992), Matsuo et al. (1988)

[161]   Taillard (1989)

[162]   Aarts / Van Laarhoven / Ulder (1991), Nakano / Yamada (1991), Yamada / Nakano (1992), Storer et al. (1991, 1992, 1992a,b)

implementations of Dell'Amico / Trubian (1993) and Nowicki / Smutnicki (1993). However, only the genetic based methods of Yamada / Nakano (1992), Dorndorf / Pesch (1993a,b), and Pesch (1993a), Taillard's parallel tabu search algorithm and the implementation of tabu search by Nowicki and Smutnicki (1993) could solve the notorious 10 job 10 machine problem optimally. Most of the current local search approaches rely on naive search neighbourhoods which fail to exploit problem specific knowledge. Applications of local and probabilistic search methods to sequencing problems are based on neighbourhoods defined in the solution space of the problem. The method of Storer et al. (1992a) is based on problem perturbation neighbourhoods, i.e. the orginal data is genetically perturbed and a neighbour is defined as a solution which is obtained when a base heuristic is applied to the perturbed problem. The obtained solution sequence for the perturbed problem is mapped to the original data, i.e. the non–perturbed operations are scheduled in the same way and the makespan of the solution to the original problem data defines the quality of the perturbed problem.[163]

The local search heuristics like simulated annealing, tabu search, and genetic algorithms are modestly robust under different problem structures and require only a reasonable amount of implementation work with relatively little insight into the combinatorial structure of the problem. Problem specific characteristica are mainly introduced via some improvement procedures the kind of representation of solutions as well as their modifications based on some neighbourhood structure.

# 2. Evolution Based Learning

We are going to consider briefly the most important solution ideas and their related neighbourhood structures leading to sophisticated branching structures for the exact methods as well as powerful and generally applicable approximation techniques, e.g. local search procedures such as simulated annealing, tabu search, or genetic algorithms.

The section is organized as follows. We describe two types of genetic based

---

[163]    Other exact and heuristic methods are described in the dissertations of Bräsel (1990)
        (and Bräsel / Werner (1989)), Hurink (1992), and Meyer (1992). Further genetic
        scheduling applications are published for instance in Männer / Manderick (1992).

learning. In each case a genetic algorithm serves as a meta–strategy to control the process of learning combinations of local job shop scheduling procedures. As local scheduling procedures we consider priority scheduling rules described in Panwalkar and Iskander (1977) as well as the shifting bottleneck procedure from Adams et al. (1988). Finally we present computational results compared to the best heuristics known up to now for the job shop scheduling problem, such as simulated annealing and the shifting bottleneck procedure.

## 2.1 Genetic Enumeration

Compared to standard heuristics "genetic algorithms are not well suited for fine–tuning structures which are very close to optimal solutions" (Grefenstette (1987b)). Therefore it is essential if a competitive genetic algorithm is desired, to incorporate (local search) improvement operators. The resulting algorithm has then been called genetic local search heuristic [164]. Putting things into a more general framework, a solution of a combinatorial optimization problem may be considered as a sequence of local decisions. For instance, in case of the traveling salesman problem a local decision might result in a choice of a city visited next. A local decision for the job shop scheduling problem might be the choice of an operation to be scheduled next. In what follows we will consider more general decision rules. In an enumeration tree of all possible decision sequences a solution of the problem is represented as a path corresponding to the different decisions from the root of the tree to some leaf. A branch and bound algorithm learns to find those decisions leading to an optimal solution with respect to the space of all decision sequences. Genetics can guide a search process in order to learn to find the most promising decisions within a reasonable amount of time. The scheme of a so called genetic enumeration algorithm is described in Figure 1, it requires further refinements in order to design a successful algorithm.

> **Initialisation:** Construct an initial population of individuals each of which is a string of local decision rules.
>
> **Assessment / Improvement:** Assess each individual in the current population introducing problem specific knowledge by special purpose heuristics (such as local search) which are guided by the sequence of local decisions.
>
> > *if* special purpose heuristics lead to a new string of local decision rules *then*
> > replace each individual by the new one, for instance a locally optimal one.

---

[164]  see Chapter II

*repeat*

> **Recombination:** Extend the current population by adding individuals obtained by unary and binary transformations (crossover, mutation) on one or two individuals in the current population.

> **Assessment / Improvement:** Assess each individual in the current population introducing problem specific knowledge by special purpose heuristics (such as local search) which are guided by the sequence of local decisions.

> > *if* special purpose heuristics lead to a new string of local decision rules *then*
> > > replace each individual by the new one, for instance a locally optimal one.

> **Selection:** Reduce the extended population to its original size according to the selection rules.

*until* some stopping criterion is met.

Figure 1.    Genetic enumeration.

The fitness value of each individual has to be defined. Usually it is the value of the objective function or some scaled version of it when all local decision rules and the improvment step have been applied. Next a new population is constructed.

The traditional genetic algorithm based on a binary string repesentation of a solution is often unsuitable for many optimization problems involving integral variables because it is very difficult to represent a solution such that substrings have a meaningful interpretation [165]. Choosing a more natural representation of solutions, however, involves more intricated recombination operators, in particular crossover operators in order to get feasible offspring [166]. To overcome these difficulties and apply the simple crossover operator one can use an interpretation of an individual solution as a sequence of decision rules as described first in Dorndorf / Pesch (1992) [167]. Each indiviual of a population is considered to be a subset of feasible schedules from the set of all feasible schedules, i.e. a solution can also be considered as a set of solutions of the problem instance.

---

[165]    cf. Whitley et al. (1989)

[166]    For the job shop scheduling problem see Aarts et al. (1991); for the traveling salesman problem see Mühlenbein et al. (1987, 1988), Gorges–Schleuter (1989), or Kolen / Pesch (1991).
The traveling salesman was also the driving force behind earlier approaches on genetic based scheduling, inter alia motivated by the close relationship of the traveling salesman and the flow shop problem, see Ablay (1987) and Stöppler / Bierwirth (1992).

[167]    the idea of which is outlined in this section

## 2.2   Heuristics for the Job Shop Scheduling Problem

**Priority rules** are probably the most frequently applied heuristics for solving (job shop) scheduling problems in practice because of their ease of implementation and their low time complexity. The algorithm of Giffler and Thompson (1960) can be considered as a common basis of all priority rule based heuristics. Let $Q(t)$ be the set of all unscheduled operations at time t. Let $r_i$ and $c_i$ denote the earliest possible start and the earliest possible completion time, respectively, of operation i. The algorithm of Giffler and Thompson assigns available operations to machines, i.e. operations which can start being processed. Conflicts, i. e. operations competing for the same machine, are solved randomly. A brief outline of the algorithm is given in Figure 2.

$t := 0; \; Q(t) := \{1, \ldots, n\text{--}1\};$
*repeat*

Among all unscheduled operations in $Q(t)$ let $j^*$ be that one with smallest completion time, i. e. $c_{j^*} = \min \{c_j \mid j \in Q(t)\}$. Let $m^*$ denote the machine $j^*$ has to be processed on.

Randomly choose an operation i from the conflict set $\{j \in Q(t) \mid j \text{ has to be processed on machine } m^* \text{ and } r_j < c_{j^*}\}$ .

$Q(t) := Q(t) \setminus \{i\};$ Modify $c_j$ for all operations $j \in Q(t)$. Set t to the next possible operation to machine assignment.

*until* $Q(t)$ is empty.

Figure 2.   The algorithm of Giffler and Thompson.

The Giffler / Thompson algorithm can generate all active schedules [168] among which are also optimal schedules. As the conflict set consists only of operations, i.e. jobs, competing for the same machine, the random choice of an operation or job from the conflict set may be considered as the simplest version of a priority rule where the priority assigned to each operation or job in the conflict set corresponds to a certain probability. Many other priority rules can be considered, for instance the total processing time of all subsequent job operations of operation i may be a criterion. We apply only those rules presented in Table 1 [169]. The first column of Table 1 contains an abbreviation and name of the rule

---

[168]   A schedule is said to be active if there is no operation which can start being processed earlier without delaying any other operation.

[169]   For an extended summary and discussion see Panwalkar and Iskander (1977) as well as Blackstone et al. (1982) and Haupt (1989).

while the last column describes which of the operations or jobs in the conflict set gets highest priority.

| | rule | description |
|---|---|---|
| 1. | SOT—rule (shortest operation time) | An operation with shortest processing time on the considered machine. |
| 2. | LOT—rule (longest operation time) | An operation with longest processing time on the machine considered. |
| 3. | LRPT—rule (longest remaining processing time) | An operation with longest remaining job processing time. |
| 4. | SRPT—rule (shortest remaining processing time) | An operation with shortest remaining job processing time. |
| 5. | LORPT—rule (longest operation remaining processing time) | An operation with highest sum of tail and operation processing time. |
| 6. | Random | The operation for the considered machine is randomly chosen. |
| 7. | FCFS—rule (first come first serve) | The first operation in the queue of jobs waiting for the same machine. |
| 8. | SPT—rule (shortest processing time) | A job with smallest total processing time. |
| 9. | LPT—rule (longest processing time) | A job with longest total processing time. |
| 10. | LOS—rule (longest operation successor) | An operation with longest subsequent operation processing time. |
| 11. | SNRO—rule (smallest number of remaining operations) | An operation with smallest number of subsequent job operations. |
| 12. | LNRO—rule (largest number of remaining operations) | An operation with largest number of subsequent job operations. |

Table 1.    Priority rules.

The **Shifting Bottleneck Heuristic** from Adams, Balas, and Zawack (1988) and recently from Balas, Lenstra, and Vazacopoulos (1992) is probably the most powerful procedure known up to now among all heuristics for the job shop scheduling problem. The idea is to solve for each machine a one machine scheduling problem to optimality under the assumption that a lot of arc directions in the optimal one machine schedules coincide with an optimal job shop schedule. Consider all operations of a job shop scheduling instance that have to be scheduled on machine m. In the (disjunctive) graph including a partial selection of opposite directed arcs (corresponding to a partial schedule) there exists a longest path of length $h_i$ from dummy operation 0 to each operation i scheduled on machine m. Processing of operation i cannot start before time (also called head) $h_i$. There is also a longest path of length $q_i$ (called tail) from i to the dummy operation n. Obviously, when i is finished it will take at least $q_i$ time units to finish the whole schedule. The one machine scheduling problem with heads and tails is also NP–complete [170], however, there is a powerful branch and bound method proposed by Potts (1980b) and Carlier (1982, 1987) which dynamically changes heads and tails in order to improve the operations' sequence [171].

The shifting bottleneck heuristic consists of two subroutines. The first one (SB1) repeatedly solves one machine scheduling problems while the second one (SB2) builds a partial enumeration tree where each path from the root to a leaf is similar to an application of SB1. In order to understand the genetic approach, we have to consider the heuristic in more detail. As the very name suggests, the shifting bottleneck heuristic always schedules bottleneck machines first. As a measure of the bottleneck quality of machine m, the value of an optimal solution of a certain one machine scheduling problem on machine m is used. The one machine scheduling problems in consideration are those which arise from the disjunctive graph model when certain machines are already sequenced. The operation orders on sequenced machines are fully determined. Hence sequencing an additional machine probably results in a change of heads and tails of those operations whose machine order is still open. For all machines not sequenced, the maximum makespan of the corresponding optimal one machine schedules, where the arc directions of the already sequenced machines are fixed, determines the bottleneck machine. In order to minimize the makespan of the job shop scheduling problem the bottleneck machine should be sequenced first. A brief statement of the shifting bottleneck procedure is given in Figure 3.

---

[170]    see Garey / Johnson (1979)

[171]    see also Larson et al. (1985) and Nowicki / Zdrzalka (1986)

Let M be the set of all machines and let $M' := \{\}$ be the set of all sequenced machines;
*repeat*
    *for* m ∈ M\M' *do*
    *begin*
        Compute head and tail for each operation i that has to be scheduled on machine m.
        Solve the one machine scheduling problem to optimality for machine m; let v(m) be the resulting makespan for this machine.
    *end;*
    Let $m^*$ be the bottleneck machine, i. e. $v(m^*) \geq v(m)$ for all m ∈ M\M' .
    $M' := M' \cup \{m^*\}$;

    /* local reoptimization */
    *for* m ∈ M' in the order of its inclusion *do*
    *begin*
        Delete all arcs between operations on m while all arc directions between operations on machines from $M' \setminus \{m\}$ are fixed.
        Compute heads and tails of all operations on machine m and solve the one machine scheduling problem.
    *end;*
*until* $M = M'$ .

Figure 3.    Outline of the SB1–heuristic.

The one machine scheduling problems, although they are NP–hard [172]    can quickly be solved using the algorithm of Carlier (1982). Unfortunately, adjusting heads and tails does not take into account a possible already fixed processing order of operations connecting two operations i and j on the same machine, whereby this particular machine is still unscheduled. So, we get one machine scheduling problems with heads, tails, and time lags (minimum delay between two operations), problems which cannot be handled with Carlier's algorithm. In order to overcome these difficulties an improved SB1 version is suggested by Dauzere–Peres and Lasserre (1993) using approximate one machine solutions. Balas et al. (1992) solved the one machine problems exactly [173].    So, there is a SB1–heuristic superior to the SB1–heuristic proposed by Adams et al. (1988). However, on the average, the results are slightly worse than those obtained by Balas et al.'s SB2–heuristic.

During the local reoptimization part of the SB1–heuristic, for each machine, the operation sequence is redetermined keeping the sequences of all other already scheduled machines untouched. As suggested by Adams et al. (1988), we go through at most three local reoptimization iterations for each set M'. Only in

---

[172]    contrary to the preemptive case, cf. Baker et al. (1983)

[173]    see also Dell'Amico (1993)

the last step, when $M = M'$, we continue until there is no improvement for a full iteration.[174]

The quality of the schedules obtained by the SB1–heuristic heavily depends on the sequence in which the one machine problems are solved and these machines included into set $M'$. Sequence changes may yield substantial improvements. This is the idea behind the second version of the shifting bottleneck procedure, i.e. the SB2–heuristic, as well as behind our second genetic algorithm approach. The SB2–heuristic applies a slightly modified SB1–heuristic to the nodes of a partial enumeration tree. A node corresponds to a set $M'$ of machines that have been sequenced in a particular way. The root of the search tree corresponds to $M' = \{\}$. A branch corresponds to the inclusion of a machine m into $M'$, thus the branch leads to a node representing an extended set $M' \cup \{m\}$. At each node of the search tree a single step of the SB1–heuristic is applied, i. e. machine m is included possibly followed by a local reoptimization. Each node in the search tree corresponds to a particular sequence of inclusion of the machines into the set $M'$. Thus, the bottleneck criterion no longer determines the inclusion into $M'$. Obviously a complete enumeration of the search tree is not acceptable. Therefore a breadth–first search up to depth $\ell$ is followed by a depth–first search. In the former case, for a search node corresponding to set $M'$ all possible branches are considered which result from inclusion of a machine $m \notin M'$. Hence the successor nodes of node $M'$ correspond to machine sets $M' \cup \{m\}$ for all $m \in M\backslash M'$. Beyond the depth $\ell$ an extended bottleneck criterion is applied, i. e. instead of $|M\backslash M'|$ successor nodes there are several successor nodes generated corresponding to the inclusion of the bottleneck machine as well as several other machines m to $M'$. In our implementation of the SB2–heuristic we kept $\ell$ equal to 3, and beyond this depth applied the bottleneck criterion, i.e. only one successor node corresponding to the inclusion of the bottleneck machine was generated.

## 2.3 Learning by Population Genetics

In this section we demonstrate how special purpose heuristics can be incorporated into a genetic framework such that population genetics serves as a strategy to determine how high quality solutions can be constructed. We consider two strategies, one is based on priority rules while the other uses ideas

---

174 For details see Adams / Balas / Zawack (1988).

of the SB2–heuristic.

Theory and evidence suggest that genetic algorithms perform better when augmented with problem–specific knowledge and advanced recombination operators that exploit the added information [175]. A genetic algorithm should be used to rapidly reduce the search space to a size that can subsequently be handled by deterministic search algorithms. The idea here is to use the genetic algorithm to identify the possible "good solution regions" in the search space, which can later be climbed by conventional algorithms. There are several possibilities to apply an improvement heuristic. For instance, as in the case of the traveling saleman problem, one can find an encoding such that each individual of a population in fact is a solution of the problem. Hence during the recombination phase an improvement step can be applied to all or several of the solutions in a population. Using 2–opt or the Lin–Kernighan heuristic for the traveling salesman problem provides a powerful genetic local search procedure [176]. Some type of an improvement heuristic may also be incorporated into the crossover operator [177]. In any case the improvement step as well as the crossover operator heavily depend on the representation of the solution [178].

Usually a simple representation requires more sophisticated recombination operators and vice versa. To overcome these difficulties we tried a completely different encoding scheme for the job shop scheduling problem. Our solution representation enables us to use the simplest type of crossover as well as to incorporate problem specific knowledge, i.e. as Davis (1985) claimed "to examine the workings of a good deterministic program in that domain", in order to be competitive with special purpose heuristics. The strategy controls a sequence of priority rules or the machine inclusion sequence for the SB1–heuristic and learns

---

[175]  cf. Davis (1985), Whitley et al. (1989), Husbands et al. (1991), and Hilliard / Liepins (1988), Nakano / Yamada (1991)

[176]  see Johnson (1990a), Ulder et al. (1991), and Kolen / Pesch (1991)

[177]  see Kolen / Pesch (1991), Yamada / Nakano (1992)

[178]  cf. Bagchi et al. (1991) and Schönburg / Heinzmann (1992)

to find best combinations in both cases.[179] Each individual of the priority rule based genetic algorithm (for short: P–GA) is a string of n – 1 entries $(pr_1, pr_2, ..., pr_{n-1})$ where n – 1 is the number of operations in the underlying problem instance. An entry $pr_i$ represents one rule of the set of twelve priority rules described in Table 1. The entry in the i–th position says that a conflict in the i–th iteration of the Giffler / Thompson algorithm should be resolved using priority rule $pr_i$. More precisely, an operation from the conflict set has to be selected by rule $pr_i$; ties are broken by a random choice. Within a genetic framework a best sequence of priority rules has to be determined. The crossover operator is straightforward. Obviously, the simple crossover, where the substrings of two cut strings are exchanged, applies and always yields feasible offspring. Heuristic information already occurs in the encoding scheme and a particular improvement step is dropped. From our experiments we recommend to use a set of priority rules which are partially complementary (e.g. SOT and LOT, SPT and LPT etc.) in order to be able (hopefully) to choose each member in a conflict set. Otherwise we could not prevent a premature convergence or exclusion of an optimal solution. The mutation operator applied with a very small probability simply switches a string position to another one, i.e. the priority rule of a randomly chosen string entry is replaced by a new rule randomly chosen among the remaining ones.[180] Besides the priority rules of Table 1 one can use other kinds of rules which fix the order of operations on the same machine. For instance, any permutation of unscheduled operations might represent a rule. However, this will result in string codes over a huge alphabet contrary to the observation that small alphabets usually accelerate a uniform convergence [181,182]

---

[179] Another genetic local search approach based on an arc solution representation is described in Aarts et al. (1991). Their ideas are stimulated by the encouraging results obtained for the traveling salesman problem (cf. Ulder et al. (1991)). A different approach was followed by Storer / Wu / Vaccari (1991, 1992a, 1992b) and Storer / Wu / Park (1992). They map the original data of the underlying problem instance to slightly disturbed and genetically controlled data representing new problem instances. The latter are solved heuristically and the solution, i.e. the operations' processing orders, are considered to be solutions of the original problem. Their idea can be considered as uncontrolled fuzzy sequencing as introduced by McCahon / Lee (1992) and Zimmermann (1990).

[180] We also tested an inversion operator (a substring swap, see Holland (1975)) but the experiments did not show a clear advantage of using inversion.

[181] see Goldberg (1989b)

[182] We did some experiments in this direction where we restricted the number of rules to k out of k! permutations if k is the number of operations that can be in conflict. The less encouraging results let us drop this approach.

Our approach to search a best sequence of decision rules for selecting operations is just in line with the ideas of Fisher and Thompson (1963) on probabilistic learning of sequences consisting of two priority rules, and Crowston et al. (1963) or O'Grady and Harrison (1985) on learning how to find promising linear combinations of basic priorities.

In our first implementation the genetic algorithm serves as a meta–strategy to optimally control the use of priority rules, whereas the genetic algorithm controls the selection of nodes in the enumeration tree of the shifting bottleneck heuristic in our second implementation, the **shifting bottleneck based genetic algorithm** (for short: **SB–GA**). Remember that the SB2–heuristic is only a repeated application of a part of the SB1–heuristic where the sequence in which the one machine problems are solved is predetermined. Up to some depth $\ell$, a complete enumeration tree is generated and a partial tree for the remaining search levels. The SB2–heuristic tries to determine the best single machine sequence for the SB1–heuristic within a reasonable amount of time. We demonstrate that this can also be achieved by a genetic strategy, even in a more effective way. Contrary to Whitley et al. (1989) scheduling of one machine alone does not in general provide a highly constrained solution space.

The length of a string representation of an individual in the population equals the number of machines in the problem which is equal to the depth of the enumeration tree in the SB2–heuristic. Hence, an individual is encoded over the alphabet from 1 to the number of machines and a partial string from the first to the k–th entry just describes the sequence in which the single machines are considered in the SB1–heuristic. As a crossover operator we can use any traveling salesman crossover; we chose the cycle crossover as described in Goldberg (1989a) or Starkweather et al. (1991). Neither a mutation nor an inversion operator is applied. In order to reduce the computation time we decreased the number of reoptimization cycles during the execution of the SB1–heuristic. Instead of immediately starting a reoptimization cycle when two machines are sequenced we initiated a reoptimization cycle when less than 6 machines are left over, i.e. not included in the partial solution. The value 6 has been empirically found to be best with respect to the 10 × 10 problem of Fisher and Thompson (1963). The difference between the shifting bottleneck heuristic and our genetic approach is that the bottleneck is no longer a decision criterion for the choice of the next machine.

## 2.4   Details of Implementation and Computational Results

Setting the control parameters for a genetic algorithm is an important decision. We chose the empirically best parameter configuration with respect to probably the 3 best known test problems for job shop scheduling: the 6 × 6, the 5 × 20, and the 10 × 10 problem the optima of which are 55, 1165, and 930, respectively [183].

Our implementation uses the algorithm of Baker (1987) for selection, the simple (P–GA) as well as the cycle (SB–GA) crossover as described in Goldberg (1989a), mutation and inversion in case of the P–GA. Near–optimal individuals with high objective function values almost have equal survival probabilities during the selection process if unscaled fitness values serve as selection criterion. While this might be a desirable effect at the beginning of the genetic search process, later on slightly better individuals should survive more often. Scaling the objective function values with respect to some scaling window will meet this requirement. A scaling window of size k means that the worst objective function value $f_w$ in the last k populations is a scaling basis, i.e. each objective function value f in the current population is replaced by $f' := f_w - f$ . For window size 0 there is no scaling. We also used an elitist strategy [184], i.e. the best individual in each population always survived.

Besides the 3 well known test instances we generated 105 random problems in order to test our parameter configuration. For each problem type we generated five instances where each job has to be processed by each machine; the machine sequence for a job is randomly generated and the randomly chosen operation processing times are uniformly distributed within the interval [5...99].

Besides our genetic strategies each problem instance (see the tables of computational results) was solved by the SB1–heuristic (SB1), our implementation of the SB2–heuristic (SB2), each of the priority rules where only the result of the priority rule performing best is mentioned (in column "PRIO"), a random choice of priority rules, i.e. whenever during the execution of the Giffler / Thompson algorithm a choice point is reached, where a next operation has to be selected, one of the 12 priority rules is chosen at random in order to select the next operation with respect to this priority rule (see column "RND"). Table 2 summarizes the set of control parameters for our genetic algorithms (consider only the first column for each of the algorithms, the second columns

---

[183]   Fisher and Thompson (1963)

[184]   see Goldberg (1989a)

are used later).

| parameter | value for P—GA | | value for SB—GA | |
|---|---|---|---|---|
| number of objective function evaluations | 60000 | 70000 | 300 | 400 |
| population size | 200 | 300 | 40 | 40 |
| crossover rate | 0.65 | 0.65 | 0.75 | 0.65 |
| mutation rate | 0.001 | 0.0 | 0.0 | 0.0 |
| inversion rate | 0.7 | 0.0 | 0.0 | 0.0 |
| window size for scaling | 4 | 3 | 4 | 3 |
| elitist strategy | yes | yes | yes | yes |

Table 2.   Genetic algorithm control parameters.

Fine tuning of these control parameters is of some influence on the quality of the solutions [185] but changing the instance most often also changes the best parameter configuration. To evaluate different parameter configurations several performance measures may be used.

The online–performance is the average value of all objective function evaluations of the individuals in all populations, while the offline–performance is the average objective function value over all best individuals in all populations. A high online–performance is reached if a high–quality search area can be found very soon, while a high offline performance needs in each population a high–quality solution. Hence the offline performance is not affected while areas of low quality solutions are searched. Other measures of performance are the average objective function value of all members in the current population or the quality of the best solution in the current population.

The whole system was implemented as a general genetic algorithm environment for different types of problems. It was written in PASCAL and all problems were solved on a DECstation 3100 under the operating system Ultrix.

Tables 3a and 3b contain the results for the well known test problems from Fisher and Thompson (1963). The makespan of the best solution found by each

---

[185]   See Schaffer et al. (1989), De Jong / Spears (1989); and Grefenstette (1986), Kakazu et al. (1992), Freisleben / Härtfelder (1993) who used a genetic algorithm for the control parameter setting.

of the heuristics and the makespan of an optimal solution for the corresponding problem instance are shown in column "f" and column "opt", respectively. Column "t sec" refers to the computation time needed for the correponding algorithm. Besides "PRIO", "SB1", and "SB2" which present the results of the best priority rule and the two shifting bottleneck versions where the enumeration depth $l$ was fixed to 4, the columns "P–GA" and "SB–GA" contain the results obtained by the two genetic learning strategies. The results for the randomly chosen priority rules in column "RND" are based on the same number of objective function evaluations (fitness evaluations) as in the P–GA algorithm. The time difference is an effect of the random number generation. Differences of the solution values obtained by the shifting bottleneck heuristics compared to the results (on a VAX 780/11) from Adams et al. (1988) – column "SB–ABZ" in Table 3b – probably stem from slightly different implementations of the reoptimization cycle part and different implementations of SB2, respectively [186].
Columns "GLS–ALU" and "SA–ALU" of Table 3b contain the best of two results obtained by two different implementations for each, a genetic local search strategy and simulated annealing, respectively [187]. These four algorithms got the same time restrictions and the part after the decimal point is truncated. Their values are average results over 5 runs on a VAX 8650. The neighbourhood structure for the simulated annealing and the genetic local search versions are the same: (i) Reversing an edge on a longest path in the graph, and (ii) reversing an edge on a longest path in the graph such that this edge is incident to an arc on the longest path from the arc set A. For details we refer to Aarts et al. (1991). In the genetic local search algorithms a local search procedure is performed after the recombination step of the genetic algorithm such that each member of the population is made locally optimal with respect to the two neighbourhood structures. The last two columns contain the results (there are no times) obtained by Nakano and Yamada (1991, 1992). Their Giffler / Thompson based crossover operator [188] solves conflicts (in the offspring) equally distributed with respect to parents' solutions. Column "TS–DT" presents the excellent tabu search results of Dell'Amico / Trubian (1992) on a 80386PC, 33 MHz, using an extension of the above mentioned neighbourhood structure [189].

---

[186] See also the worse results obtained by Applegate and Cook (1991) from their shifting bottleneck implementation.

[187] see Aarts et al. (1991)

[188] see Yamada / Nakano (1992)

[189] cf. the simulated annealing approach of Van Laarhoven et al. (1992)

In particular for the famous 10 × 10 problem we wished to have more information about the behaviour of both genetic algorithm strategies. Therefore we increased the number of fitness evaluations substantially. Figure 4 and Figure 5 show the performance of the P–GA over 250000 fitness evaluations, and the SB–GA over 800 fitness evaluations, respectively. In both figures the vertical axis shows the quality of the performance measure, i. e. the online, the offline, the average performance, and the best objective function value obtained in each population. In both cases the best solution converges to a suboptimum where it remains stable. In the SB–GA the speed of convergence is conspicuous due to the fact that the incorporated heuristic – the shifting bottleneck procedure – is much more powerful than priority rule based heuristics. In both figures the average and the offline performance become almost the same in later steps of the genetic algorithms.

Besides the standard test problems we considered 21 problem types, each consisting of 5 randomly generated problem instances. We applied the same set of heuristics as described for Table 3a to each of these 105 problem instances, the only difference being that because of memory restrictions we chose an enumeration depth $\ell = 3$ in the SB2–heuristic. Carlier's (1982) one machine scheduling branch and bound algorithm provides a lower bound on the makespans for each instance. Table 4 contains for each problem type (column "type", machines × jobs) the average relative deviation over 5 instances from the lower bound in percent. The computation times on average over 5 instances on a DECstation 3100 can be found in Table 5. A dash entry in the SB2 column indicates that the breadth–first search of the SB2–heuristic caused a memory problem (excessive paging).

| problem | PRIO | | RND | | P-GA | | SB1 | | SB2 | | SB-GA | | OPT |
|---|---|---|---|---|---|---|---|---|---|---|---|---|---|
| | f | t sec | f | t sec | f | t sec | f | t sec | f | t sec | f | t sec | |
| 6 x 6 | 58 | 0.0 | 58 | 12.1 | 55 | 11.4 | 57 | 0.1 | 55 | 3.6 | 55 | 19.7 | 55 |
| 5 x 20 | 1489 | 0.2 | 1374 | 1992.5 | 1249 | 1609.6 | 1274 | 0.4 | 1240 | 10.8 | 1178 | 95.7 | 1165 |
| 10 x 10 | 1191 | 0.1 | 1088 | 1498.1 | 960 | 932.6 | 1031 | 0.5 | 951 | 186.5 | 938 | 106.7 | 930 |

| problem | RND | | P-GA | | SB-ABZ | | SB-GA | | GLS-ALU | | SA-ALU | | TS-DT | | GA-NY | GA-YN | |
|---|---|---|---|---|---|---|---|---|---|---|---|---|---|---|---|---|---|
| | f | t sec | f | t sec | f | t sec | f | t sec | f | t sec | f | t sec | f | t sec | f | f | OPT |
| 6 x 6 | 58 | 12.1 | 55 | 11.4 | 55 | 1.5 | 55 | 19.7 | 55 | 9.4 | 55 | 9.4 | 55 | 2.4 | 55 | 55 | 55 |
| 5 x 20 | 1374 | 1992.5 | 1249 | 1609.6 | 1178 | 80 | 1178 | 95.7 | 1294 | 88.4 | 1216 | 88.4 | 1165 | 160.1 | 1235 | 1184 | 1165 |
| 10 x 10 | 1088 | 1498.1 | 960 | 932.6 | 930 | 851 | 938 | 106.7 | 978 | 99.4 | 969 | 99.4 | 935 | 155.8 | 965 | 930 | 930 |

Table 3a (top) and Table 3b (bottom).   The 3 Fisher / Thompson standard problems.

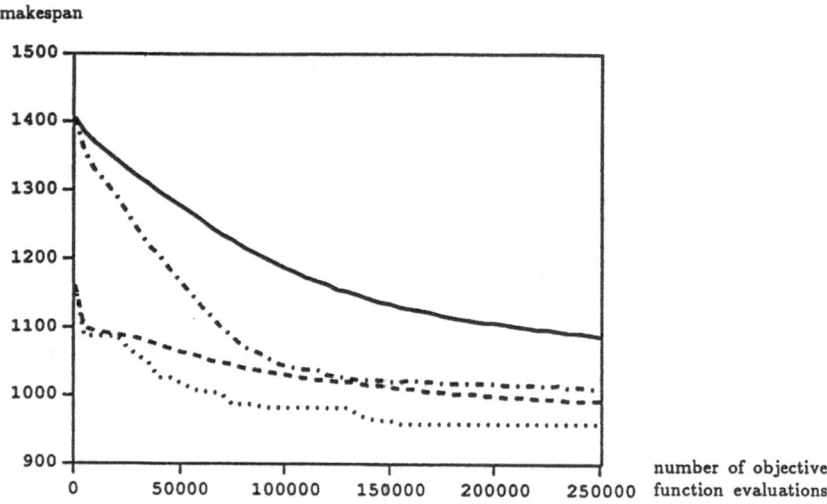

Figure 4.     Performance measures of the P–GA for the 10 × 10 problem.

————— online performance
—·—·— average performance
— — — offline performance
· · · · · · best fitness

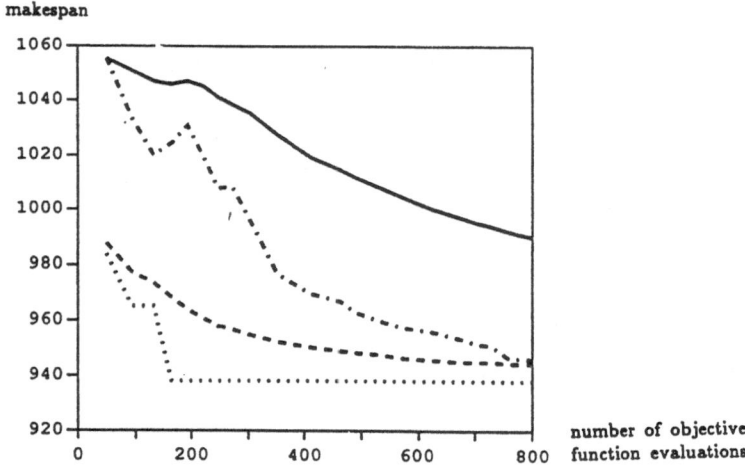

Figure 5.     Performance measures of the SB–GA for the 10 × 10 problem.

————— online performance
—·—·— average performance
— — — offline performance
· · · · · · best fitness

| type | PRIO | RND | P–GA | SB1 | SB2 | SB–GA |
|------|------|-----|------|-----|-----|-------|
| 5 x 10 | 18.7 | 13.9 | 4.5 | 4.1 | 3.8 | 3.2 |
| 5 x 15 | 16.2 | 6.8 | 0.6 | 1.3 | 0.2 | 0.2 |
| 5 x 20 | 15.7 | 8.0 | 1.0 | 0.1 | 0.0 | 0.0 |
| 10 x 5 | 15.2 | 3.3 | 3.2 | 9.6 | 5.3 | 4.6 |
| 15 x 5 | 7.8 | 3.1 | 3.4 | 4.6 | 3.1 | 3.1 |
| 20 x 5 | 8.2 | 3.3 | 1.8 | 3.8 | - | 3.1 |
| 10 x 10 | 36.0 | 24.2 | 15.2 | 22.3 | 13.2 | 12.0 |
| 10 x 15 | 31.2 | 28.3 | 13.6 | 13.5 | 8.9 | 7.2 |
| 10 x 20 | 21.6 | 21.7 | 7.4 | 7.4 | 2.0 | 1.7 |
| 15 x 10 | 28.1 | 20.5 | 13.2 | 15.4 | 10.5 | 10.3 |
| 20 x 10 | 26.2 | 21.9 | 13.8 | 17.6 | - | 11.6 |
| 5 x 5 | 16.8 | 7.7 | 7.7 | 9.7 | 8.2 | 8.2 |
| 6 x 6 | 20.3 | 8.5 | 7.8 | 12.5 | 7.2 | 7.1 |
| 7 x 7 | 22.7 | 9.8 | 8.7 | 12.7 | 7.7 | 6.9 |
| 8 x 8 | 29.8 | 17.3 | 13.3 | 17.6 | 11.6 | 10.7 |
| 9 x 9 | 34.8 | 23.0 | 15.4 | 19.0 | 12.2 | 11.4 |
| 11 x 11 | 31.2 | 26.6 | 15.0 | 16.6 | 11.5 | 10.0 |
| 12 x 12 | 34.6 | 24.3 | 11.7 | 12.1 | 8.7 | 8.1 |
| 13 x 13 | 35.4 | 29.0 | 16.1 | 19.5 | 12.9 | 11.5 |
| 14 x 14 | 38.4 | 35.8 | 18.3 | 21.3 | 15.7 | 15.0 |
| 15 x 15 | 38.3 | 32.0 | 18.1 | 19.9 | 13.6 | 13.6 |

Table 4.   Performance comparison of 6 algorithms: average relative deviation from the lower bound in percent.

| type | PRIO | RND | P–GA | SB1 | SB2 | SB–GA |
|------|------|-----|------|-----|-----|-------|
| 5 x 10 | 0.05 | 195.0 | 186.9 | 0.1 | 1.2 | 21.4 |
| 5 x 15 | 0.1 | 324.7 | 304.1 | 0.1 | 1.6 | 34.3 |
| 5 x 20 | 0.2 | 482.7 | 449.5 | 0.2 | 2.1 | 59.1 |
| 10 x 5 | 0.05 | 166.1 | 158.2 | 0.3 | 37.1 | 48.2 |
| 15 x 5 | 0.05 | 236.9 | 212.2 | 0.7 | 350.1 | 107.5 |
| 20 x 5 | 0.1 | 307.1 | 263.8 | 1.6 | - | 193.5 |
| 10 x 10 | 0.1 | 365.8 | 324.4 | 0.6 | 75.5 | 101.7 |
| 10 x 15 | 0.2 | 628.5 | 559.3 | 1.0 | 118.4 | 169.2 |
| 10 x 20 | 0.4 | 955.2 | 852.5 | 1.6 | 116.3 | 253.8 |
| 15 x 10 | 0.2 | 538.5 | 459.9 | 1.7 | 696.7 | 241.4 |
| 20 x 10 | 0.2 | 716.6 | 597.6 | 4.1 | - | 442.5 |
| 5 x 5 | 0.05 | 97.5 | 110.8 | 0.05 | 0.8 | 12.6 |
| 6 x 6 | 0.05 | 131.1 | 136.0 | 0.1 | 3.0 | 20.3 |
| 7 x 7 | 0.05 | 175.3 | 171.2 | 0.1 | 7.7 | 32.8 |
| 8 x 8 | 0.05 | 227.8 | 212.9 | 0.2 | 19.6 | 48.9 |
| 9 x 9 | 0.1 | 290.5 | 264.2 | 0.4 | 33.1 | 71.9 |
| 11 x 11 | 0.1 | 451.0 | 394.5 | 0.7 | 136.7 | 147.3 |
| 12 x 12 | 0.2 | 554.6 | 482.3 | 1.3 | 244.7 | 189.4 |
| 13 x 13 | 0.2 | 664.8 | 542.4 | 1.9 | 484.6 | 250.5 |
| 14 x 14 | 0.3 | 793.9 | 634.1 | 2.4 | 845.9 | 319.9 |
| 15 x 15 | 0.3 | 940.4 | 822.3 | 3.3 | 1257.8 | 414.5 |

Table 5.   Average running times for the results of Table 4 on a DECstation 3100.

The worst solutions are those obtained by the priority rules (column "PRIO"). Among all 105 runs the LOT–rule, the LNRO–rule, the LORPT–rule, the LPT–rule, and the Random rule yielded the best solution 45, 30, 28, 1, and 1 times, respectively; the other priority rules never dominated. A random choice of a priority rule in order to detect the next operation to be scheduled turned out to yield better solutions than a fixed choice of a priority rule. However the running times are pretty high, even higher than in P–GA, due to the elaborative random number generation. In general the SB2–heuristic dominates P–GA, however, for a large number of machines the conflict sets from Giffler / Thompson's algorithm tend to be small so that the P–GA could be superior to SB2. If the performance measure is put into relation to the running time then SB1 may become the most favorite algorithm. As expected, if the problem sizes turn out to be large enough SB–GA dominates all others, for both, the quality of the obtained solutions as well as the running time. A main aspect to consider SB–GA as a favorite is also its amount of memory in use; SB–GA uses very little memory compared to the SB2–heuristic. The latter, however, can be implemented similarly at the expense of its running time.

The encouraging results for the 105 randomly generated problems and the 3 famous instances as well as the desire to compete with alternative heuristics reported in the literature led us to consider the behaviour of our algorithms for a well known testbed of 40 problem instances. It is the same set of instances underlying the shifting bottleneck experiments of Adams et al. (1988), the genetic local search experiments of Aarts et al. (1991), as well as the simulated annealing experiments of van Laarhoven et al. (1992). The optimum solution of most of the instances is known [190].

We avoided to spend hours of computation time on fine parameter tuning and chose our parameters with respect to the outcomes of our former experiments. Different parameter settings for the different problem sizes probably would lead to slight performance improvements in particular for the larger and the smaller problem instances, but we adopted the same overall control parameters that seemed to be reasonable for middle sized instances. The parameters correspond to the entries of the second columns in Table 2 for each of the two algorithms.

Table 6 contains our results compared to those obtained by alternative approximation algorithms. The numbering of the problems in the first column is identical to that one in Adams et al. (1988). We dropped the instances 31 to 35 because the SB1 implementation already solved them to optimality in a fraction

---

[190]    cf. Brucker / Jurisch / Sievers (1991, 1992)

of the time needed by the SB–GA. Column "SB–GA (40)" corresponds to the parameter setting as described in Table 2. Additionally, we also ran SB–GA for the larger problems from 16 to 40 on population sizes of 60 limited to 800 function evaluations (see column "SB–GA (60)"). The results reported for SB–GA are average results over two runs while P–GA was run only once. The initial population was always randomly generated. Column "SB–ABZ" contains the makespans that Adams et al. (1988) obtained from their shifting bottleneck implementation. Columns "f–GLS" and "f–SA" contain the best of two results obtained by two different implementations for each, a genetic local search strategy and simulated annealing, respectively. These four algorithms got the same time restrictions from column "t sec". The last three columns are headed by the name "ALU" indicating that their values are taken from Aarts et al. (1991) and the part after the decimal point is truncated. Finally the last column "opt" presents the smallest makespan published in Carlier / Pinson (1990), Brucker et al. (1991), Applegate and Cook (1991), or Dell'Amico / Trubian (1993), for the particular problem instance. An asterisk (*) says that the value is proven to be optimal.

The values in the "ALU" column are average results over 5 runs on a VAX 8650. The neighbourhood structure for the simulated annealing and the genetic local search versions are the same: (i) Reversing an edge on a longest path in the graph, and (ii) reversing an edge on a longest path in the graph such that this edge is incident to an edge of the arc set A. For details we refer to Aarts et al. (1991). In the genetic local search algorithms a local search procedure is performed after the recombination step of the genetic algorithm such that each member of the population is made locally optimal with respect to the two neighbourhood structures.

Although P–GA could even find an optimal solution for several problems, and probably can do better for larger population sizes (for instance: For a population size of 500 we got the makespans 987, 807, 855, 863, 928, for the five instances from 16 to 20, respectively, within an average running time of 180 seconds for 70000 function evaluations.), the P–GA is never a serious competitor to other heuristics because of its running time compared to the quality of the results. However, for the largest instances the relation between running time and quality is quite good.

| problem | P-GA | | SB-GA (40) | | SB-GA (60) | | SB-ABZ | ALU | | | | opt |
|---|---|---|---|---|---|---|---|---|---|---|---|---|
| | f | t sec | f | t sec | f | t sec | f | f-GLS | f-SA | t sec | | opt |
| **5 machines, 10 jobs** | | | | | | | | | | | | |
| 01 | 666 | 106 | 666 | 15.6 | | | 666 | 666 | 666 | 17 | | 666*BJS |
| 02 | 681 | 108 | 666 | 16.4 | | | 669 | 659 | 658 | 18 | | 655*CP |
| 03 | 620 | 104 | 604 | 11.3 | | | 605 | 609 | 616 | 21 | | 597*BJS |
| 04 | 620 | 106 | 590 | 11.1 | | | 593 | 594 | 590 | 16 | | 590*BJS |
| 05 | 593 | 106 | 593 | 14.8 | | | 593 | 593 | 593 | 13 | | 593*BJS |
| **5 machines, 15 jobs** | | | | | | | | | | | | |
| 06 | 926 | 185 | 926 | 25.2 | | | 926 | 926 | 926 | 25 | | 926*CP |
| 07 | 890 | 182 | 890 | 27.6 | | | 890 | 890 | 890 | 42 | | 890*BJS |
| 08 | 863 | 184 | 863 | 26.1 | | | 863 | 863 | 863 | 39 | | 863*BJS |
| 09 | 951 | 186 | 951 | 23.6 | | | 951 | 951 | 951 | 35 | | 951*BJS |
| 10 | 958 | 185 | 958 | 24.3 | | | 959 | 958 | 958 | 17 | | 958*BJS |
| **5 machines, 20 jobs** | | | | | | | | | | | | |
| 11 | 1222 | 290 | 1222 | 39.3 | | | 1222 | 1222 | 1222 | 59 | | 1222*BJS |
| 12 | 1039 | 292 | 1039 | 35.9 | | | 1039 | 1039 | 1039 | 50 | | 1039*BJS |
| 13 | 1150 | 289 | 1150 | 33.5 | | | 1150 | 1150 | 1150 | 50 | | 1150*BJS |
| 14 | 1292 | 289 | 1292 | 34.9 | | | 1292 | 1292 | 1292 | 21 | | 1292*BJS |
| 15 | 1237 | 282 | 1207 | 48.8 | | | 1207 | 1207 | 1207 | 76 | | 1207*BJS |
| **10 machines, 10 jobs** | | | | | | | | | | | | |
| 16 | 1008 | 191 | 961 | 76.3 | 961 | 160.3 | 978 | 976 | 969 | 88 | | 945*CP |
| 17 | 809 | 192 | 787 | 77.7 | 784 | 163.3 | 787 | 791 | 785 | 91 | | 784*CP |
| 18 | 916 | 190 | 848 | 76.2 | 848 | 165.8 | 859 | 856 | 856 | 96 | | 848*BJS |
| 19 | 880 | 191 | 863 | 77.4 | 848 | 161.3 | 860 | 859 | 854 | 93 | | 842*BJS |
| 20 | 928 | 188 | 911 | 80.4 | 910 | 158.5 | 914 | 913 | 911 | 107 | | 902*BJS |
| **10 machines, 15 jobs** | | | | | | | | | | | | |
| 21 | 1139 | 352 | 1074 | 134.8 | 1074 | 292.8 | 1084 | 1084 | 1078 | 243 | | 1050 BJS |
| 22 | 998 | 350 | 935 | 136.1 | 936 | 290.0 | 944 | 944 | 950 | 254 | | 927*BJS |
| 23 | 1072 | 353 | 1032 | 139.6 | 1032 | 295.6 | 1032 | 1032 | 1032 | 242 | | 1032*CP |
| 24 | 1014 | 352 | 960 | 137.3 | 957 | 289.0 | 976 | 970 | 960 | 234 | | 935*AC |
| 25 | 1014 | 350 | 1008 | 134.2 | 1007 | 228.9 | 1017 | 1010 | 1003 | 254 | | 977*AC |
| **10 machines, 20 jobs** | | | | | | | | | | | | |
| 26 | 1278 | 569 | 1219 | 227.3 | 1218 | 442.8 | 1224 | 1236 | 1221 | 469 | | 1218*BJS |
| 27 | 1378 | 565 | 1272 | 242.5 | 1269 | 446.2 | 1291 | 1300 | 1275 | 492 | | 1242 DT |
| 28 | 1327 | 569 | 1240 | 221.8 | 1241 | 463.5 | 1250 | 1264 | 1242 | 455 | | 1216*DT |
| 29 | 1336 | 570 | 1204 | 241.0 | 1210 | 453.1 | 1239 | 1260 | 1225 | 471 | | 1182 DT |
| 30 | 1411 | 568 | 1355 | 230.5 | 1355 | 455.6 | 1355 | 1386 | 1355 | 441 | | 1355*CP |
| **15 machines, 15 jobs** | | | | | | | | | | | | |
| 36 | 1373 | 524 | 1317 | 335.6 | 1317 | 688.1 | 1305 | 1310 | 1307 | 602 | | 1268*CP |
| 37 | 1498 | 520 | 1484 | 350.5 | 1446 | 665.9 | 1423 | 1449 | 1440 | 636 | | 1402 CP |
| 38 | 1296 | 525 | 1251 | 335.7 | 1241 | 665.9 | 1255 | 1283 | 1235 | 635 | | 1203 DT |
| 39 | 1351 | 525 | 1282 | 327.2 | 1277 | 687.5 | 1273 | 1279 | 1258 | 592 | | 1233*BJS |
| 40 | 1321 | 526 | 1274 | 348.0 | 1252 | 698.4 | 1269 | 1260 | 1254 | 596 | | 1222*AC |

Table 6.
Computational results on a DECstation 3100 for 35 well known problem instances. Comparison to the results of the shifting bottleneck procedure from Adams et al. (1988) ("SB-ABZ") on a VAX 780/11 and to the results from Aarts et al. (1991) ("ALU") on a VAX 8650 for two genetic local search ("f-GLS") and two simulated annealing ("f-SA") approaches. Column "opt" contains the value of an optimal (*) or best known solution, which results from the exact algorithms of Brucker et al. (1991) (BJS), Applegate and Cook (1991) (AC), Carlier and Pinson (1990) (CP), or from tabu search by Dell'Amico and Trubian (1993) (DT).

For the smaller problems (from 1 to 15) GLS, SA, and SB–GA are almost equal in their solution quality; the optimum could be found in nearly all runs. For these problem instances the pure shifting bottleneck implementation is the winner in both, quality of the solution and the time. For the larger instances from 16 to 30, where the running time of the shifting bottleneck implementation is still the lowest, SB–GA performs better than the pure shifting bottleneck alternative as well as the GLS and SA algorithms (even if we consider the running times to the latter as comparable). For the last five instances the performance of simulated annealing and SB–GA is quite comparable. (Remember that columns "f–GLS" and "f–SA" contain always the best of two values.)

We would like to mention that for the famous $5 \times 20$ and $10 \times 10$ problem the makespans published in Aarts et al. (1991) are 1294 (GLS) / 1216 (SA) and 978 (GLS) / 969 (SA) found in 88 and 99 seconds, respectively. Hence, under fine parameter tuning SB–GA clearly dominates (see Table 3b).

Although P–GA is less powerful than the other approaches it has its advantages: Ease of implementation, also in a more general framework, as well as robustness to problem changes (SB–GA seems to be much more sensitive) can give P–GA the user's preference, even to simulated annealing.

## 2.5  Conclusions

We presented a special type of probabilistic learning by population genetics. Learning to find a best sequence of priority rules for solving job shop scheduling problems did not work best because of the amount of computation time needed. This is not a big surprise as we do not know if small makespan schedules can be produced by the fixed set of priority rules from Table 1. Moreover, priority rules only make little use of problem specific knowledge. In contrast, the solution of a one machine problem (with respect to some former decisions) involves a lot of problem specific knowledge, the use of which may be the main reason for the success of the genetically guided shifting bottleneck procedure. Introducing problem specific knowledge makes a genetic strategy a promising alternative. This knowledge might be incorporated by the application of a heuristic performing well on the particular problem, such as the shifting bottleneck procedure for the job shop problem and the Lin–Kernighan algorithm for the traveling salesman problem. We may also think of local search algorithms such as tabu search or simulated annealing as "special purpose" heuristics. So a simple genetic algorithm can do its work very well if an individual is considered to be a sequence of local decisions on the use of the knowledge provided by special purpose heuristics or parts of them.

The "general problem solver" nature of a genetic algorithm is maintained and the interface to the specific problem is handled via the encoding scheme, i.e. sequences of local decisions. The decoding routine for these representations was a simulation of the job shop, assuming that at any choice point of the simulation, a machine would perform the first or only allowable operation determined by a local decision rule which might be also a preference list.[191]

# 3.  Learning by Constraint Propagation

## 3.1  Constraint Propagation and Backtrack Search

Knowledge based systems [192],   which also include expert systems, basically consist of two important components: the knowledge base and the inference engine [193].   Knowledge and inference are the two main elements of intelligent systems, where, in particular, inference makes the big difference compared to regular systems, such as text processing or operating systems. Obviously, inference is also some knowledge, namely the knowledge how to reason on the existing knowledge base in order to increase it or in order to make implicitly existing knowledge more explicit.

The structur of a knowledge base is declarative with respect to the underlying problem, for instance, in the form of frames, rules, semantic networks, or

---

[191]   A broader application that considers scheduling decisions based on priority sequencing in general job shops (cf. Wein / Chevalier (1992), Berkley / Kiran (1991), Adam / Sukis (1980), Bowden (1969), Sen / Gupta (1984)) under different cost structures such as the cost of idle machines, cost of work–in–process inventory, cost of long production lead times, cost of missed due dates, etc. (Jones (1973)), might learn attractive solution procedures probably superior to previous simulations (see Hershauer / Ebert (1975), Green / Appel (1981), Baker (1984), Rochette / Sadowski (1976), Arumugam / Ramani (1980), Scudder / Hoffmann (1985). This is in particular true under certain stochastic assumptions (cf. Elver / Taube (1983).

[192]   Schefe (1986), Winston (1987), Charniak / McDermott (1985), Sundermeyer (1991), Schneeweiß (1992), Kurbel (1989)

[193]   Finlay (1990)

constraints. It is only descriptive without any rules for its use contrary to the procedural knowledge applied in common computer programs. Hence, knowledge based systems can be easily build and modified, or even certain entries dropped (e.g. rules or frames). In fact, this is a major idea, to be able to decompose the knowledge of a knowledge base into independent and comprehensible pieces which can be extracted. These parts are supposed to activate or inactivate each other in order to generate new knowledge while the problem is undergoing its resolution. Clearly, the inference engine need not work on a complete knowledge base immediately, additional knowledge such as constraints or rules can be introduced lateron without disturbing the solution process.

Unfortunately there are many problems that permit a problem representation in a reasonable small knowledge base consisting of only a small set of constraints but the number of solutions to the problem is exponentially increasing in its size. Most of the problems in combinatorial optimization belong to this set and the only way of attack to get feasible or even optimal solutions is by means of backtracking. Consistency checks as a part of the inference engine enriched by procedures such as intelligent backtracking, selective backtracking, dependency–directed backtracking, backjumping, or backmarking are in the recent past intensively applied in order to keep the search tree small [194]. Although there are retrospective and prospective techniques, relaxations, and heuristic search which can cut the enumeration tree substantially but they are not sufficient in order to solve NP–hard combinatorial optimization problems in a reasonable amount of time. Neither logic based nor procedural programming languages can overcome these difficulties. The former, however, provide. because of their declarative nature, implicitly a problem independent enumeration of the search tree. The user is free of details of implementation and only has to guide the generation and the structure of the search tree. Logic based programming languages provide a generic strategy of searching in the problem space, but, because of their nature as a general problem solver, some of them are only capable to solve easy or small problem instances. Backtrack search became increasingly important because of its use in PROLOG [195], in truth maintenance systems [196]    and as a way to solve constraint satisfaction problems [197].

---

[194]    Tate (1985), Bitner / Reingold (1975), Freuder / Wallace (1992), Bruynooghe / Pereira (1984)

[195]    Cox (1984), Clocksin / Mellish (1984)

[196]    Doyle (1979), De Kleer (1986)

[197]    De Kleer (1989)

PROLOG II [198]   provides the necessary concepts to smooth the way for what is nowadays called "logic programming with constraints" (Cohen (1990)). It is an important step compared to those languages operating on knowledge bases of equations [199].   PROLOG III [200]   is still another step further and offers the chance to define constraints on different data types, hence expressiveness and performance now allow applications which are difficult to handle under PROLOG [201].

The development of logic based programming languages, partially motivated by some work on automated theorem proving [202],   dates back to the early work of Sussman / Steele (1980) and Steele (1980) as well as Borning (1981) and is suggested in their programming languages CONSTRAINTS and ThingLab, respectively. Nowadays the most famous representatives among the logic programming languages are PROLOG III and the simultaneously developed language CHIP [203].

Lauriere (1978) realized with the system ALICE one of the first successful generic problem solving systems including a check of consistency with respect to the problem defining constraints and the variable domains. From a users point of view it is a black box without any chance to intervene into the search process. Although Lauriere (1978) states that ALICE "has been really utilized to solve real life problems with an efficiency better than that of specific classical operational research codes" just the development of CHIP provides a logical programming environment in order to solve real combinatorial optimization problems. Van Hentenryck (1989b) says that "CHIP has been applied to many real–life problems in Operations Research and hardware design", such as graph coloring, scheduling, VLSI design, warehouse location and blending problems, "with an efficiency comparable to specific programs written in procedural languages", see Habraken et al. (1991). The ideas of CHIP are taken a step further in the language cc(FD). It is a language capable to exploit the intrinsic parallelism that is implicitly provided in the knowledge base. A comparison of CHIP and cc(FD) can be found in Van Hentenryck / Simonis / Dincbas (1992).

---

[198]   Pique (1988), Sterling / Shapiro (1986)

[199]   Leler (1988)

[200]   Colmerauer (1987, 1990)

[201]   Jost / Skuppin (1989), Krautter / Steinert (1988)

[202]   Bibel (1987)

[203]   constraint handling in Prolog, cf. Dincbas et al. (1988, 1990), Van Hentenryck (1987, 1989a,b)

However, Young et al. (1992) still feel "that the primary disadvantage is that they (constaint based logic programming languages) were mostly develoжed for fairly narrow application areas and they are therefore not suitable for application to the side domain of concurrent engineering". They introduced the language SPARK in order to interactively develop constraint networks. New object oriented constraint based logic programming languages were created, e.g. Pecos and ConstraintLisp [204], both of which are based on Lisp. Van der Wal / Kolen (1992) describe a general backtrack search including a check on consistency; it is written in C++. Major applications of these languages are planning and configuration tasks, timetabling and scheduling.

A comprehensive comparison of a number of logic based languages (e.g. PROLOG III, CHIP) and algorithms (e.g. a branch and cut approach) in order to solve the satisfiability problem in conjunctive normal form is published by Mitterreiter / Radermacher (1991).

PROLOG and CHIP are the most prominent representatives among all programming languages which provide a user guided backtrack search and check on consistency procedures. Pretty comparable to CHIP is the language CHARME [205], a general problem solver that permits a controlled backtrack search via a set of parameters. It includes local constraint propagation [206] in the form of arc consistency [207].

Knowledge based scheduling systems have been build by various people using various techniques [208]. Some of them are rule based systems others are based on frame representations. Some of them only use heuristic rules to construct a schedule others conduct a constraint directed state space search. ISIS [209] is a constraint directed reasoning system for the scheduling of factory job shops. The main feature is that it formalizes various scheduling influences as constraints on the system's knowledge base and uses these constraints to guide the search in order to generate heuristically the schedule. In each scheduling cycle it first selects an order to be scheduled according to priority rules and then proceeds through a level of analysis of existing schedules, a level of constraint directed search and a level of detailed assignment of resources and time intervals for each

---

[204]   Liu (1989, 1992), Liu / Ku (1992b)

[205]   Bull 1990a,b

[206]   Güsgen / Hertzberg (1988)

[207]   cf. Pesch / Drexl / Kolen (1993)

[208]   Kusiak / Chen (1988), Mertens (1988), Zelewski (1990a,b)

[209]   Fox (1987), and Fox / Smith (1984)

operation in order. A large amount of work done by ISIS actually involves the extraction and organization of constraints that are created specifically for the problem under consideration. Scheduling relies only on order based problem decomposition. The system OPIS [210]  which is a direct descendant of ISIS attempts to make some progress by concentrating more on bottlenecks and scheduling under the perspective of resource based decomposition [211]. Bottleneck machines get a higher priority for being scheduled first. Decision making is based on heuristic rules.[212]

Scheduling is a complicated problem of both resource allocation and constraint satisfaction [213]. The approaches developed in the past have had only limited success. Most existing knowledge based scheduling systems are only capable of incorporating a small fraction of scheduling knowledge. As a result the schedules produced bear little resemblance to the actual state of the factory and can only be used to provide a high level guideline for human schedulers.

Encouraged by this little success and the progress that is made in the development of general problem solvers in form of constraint based logic programming languages we are going to restrict ourself to a less complex scheduling problem, namly standard job shop scheduling under makespan minimization. Learning about this problem and to recognize some typical features probably could enormously increase the power of general problem solvers in order to solve optimization, and in particular constraint satisfaction problems.

A constraint satisfaction problem (CSP) consists of a set of n variables $X_1, ..., X_n$, their domains $D_1, ..., D_n$, respectively, and a set of constraints of these variables. An n–ary relation or constraint on the variables $X_1, ..., X_n$ is a subset of the cartesian product $D_1 \times D_2 \times ... \times D_n$ of the domains. A solution is a value assignment of the variables such that all constraints are satisfied. As a special case a binary CSP consists only of constraints on two variables [214]. The CSP is NP–complete [215].

---

[210]   Ow / Smith (1988), Smith et al. (1986)

[211]   cf. Adams et al. (1988) and Chu et al. (1992)

[212]   There are other systems, e.g. OPAL (Bensana et al. (1988)), SOJA (Lepape (1985)), FURNEX (Slotnick et al. (1992)) and some rule based systems developed by Bruno et al. (1986), Subramanyam / Askin (1986), Svestka / Jiang (1992) and Shaw (1987), etc.

[213]   Minton et al. (1990, 1992)

[214]   Dechter / Pearl (1988), Meseguer (1989), Shanahan / Southwick (1989), Kumar (1992), Mackworth (1987)

[215]   Garey / Johnson (1979)

Obviously, optimization problems can be considered as constraint satisfaction problems whereby the objective function is included into the set of constraints [216]. Backtracking [217] has been the standard way of solving a CSP. In its most simple form backtrack search fixes a variable order at the beginning of the search and also determines a fixed ordering of the values for each domain. Variable selection for value assignment during the search process is only possible in the predetermined orders. Hence, if there is no feasible value left over in the domain of the current variable then the latest value to variable assignment is canceled and the value assignment of the last but one variable in the order preceds. Backtrack search can be demonstrated using the following two procedures FORWARD and BACKWARD [218] from Figure 6. While the first procedure tries to extend the current partial solution by a value assignment to a next variable, the latter is responsible for jumping backward. TOP and POP are two subroutines which modify the list. The first one delivers the first or the top list element and the latter deletes this first entry from the current list. The initialization step is the first call of FORWARD for i = 0. The complexity is of the order $O(a^n)$, where a is a maximum number of elements in a variable domain [219]. Hence, backtrack search can be very inefficient and there are particular pathological cases reported in Sussman / McDermott (1972) and Gaschnig (1974). Nadel (1988) reports about the behaviour quality and run time of different backtrack search algorithm like "Generate and Test", "Simple Backtracking", "Forward Checking", "Partial Lookahead", "Full Lookahead", and "Really Full Lookahead"; he also explains the differences. Backtracking in its simple form chooses the variables in a previously fixed order for instantiation. Instantiation also obeys a previously fixed order of the domain entries. A constraint is tested after its instantiation of its variables, i.e. the value assignment precedes the feasibility check. If the constraint is satisfied then the process of instantiation and test can continue, otherwise the search process fails and backtracking to the latest instantiated variable with a nonempty domain of possible value assignments is the consequence. Thus, the simple form of backtracking fixes the number and sequence of backtracks already in advance and a modification during the search is impossible. Undoubtly the procedure has also some advantages, the ease of implementation and the small amount of

---

[216]   Dechter et al. (1990)

[217]   Golumb / Baumert (1965), Lawler / Wood (1966), Bitner / Reingold (1975)

[218]   Dechter / Pearl (1988)

[219]   Nudel (1983), Haralick / Elliott (1980)

FORWARD $(X_1, ..., X_n)$
*begin*
    *if* i = n *then* STOP with the output of the current value assignment
    *else*
        $C_{i+1}$ := list of feasible values for $X_{i+1}$ with respect to $X_1, ..., X_{i-1}$;
        *if* $C_{i+1}$ ≠ empty list *then*
            $X_{i+1}$ := TOP $(C_{i+1})$;
            POP $(C_{i+1})$;
            FORWARD $(X_1, ..., X_i, X_{i+1})$
        *else*
            BACKWARD $(X_1, ..., X_i)$
*end;*


BACKWARD $(X_1, ..., X_i)$
*begin*
    *if* i = 0 *then* STOP "there is no feasible solution"
    *else*
        *if* $C_i$ ≠ empty list *then*
            $X_i$ := TOP $(C_i)$;
            POP $(C_i)$;
            FORWARD $(X_1, ..., X_i)$
        *else*
            BACKWARD $(X_1, ..., X_{i-1})$
*end;*

Figure 6.    Backtrack search.


memory in use (about $O(\log n)$). These, however, is on the cost of an in the problem size exponentially increasing amount of time. So, more efficient backtrack search procedures are desired.

There are a number of ways to reduce the enumeration tree either at the beginning or dynamically during searching [220].

## Methods to cut the enumeration tree in advance

A graph may serve as an illuminating representation of constraint satisfaction problems. In the dual representation each vertex of the graph corresponds to a constraint and vertices are adjacent if the vertices representing constraints have at least one variable in common. In the primal representation – the only one we

---

[220]    Dechter (1989/90), Bruynooghe (1981)

are going to consider furtheron – each vertex of the graph corresponds to some variable of the CSP. An edge is a subset of the vertex set. The edge represents precisely those constraints which constitute of these and only these variables represented by the edge defining vertex set. Hence, an edge implicitly is defined by the set of all feasible tuples of variable instantiations of the edge defining constraints. The resulting graph is a hypergraph [221]. The situation is much simpler in case of a binary CSP, i.e. a constraint $R_{ij}(X_i,X_j)$ contains at most two variables $X_i$ and $X_j$, and corresponds to a subset of the cartesian product $D_i \times D_j$. An edge connecting the vertices of the variables $X_i$ and $X_j$ corresponds to all binary relations on these two variables. The resulting graph is said to be the constraint graph of the underlying CSP. A universal relation between any two variables $X_i$ and $X_j$, i.e. constraints which are satisfied by all tuples of the cartesian product $D_i$ and $D_j$, is not included into the graph. Universal constraints do not deliver any information. Obviously, a binary CSP can equivalently be transformed into a constraint graph but not each CSP can polynomially be transformed to a binary one [222].

Consider the following example consisting of five variables $X_1$ , $X_2$ , $X_3$ , $X_4$ , $X_5$ and the domains $D_1 = \{1,2,3,4\}$, $D_2 = \{1,2,3\}$, $D_3 = D_2$, $D_4 = D_1$, and $D_5 = D_1$. Assume that each variable $X_i$ represents after instantiation the start of processing of operation i on a machine for one time unit. There are several sequencing constraints $R_{ij}(X_i,X_j)$ among the operations i and j, i.e. operation i has to be processed before operation j on the machine.

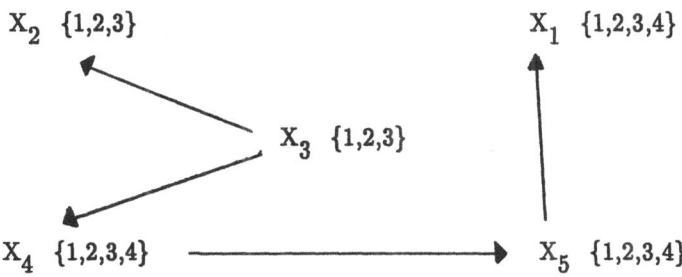

Figure 7.   Constraint Graph.

These precedence constraints are indicated by the orientation of an $X_i$ and $X_j$

---

[221]   Berge (1973), Mohr / Henderson (1986)

[222]   Montanari (1974)

connecting edge in the constraint graph of Figure 7, although edge directions are in general superfluous.

Naive backtracking with the fixed sequence $X_1$ , $X_2$ , $X_3$ , $X_4$ , $X_5$ of variable instantiations and the fixed sequence of value assignments as indicated yields the following enumeration tree of Figure 8 for each of the four possible instantiations of $X_1$. Actually the enumeration tree is four times as big as demonstrated in Figure 8 and every of its four subtrees contains exactly the same information with the only exception that the value for $X_1$ (here indicated by a *) is the only changing one. Some simple consistency checks at the beginning of the search can drastically reduce the size of this extensive search tree. These tests of consistency have the advantage that the constraint graph becomes more explicit, i.e. hidden constraints on variables currently not adjacent in the constraint graph get visible and new edges may be introduced into the graph.

Hereby we say that a set of variables is k–consistent if it is k–1–consistent and for any subset of k–1 variables and any instantiation of these k–1 variables satisfying all constraints there exists a value in the domain of the remaining variable such that all constraints on all k variables are satisfied. (0–consistency by definition is always fulfilled.) A set of variables is arc consistent if it is 2–consistent (this is said in relation to the constraint graph). A pair of variables $X_i$ and $X_j$ is path–consistent if for any feasible, i.e. $R_{ij}(X_i,X_j)$ respecting, instantiation $a_i$ and $a_j$ of $X_i$ and $X_j$ and any sequence of edges $(R_{ii_1} , R_{i_1i_2} , ... , R_{i_mj})$ in the constraint graph there is an instantiation $a_{i_1} \in D_{i_1}$ , $a_{i_2} \in D_{i_2}$ ,..., $a_{i_m} \in D_{i_m}$ of variables $X_{i_1}$ , $X_{i_2}$ ,..., $X_{i_m}$ such that all constraints $R_{ii_1}(a_i,a_{i_1})$ , $R_{i_ki_{k+1}}(a_{i_k},a_{i_{k+1}})$ and $R_{i_mj}(a_{i_m},a_j)$ , $k = 1,...,m–1$ , are satisfied. Thus path–consistency means that for any pair of variables and any explicitly feasible value pair there is also a feasible variable–value assignment on each path (edge sequence) connecting this variable pair in the constraint graph (including the universal relation). The constraint graph is arc– or path–consistent if any pair of variables is arc– or path–consistent, respectively. Arc–consistency requires an $\mathcal{O}(ea^2)$ effort while path–consistency can be reach with an effort of $\mathcal{O}(n^3a^3)$ where e is the number of edges in the constraint graph, a is the maximum number of elements in a domain, and n is the number of variables [223]. Both, arc– and path–consistency

---

[223]    Mohr / Henderson (1986); Han / Lee (1988), Mackworth (1977), Mackworth / Freuder (1985), Cooper (1989/90), Chen (1991), Dechter (1992) and some domain specific special cases in Van Hentenryck / Deville / Teng (1992), Cooper / Swain (1992), Liu / Ku (1992a)

Figure 8. Naive backtracking for the constraint graph of Figure 7.

already imply 3–consistency (Mackworth (1977)). Furthermore it suffices to consider only paths of length 3 in order to imply path–consistency (Mackworth (1977)).

As mentioned earlier consistency tests yield a more explicit constraint graph. This is quite comparable to what happens during backtrack search. The enumeration only takes those constraints into account which are explicitly contained in the constraint graph, while implicitly existing edges become just visible during the search process. Finding implicitly existing constraints means to generate new knowledge in the knowledge base, it is called constraint propagation and dates back to an early idea of Waltz (1975). Clearly, as higher the level of consistency as more constraints become visible. That means, to reach arc– path or 3–consistency can be considered as a local constraint propagation. As a remark, this is just in line with the notion of a schema or a building block of an individual in a genetic algorithm. Only a few entries fixed as a building block constituting hopefully good solutions, while the remaining entries are undecided. A building block or a schema just describes a locally feasible partial solution. Its entries are consistent with each other.

In order to reach a backtrack–poor search it is necessary to make as many constraints explicit as possible, because variable instantiations which violate implicitly existing constraints usually are detected much later during the search process. Such inconsistencies then are leading to a backtracking. In order to reach a backtrackfree search it is indispensible to make all implicitly existing constraints explicit. Montanari (1974) called that the "central problem" obviously an NP–hard one. For some special cases it is possible to require sufficient conditions in order to reach a backtrackfree [224] or a backtrack–poor search [225]. For instance, arc–consistency with respect to a constraint graph which is a tree guaranties a backtrackfree search. Let us consider again the example illustrated in Figure 7. Immediately the implicitly contained constraint $R_{31}(X_3, X_1)$ becomes visible via the path containing the variables $X_3$, $X_4$, $X_5$, $X_1$ and the constraints $R_{34}(X_3, X_4)$, $R_{45}(X_4, X_5)$, $X_{51}(X_5, X_1)$. Adding this new constraint explicitly to the constraint graph at the beginning of the search, yields a much smaller enumeration tree, in spite of the simplicity of the type of backtracking. The new enumeration tree contains only 172 nodes that one of Figure 8 contains 228 nodes.

Local constraint propagation in the sense of arc–consistency will change the

---

[224]   Freuder (1982) and Dechter / Pearl (1988, 1989)

[225]   Freuder (1985) and Dechter / Pearl (1989)

variable domains as follows. Since $R_{32}(X_3, X_2) = \{(1,2), (1,3), (2,3)\}$ the domains of $X_2$ and $X_3$ reduce to $\{2,3\}$ and $\{1,2\}$, respectively. Afterwards a consistency test of the variable pair $X_3$ and $X_4$ yields because of $R_{34}(X_3, X_4) = \{(1,2), (1,3), (1,4), (2,3), (2,4)\}$ that $D_4 = \{2,3,4\}$. Since $R_{45}(X_4, X_5) = \{(2,3), (2,4), (3,4)\}$ $D_4$ reduces to $\{2,3\}$, $D_5$ to $\{3,4\}$, and furthermore $R_{51}(X_5, X_1)$ becomes $\{(3,4)\}$. Now the variables $X_1$ and $X_5$ immediately become instantiated to the only remaining value of their domains, namly 4 and 3, respectively. Then $D_4$ reduces to $\{2\}$, i.e. $X_4$ is set to 2. This implies that $X_3$ will be instantiated to 1. Values 2 and 3 remain to be in domain $D_2$. It turns out that the constraint graph is arc–consistent now and a backtrackfree search yields the two solutions (4,2,1,2,3) or (4,3,1,2,3) for all five variables (only if the machine can handle two operations simultaneously). This example could demonstrate very clearly the advantages of (automatic) knowledge aquisition by introduction of additional edges into the constraint graph.

### Dynamic reduction of the enumeration tree

Constraint propagation, i.e. local consistency checks, can reduce the enumeration procedure before searching substantially but also during the search process it may happen that the modification of a variable domain reduces the domains of other variables, of those which are connected to the former by some constraints. That may lead to a cutting of some search tree branches. Certain methods are described in the literature in order to reduce the search tree. They are characterized as lookahead and lookback methods. Lookahead methods are, for instance, constraint propagation and some types of decomposition such as cycle–cutset decomposition of the constraint graph in order to yield backtrackfree subproblems [226]. Furthermore the following questions have to be answered:

What variable should be selected next for instantiation?

What is the best value for this variable?

A possible choice could be to select the mostly restricted variable, and the smallest or largest value of its domain for instantiation.

Among the main questions for lookback methods are: How far to backtrack? What is actually the reason for backtracking?

Backtracking to the latest instantiated variables is most common although this variable may not be reponsible for the necessary backtracking. Certain methods

---

[226]    Dechter (1989/90)

are described such as dependency directed backtracking [227]   or graph–based backjumping [228].   They rely on the common constraint variables.

## 3.2   The Job Shop Constraint Satisfaction Problem

Job shop scheduling is the assignment of time bounds to specific manufacturing operations with respect to a finite set of resources such as machines, tools, etc. In other words, it is a problem of proper allocation of these resources such that the manufacturing can be carried out in a timely and cost affective fashion. Much of the difficulty comes from the need to allocate limited resources to attend to a large and diverse set of objectives, requirements and preferences. The conflicts between these scheduling influences are often not clear. Hence, a good schedule must reflect a satisfactory compromise among these competing influences. The schedule influences or constraints can be partitioned into objective constraints (meeting due dates, minimizing work–in–process, maximizing recource utilization, etc.), local constraints (machine preferences, setups, job urgency, etc.), and technical constraints (resourcse requirements, precedence constraints, etc.) [229].

For a long time priority rules for the minimum makespan problem of job shop scheduling were the only possible way to tackle job shops of at least 100 operations. Recently, generally applicable approximation procedures such as tabu search, simulated annealing or genetic algorithm learning strategies became very attractive and successive solution strategies. Their general idea is to modify current solutions in a certain sense, where the modifications are defined by a neighbourhood operator, such that new feasible solutions are generated, the so called neighbours, which hopefully have improved or at most limited deterioration of their objective function value. In order to reach this goal problem specific knowledge, incorporated by problem specific heuristics, has to be introduced into the local search process of the general problem solvers. We are going to use different inference techniques – propagation techniques in order to reach a certain level of consistency – as a preprocessing step to accelerate exact solution methods as well as local search procedures. Model based local reasoning over the constraint set makes problem specific knowledge (which is

---

[227]   Petrie (1987)

[228]   Doyle (1979), Gaschnig (1979) and Dechter (1986)

[229]   Liu (1989)

implicitly contained in the model description) explicitly available for branching or bounding.

A solution of the job shop scheduling problem, i.e. a schedule, means to drop one arc from each disjunctive arc pair in the disjunctive graph representation such that the resulting graph G is acyclic, i.e. there are no precedence conflicts between operations. Obviously, the length of a maximum weight or longest path in G connecting vertices 0 and i equals the earliest possible starting time $t_i$ of operation i; the makespan of the schedule is equal to the length of the critical path, i.e. the weight of the longest path from start vertex 0 to end vertex n. Any arc (i,j) on a critical path is said to be critical; if i and j are operations from different jobs then (i,j) is called a disjunctive critical arc, otherwise it is a conjunctive critical arc. Clearly, a feasible schedule is a selection of disjunctive arcs from each set $E_k$ of disjunctive arcs between operations on machine k.

One of the main drawbacks of all branch and bound methods is the lack of strong lower bounds in order to cut branches of the enumeration tree as early as possible.[230] The most prominent bounding procedure has been described by Carlier (1982) and Potts (1980b). Consider any operation i in the job shop or in the disjunctive graph possibly already including a partial selection. Then there is a longest path from the artifical source 0 to i of length $r_i$ as well as a longest path of length $q_i$ connecting the end of operation i to the last one, the dummy operation n. Operation i cannot start to be processed earlier than its release date $r_i$ (also called head) and its processing has to be finished latest until its due date $t_n - q_i$ in order to cause no schedule delay. The time $q_i$ is said to be the tail of operation i. Hence there exist m one machine lower bounds for the optimal makespan of the job shop scheduling problem where each bound is obtained from the exact solution of a one machine scheduling problem with release dates and due dates. Although this problem is NP–complete Carlier's algorithm quickly solves the one machine problems optimally for all problem sizes in the job shop under consideration. Balas, Lenstra, and Vazacopoulos (1992) describe an even better branch and bound procedure that can yield improved lower bounds. Their method additionally takes minimum delays between pairs of operations into account. Carlier's algorithm extensively makes use of the fact that the shortest makespan of a one machine schedule cannot fall below

---

[230] Several types of lower bounds are applied in the literature, for instance, bounds based on Lagrangian relaxation (Van de Velde (1991)), bounds based on the optimal makespan of a subproblem consisting of only two or three jobs and all machines (Akers (1956) and Brucker (1988), or Brucker / Jurisch (1993)).

$$LB1(C) := \min \{r_i \mid i \in C\} + \sum_{i \in C} p_i + \min \{q_i \mid i \in C\}$$

for any subset C of all operations which have to be scheduled on a particular machine, where $p_i$ is the processing time of operation i.

Carlier / Pinson (1989, 1990), Brucker et al. (1991, 1992), and Applegate / Cook (1991) applied a lot of additional inference rules – several are summarized in the sequel – in order to cut the enumeration tree during a preprocessing or the search phase.

Consider a one machine scheduling problem consisting of the set N of operations, release dates $r_i$ and tails $q_i$ for all $i \in N$. Let $\mathscr{C}_{max}$ the maximum completion time of a feasible schedule, i.e. $\mathscr{C}_{max}$ is an upper bound for the makespan of an optimal one machine schedule. Let $E_C$, $S_C$ and C be subsets of N such that $E_C$, $S_C \subseteq C$ and any operation $j \in C$ also belongs to $E_C$ ($S_C$) if there is an optimal one machine schedule such that j is first (last) processed among all operations in C. Then the following conditions hold for any operation k of N :

(1)      If $r_k + \sum_{i \in C} p_i + \min \{q_i \mid i \in S_C, i \neq k\} > \mathscr{C}_{max}$ then $k \notin E_C$.

(2)      If $\min \{r_i \mid i \in E_C, i \neq k\} + \sum_{i \in C} p_i + q_k > \mathscr{C}_{max}$ then $k \notin S_C$.

(3)      If $k \notin E_C$ and $LB1(C-\{k\}) + p_k \geq \mathscr{C}_{max}$ then $k \in S_C$.

(4)      If $k \notin S_C$ and $LB1(C-\{k\}) + p_k \geq \mathscr{C}_{max}$ then $k \in E_C$.

(1) and (2) describe sufficient conditions under which there is no one machine schedule such that an operation k is processed before or after all operations belonging to set C. Conditions (3) and (4) are sufficient in order to conlude that there exists an optimal one machine schedule such that operation k is processed after or before all operations of set C.

The preceding results tell us that, if C contains only two operations i and k such that $r_k + p_k + p_i + q_i \geq \mathscr{C}_{max}$ then operation i is processed before operation k, i.e. from the disjunctive arc pair connecting i and k arc (i,k) is selected. Moreover (3) and (4) can be used in order to adapt heads and tails within the branch and bound process, see Carlier (1982) and Carlier / Pinson (1989). If (3) or (4) holds then one can fix all arcs (i,k) or (k,i), respectively, for all operations

$i \in C$, $i \neq k$. Application of (1) to (4) guarantees an immediate selection of certain arcs from disjunctive arc pairs before branching into a subtree. There are problem instances such that conditions (1) to (4) cut the enumeration tree substantially. The branching structure of Carlier / Pinson (1989) is based on the disjunctive arc pairs which define exactly two subtrees. Let i and j be such a pair of operations which have to be scheduled on the same machine. Then, roughly speaking, according to Bertier and Roy (1965), both sequences of the two operations are checked with respect to their regrets if they increase the best lower bound LB. Let

$$d_{ij} := \max \{0, r_i + p_i + p_j + q_j - LB\}$$

$$d_{ji} := \max \{0, r_j + p_j + p_i + q_i - LB\},$$

$$a_{ij} := \min \{d_{ij}, d_{ji}\} \quad \text{and} \quad b_{ij} := |d_{ij} - d_{ji}|$$

Among all possible candidates of disjunctive arc pairs that one is chosen that maximizes $b_{ij}$ and, in case of a tie, the pair is chosen with the maximum $a_{ij}$. Branching in the algorithm of Brucker, Jurisch, and Sievers (1991) is restricted to moves of operations which belong to a critical path. For a block, i.e. successively processed operations on the same machine, that belongs to a critical path new subtrees are generated if an operation is moved to the very beginning or the very end of this block. In any case the critical path is modified. In order to keep the generated subproblems nonintersecting it is necessary to fix some additional arcs. Promising subproblems are heuristically detected.

Consider the model of a disjunctive graph for the job shop scheduling problem [231]. Such a disjunctive graph is a particular form of a binary constraint graph. An operation representing vertex corresponds to a variable in the constraint satisfaction model. A variable's domain consists of all possible starting times of the operation. Conjunctive arcs (arcs between operations of the same job) describe the machine requirements for a job and the orientation of the arc defines certain time dependencies. A disjunctive arc pair connecting two operations i, j which have to be scheduled on the same machine may be replaced by an undirected edge connecting operations i and j, or variables $X_i$ and $X_j$, respectively. Such an edge defines the constraint $R_{ij}(X_i, X_j) = R_{ji}(X_j, X_i)$ corresponding to $X_i + p_i \leq X_j$ or $X_j + p_j \leq X_i$, either operation i is scheduled

---

[231] see Figure 5 of Chapter I

before operation j or the other way round.

As an example consider three operations which have to be scheduled completely and non–overlapping on one machine within a time interval $[0...7]$. Assume the operations have the following processing times, release dates and tails: $p_1 = 1$, $p_2 = 2$, $p_3 = 4$, $r_1 = 2$, $r_2 = r_3 = 0$, $q_2 = 2$, $q_1 = q_3 = 0$. A 1–consistency (also called node–consistency) test yields the following domains for all three variables $X_1$, $X_2$, and $X_3$: $D_1 = [2...6]$, $D_2 = [0...3]$, and $D_3 = [0...3]$. Before reaching arc–consistency we get the following tuples associated to each edge of the constraint graph. Regard, that we are using a matrix constraint notation where a 1–entry at the matrix position $(R_{ij})_{t1,t2}$ means that $(t1,t2)$ is an edge connecting $X_i$ to $X_j$ associated tuple. So we get:

| $R_{12}$ | 0 | 1 | 2 | 3 |
|---|---|---|---|---|
| 2 | 1 | . | . | 1 |
| 3 | 1 | 1 | . | . |
| 4 | 1 | 1 | 1 | . |
| 5 | 1 | 1 | 1 | 1 |
| 6 | 1 | 1 | 1 | 1 |

| $R_{13}$ | 0 | 1 | 2 | 3 |
|---|---|---|---|---|
| 2 | . | . | . | 1 |
| 3 | . | . | . | . |
| 4 | 1 | . | . | . |
| 5 | 1 | 1 | . | . |
| 6 | 1 | 1 | 1 | . |

| $R_{23}$ | 0 | 1 | 2 | 3 |
|---|---|---|---|---|
| 0 | . | . | 1 | 1 |
| 1 | . | . | . | 1 |
| 2 | . | . | . | . |
| 3 | . | . | . | . |

Performing local constraint popagation in order to reach arc–consistency yields:

| $R_{12}$ | 0 | 1 | 2 | 3 |
|---|---|---|---|---|
| 2 | 1 | . | . | . |
| 3 | . | . | . | . |
| 4 | . | . | . | . |
| 5 | . | . | . | . |
| 6 | 1 | 1 | . | . |

| $R_{13}$ | 0 | 1 | 2 | 3 |
|---|---|---|---|---|
| 2 | . | . | . | 1 |
| 3 | . | . | . | . |
| 4 | . | . | . | . |
| 5 | . | . | . | . |
| 6 | . | . | 1 | . |

| $R_{23}$ | 0 | 1 | 2 | 3 |
|---|---|---|---|---|
| 0 | . | . | 1 | 1 |
| 1 | . | . | . | 1 |
| 2 | . | . | . | . |
| 3 | . | . | . | . |

A path–consistency check finally infers that the tuple $(6,1)$ associated to constraint $R_{12}(X_1,X_2)$ is infeasible, and we get:

| $R_{12}$ | 0 | 1 | 2 | 3 |
|---|---|---|---|---|
| 2 | 1 | . | . | . |
| 3 | . | . | . | . |
| 4 | . | . | . | . |
| 5 | . | . | . | . |
| 6 | 1 | . | . | . |

| $R_{13}$ | 0 | 1 | 2 | 3 |
|---|---|---|---|---|
| 2 | . | . | . | 1 |
| 3 | . | . | . | . |
| 4 | . | . | . | . |
| 5 | . | . | . | . |
| 6 | . | . | 1 | . |

| $R_{23}$ | 0 | 1 | 2 | 3 |
|---|---|---|---|---|
| 0 | . | . | 1 | 1 |
| 1 | . | . | . | . |
| 2 | . | . | . | . |
| 3 | . | . | . | . |

Now, we can conclude that $(X_1, X_2, X_3) \in \{(2,0,3), (6,0,2)\}$ are the only possible solutions.

Unfortunately, in the worst case the representation of any relation $R_{ij}$ needs a $|D_i| \times |D_j|$ – matrix. If there are n operations which have to be processed on m different machines $\mathcal{O}(m \cdot n^2 \cdot a^2)$ entries have to be handled (without the temporary use of memory for the path–consistency tests) where a is the maximum size of a domain. This leads even for small sized problems to an enormous amount of memory what is not at all comparable to the increase of efficiency that can be reached. Therefore we are going to represent a domain as an interval of integers of possible processing starts, but only the interval defining endpoints are considered as domain values [232]. Thus, an inconsistency of a tuple is not realized if one of the tuple entries belongs to the interior of an interval. Such an inconsistency is only excluded from the interval if the value of the tuple responsible for the inconsistency, will become an interval endpoint at a certain time. Consider again any constraint $R_{ij}(X_i, X_j)$. Then the left bound (with respect to arc–consistency on constraint $R_{ij}$) of the domain interval $D_i$ is defined by the earliest possible starting time of operation i such that there is a possible starting time of operation j within its domain $D_j$. The right bound (with respect to arc–consistency on constraint $R_{ij}$) of the domain $D_i$ is defined as the latest possible starting time of operation i such that there is a possible starting time of operation j in $D_j$ and both operations can be processed. Regard, the order in which i and j are processed is not fixed. An arc–consistent domain of operation i means that $D_i$ is bounded to the left by the maximum of all such earliest possible starting times with respect to all operations j on the same machine. The right bound is given by the minimum of all such latest possible starting times of operation i with respect to all operations j on the same machine. Arc–consistency of an interior point of the interval is not guarantied unless it becomes an interval endpoint at some time. Consider the above example once more. Operation 1 is arc–consistent to operations 2 and 3 in the interval [2...6]. Operation 2 is arc–consistent to operations 1 and 3 in the interval [0...3] and [0...1], respectively. Finally, operation 3 is arc–consistent to 1 and 2 in [0...3] and [2...3], respectively. So, we may conclude that operations 1, 2, and 3 are arc–consistent in the interval [2...6], [0...1], and [2...3], respectively. Consider the disjunctive graph model. Then the longest path from 0 to operation i, which defines the release date $r_i$ of i, is a lower bound of the left bound of an

---

arc–consistent interval. Correspondingly, if we substract the tail $q_i$ from the makespan then we obtain an upper bound for the right endpoint of the interval. Hence computing heads and tails generally yields only node–consistent (1–consistent) domains of possible starting points for all operations but, if one arc is selected from each disjunctive arc pair, i.e. we have an acyclic graph defining a feasible schedule then the computation of heads and tails provides an arc–consistent solution. Even more, without backtracking a feasible solution can be found.

A test of path–consistency means that for any two path–consistent operations i and j, where i may be processed before j, and any third operation k either k, i, j is a possible processing sequence or i, k, j or i, j, k. All three possibilities are checked and the intervals respectively modified. This corresponds to a restricted check of (1) to (4) with respect to three operations. Arc–consistency is limited to only two operations and constraints (1) and (2).

### 3.3    An Immediate Selection Heuristic

The algorithm of Carlier / Pinson (1989) is based on the above mentioned results (1) to (4) and hence it heavily depends on good upper bounds on the makespan of an optimal schedule in order to select an arc from a disjunctive arc pair in advance. The better the upper bound the more arcs can be selected. Obviously, (1) to (4) can also be applied to any "upper" bound even without the assertion that the applied bound is actually an upper bound of a feasible schedule. An order k–consistency algorithm will recognize a bound which is chosen too small for a feasible schedule, but arc– or path–consistency probably is not sufficient to find conflicting operations resulting from a bound too small for feasibility. However, it might occur that the arc– or path–consistency checks lead to the selection of arcs from disjunctive arc pairs, which are not in any optimal schedule. On the other hand the selection of arcs, even if they are illegal for an optimal schedule, might lead to an increase of the best lower bound which also may not be any longer a feasible lower bound. If this new "lower" bound exceeds the "upper" bound then it is obvious that the "upper" bound actually is a true lower bound of the problem. Unfortunately there are many problems such that any feasible upper bound (even the makespan of an optimal schedule) yields only a few arc selections (which would also occur in all optimal schedules). Most of these arc selections are trivial, so they cannot accelerate a branch and bound procedure.

Therefore the basic idea is as follows. Consider an optimal solution $\mathscr{A}$ of a job shop scheduling problem, then we can extract the schedule part of an arbitrary subproblem $\mathscr{S}$ consisting of the operations to be processed on several machines or of several jobs. Now remove all operations from $\mathscr{A}$ that do not belong to $\mathscr{S}$, i.e. all vertices corresponding to operations not in $\mathscr{S}$ and all arcs incident to any of these vertices are deleted from the disjunctive graph. All remaining arc directions from the optimal schedule $\mathscr{A}$ in the disjunctive graph are kept. Thus we get a new acyclic graph corresponding to a feasible schedule of the subproblem $\mathscr{S}$, actually it is a restriction of $\mathscr{A}$ to $\mathscr{S}$ where all operations are starting as early as possible (in a Gantt diagram operations are shifted left if possible). Let $C_{max}(\mathscr{A}_{|\mathscr{S}})$ be the makespan of the feasible schedule. Consider again the complete problem and a set $\mathscr{P}$ of subproblems $\mathscr{S} \in \mathscr{P}$. If it would be possible to know all makespans $C_{max}(\mathscr{A}_{|\mathscr{S}})$ of all subproblems $\mathscr{S} \in \mathscr{P}$ in an optimal schedule $\mathscr{A}$ of the complete problem then we could easily reach arc- and path-consistency of the subproblem $\mathscr{S}$ with respect to its makespan $C_{max}(\mathscr{A}_{|\mathscr{S}})$, for all $\mathscr{S} \in \mathscr{P}$. Probably we can select much more arcs from disjunctive arc pairs than we could for the complete problem (with respect to constraint propagation) and all of these selected arcs would be feasible for at least one optimal schedule, namely $\mathscr{A}$, of the complete problem. The introduction of these new arcs could enormously accelerate branch and bound solution procedures. However one handicap remains. Usually it is impossible to compute the exact values $C_{max}(\mathscr{A}_{|\mathscr{S}})$. So we considered a slightly different strategy that probably gives us improved lower bounds for the complete problem but might expel all optimal solutions. In order to find a good approximation of $C_{max}(\mathscr{A}_{|\mathscr{S}})$ we are solving the multiple job subproblems $\mathscr{S} \in \mathscr{P}$ to optimality, a task that can easily be achived by the branch and bound algorithm of Carlier / Pinson (1989) or Brucker et al. (1991). The subproblems are reasonable small, so the computation of even an optimal schedule does not cause any time problems. Let $C_{max}(\mathscr{S})$, for all $\mathscr{S} \in \mathscr{P}$, be the makespan of an optimal subproblem schedule. Replacing in the above constraint propagation procedure $C_{max}(\mathscr{A}_{|\mathscr{S}})$ by $C_{max}(\mathscr{S})$ would lead to a lot of wrongly selected edges for the complete job shop problem. Much better results could be obtained when the tight bound $C_{max}(\mathscr{S})$ is relaxed, i.e. we added a certain percentage p% of $C_{max}(\mathscr{S})$ and in the constraint propagation procedure the bound $C_{max}(\mathscr{A}_{|\mathscr{S}})$ is replaced by $C_{max}(\mathscr{S}) \cdot (1 + p/100)$. In our computational results we will show that the right choice of p can lead to almost optimal solutions in a reasonable amount of time.

The performance of the resulting heuristic combining a constraint propagation and a branch and bound strategy can lead to effective heuristics even faster than many local search procedures. Moreover, all values $C_{max}(\mathcal{S})$ are lower bounds for the makespan of an optimal schedule of the complete problem. It is not necessary to use an exact method in order to solve the subproblems and the finally resulting restricted complete problem. Fast heuristics, such as local search procedures [233] would do an almost equally effective job. In this study however we focused on the use of exact methods in order to learn about the problem structure and to reach solutions as close to the optimum as possible.

A job shop scheduling problem of n jobs contains $\binom{n}{k}$ subproblems consisting of k out of these n jobs. For large n and k = n/2 there are too many subproblems to solve them in a reasonable amount of time, and for k close to n the subproblem optimal schedules become very similar to the corresponding part in the complete problem optimal schedule. In both cases it becomes too time consuming to solve all such subproblems. If k is very small the optimal schedules of the subproblems possibly differ substantially from an overall optimal schedule. Moreover, several subproblems should overlap. So, it may happen that after the constraint propagation phase on overlapping subproblems opposite directed arcs or cycles are created, in particular for small values of k. It seems to be impossible to judge for the complete problem which of these contradicting arcs are the right ones in an optimal schedule. Therefore a test on feasibility is necessary and all contradicting non–conjunctive arcs, those which belong to a cycle and all disjunctive selected arc pairs are deleted. We examine the best values for k, p, and the number of solved subproblems in order to reflect the overall problem structure as far as possible within the subproblems. So, we will have the following phases:

(1)        Choose the number and size of the subproblems and solve all chosen subproblems to optimality.

(2)        Constraint propagation with respect to the subproblems. Include the obtained arcs into the whole problem and remove infeasibilities.

(3)        Solve the complete problem optimally under the additional arc constraints of step (2).

---

[233]    for instance the tabu search implementation of Dell'Amico / Trubian (1993) or Nowicki / Smutnicki (1993)

### 3.4 Computational Results

As a testbed for the proposed heuristic we considered the well–known benchmark problem of Fisher and Thompson (1963), the 10×10–job shop scheduling problem. Besides this famous problem instance we considered 10 benchmark problems which also have been used by Adams et al. (1988) and by Brucker et al. (1991). In order to distinguish these problems from literature we adopted the naming from Brucker et al. (1991) respectively Adams et al. (1988). So, the 10×10–problem is indicated T1–P02. Among the others are seven 10 jobs 10 machines problems T1–P04, T1–P05, T2–P16, T2–P17, T2–P18, T2–P19, T2–P20, and three problems of a size 15 jobs and 10 machines, namly T2–P22, T2–P23, and T2–P24. We were particularly interested in a study of the well–known 10×10–problem and wondered what makes this instance so difficult to solve. Therefore we randomly generated 50 job shop scheduling problems consisting of 10 jobs and 10 machines numbered from 1010–01 to 1010–50. The processing time of each operation was randomly chosen from the interval [1...99], the machine sequence of each job was also generated at random under the assumption that each job is supposed to be processed on each machine exactly once. In all tables below we used the notation f_k.p with respect to a particular problem of size n×m indicating that we considered f subproblems $\mathscr{I}$ consisting of randomly chosen k out of n jobs. Then an upper bound $C_{max}(\mathscr{I}) \cdot (1 + p/100)$ serves as a bound for the constraint propagation phase on $\mathscr{I}$ where $C_{max}(\mathscr{I})$ is the makespan of an optimal schedule for $\mathscr{I}$. Every of the n jobs has the same chance to be chosen a member of $\mathscr{I}$. A value p = 0 is skipped. The branch and bound algorithm of Carlier and Pinson (1989) was substantially slower than that one of Brucker / Jurisch / Sievers (1992), for instance it took more than four times the running time of the Brucker et al. (1991) algorithm to solve the benchmark 10×10 to optimality, starting from an upper bound 942. Therefore we solved all subproblems as well as the complete problem containing a partial selection, unless otherwise stated, always by the branch and bound procedure of Brucker, Jurisch, and Sievers (1992). The algorithms are implemented in PASCAL and C. We used our own implementation for the Carlier / Pinson branch and bound procedure, and obtained the implementation of Brucker, Jurisch and Sievers algorithm from them. The latter was modified for the overall optimization part starting from a partial selection. All runs were performed on two comparable PCs (a 80386 processor and an IBM PS/70 both under 25 MHz).

In Tables 7 and 8 we can see the obvious fact that if p is increased then the running time is also increased and the quality of the solutions is improved (keeping f and k constant). Clearly, if p is increased then the upper bounds for the subproblems are less tight leading to a smaller set of (possibly wrongly) selected disjunctive arcs. Hence, the overall optimization step finally will require more branches. If the number of jobs k is increased then the quality of the solutions is higher on the cost of an increased running time. There seems to be a tradeoff between the number f of subproblems solved and the quality of the final outcome in relation to the running time. Exceeding a certain number of subproblems does not contribute to the overall outcome, it only rises the running time. A large number of subproblems leads to a large set of selected arcs, some of which are not existing in an optimal solution of the complete problem. The final outcome turns to become worse. It might happen that only one wrongly chosen arc among a larger set of selected arcs may lead to even worse results than several wrongly chosen arcs in a small set of selected ones. The reason is that in the first case there is almost no chance for the final branch and bound procedure to compensate the wrong choice while this compensation is still possible in the second case.

Considering the randomly generated 10 jobs 10 machines problem instances we can see that the best results with respect to quality of the outcome are obtained in column 10_7.10 of Table 7. Only one instance, 1010–48 could not be solved to optimality and on average the makespans are only 0.005 % above the optimal average makespan. The time to get the results is nearly the same as that one needed by the branch and bound method of Brucker et al. (1991). Columns 10_6.10 and 20_6.10 show results which are on an average only 0.178 % and 0.271 %, respectively, above the optimal values and the computation time is only about 2/3 of the branch and bound running time. Columns 10_5.5, 10_5.10, and 20_5.10 present results of about 1% above the optimum (on average) requiring only about 30% to 50% of the running time of the exact method. The best results are obtained if the subproblems are about half the size of the complete problem, the number of considered subproblems is about the number of jobs, and the optimal makespan of the subproblems is increased by about 10% in order to get a reasonable upper bound for the constraint propagation phase. Surely, smaller upper bounds lead to a much faster algorithm with a running time of less than 10% of the exact one, but on the cost that the outcome is also about 10% above the optimum value. Differences between the structure of an optimal schedule of the complete problem and the optimal schedules of the subproblems are more serious. Also if the probably best configurations are chosen, see columns 10_5.10 and 10_6.10, still several problems remain to be unsolved to optimality. So we can conclude that in such a case wrong arcs of disjunctive arc pairs are chosen. That is true in particular for the more difficult instances. Hence it might be save to say, that a criterion of a more difficult problem is the value

$$\max \{C_{max}(\mathscr{A}_{|\mathscr{S}}) - C_{max}(\mathscr{S}) \mid \mathscr{S} \text{ is subproblem of } \mathscr{A}\},$$

i.e. the maximum difference between the makespan of a subproblem $\mathscr{S}$ in an optimal schedule of the complete problem $\mathscr{A}$, and the optimal makespan of $\mathscr{S}$.
That might also be the reason why the famous T1–P02 problem is so difficult to solve. Table 12 shows that we had to add up to 28% to the makespan of an optimal subproblem consisting of 5 jobs in order to reach the optimum 930.
The computational results for the problems from literature (except T1–P02) are shown in Table 8. The best results are obtained for 10_5.10, 10_6.10, 20_6.10, 20_6A.10, and 10_7.10 with on an average a time consumption of 1/3, 1/2, about 1/3, 1/3, and 2/3, respectively, of the running time needed for the exact algorithm. The quality of the solutions is far less than 1/2% above the optimum. Hereby the problem 20_6A.10 (and 20_6A) indicates that twenty subproblems each consisting of 6 jobs are considered and for any two jobs the number of occurrencies of these jobs in subproblems differs at most by one. Again tight upper bounds can yield a much faster heuristic, the running time of which may decrease to less than 2% of the optimal procedure, but the quality of the solutions will slow down to almost 9% above the optimum value (column 10_5). Again a subproblem size of slightly larger than half the problem size in the number of jobs and a number of subproblems comparable to the number of jobs in the complete problem worked best. A value $p = 10$ seems to be the best choice. We did computations for larger $p$ without improving the relation of quality to running time. For these problems from literature it turned out that the subproblems should be constituted such that each job is in almost the same number of subproblems. Hence, it is not sufficient to choose the jobs with the same chance if the number of subproblems is small, it is also recommended to make sure that jobs are actually chosen almost equally often.
A similar behavior seems to show the bigger examples of 15 jobs and 10 machines, see Table 9. We only compared T2–P22 and T2–P23 to the results of the exact method of Brucker et al. (1991). Problem T2–P24 could not be solved satisfactory (indicated by a dotted line) neither by Brucker, Jurisch and Sievers

(1992), nor with our method using weak subproblem upper bounds, within the given time limit. The stratagy 15_8.10 seems to perform best which is completely in line with our results obtained for the set of 10×10–problems.

Tables 10 and 11 present the results were the branch and bound algorithm of Brucker et al. (1992) (in Table 11) is replaced by our implementation of the branch and bound algorithm of Carlier and Pinson (1989) (in Table 10). We restricted our attention to only nine randomly chosen problems out of fifty. The algorithm of Carlier and Pinson is used to solve the subproblems to optimality and to solve the complete problem to optimality under the additional assumption that already some of the disjunctive arcs are selected. The time savings are much higher than those compared to the algorithm of Brucker et al. We also included the results for those computations where we increased the makespan of the optimally solved subproblems by $p = 15\%$. It shows that we could always reach the optimal value saving 60% of the time for 10_5.15 and 19% of the time for 20_7.15, with respect to the algorithm of Carlier and Pinson.

If we compare our results from Table 8 to those obtained by Balas et al. (1988) with the famous shifting bottleneck heuristic (SB2) then SB2 did a worse job on the instances of Table 8, see Table 14. The shifting bottleneck heuristic needed on a VAX 780/11 1 hour 3 minutes and 10 seconds to solve the seven problem instances. Sunmming up all makespans gives a value of 6580. However, the strategies 10_5.10, 20_5.10, 10_6.10, and 20_6.10 did a better job. The makespan sums are 6524, 6566, 6504, and 6517, respectively, i.e. in all four cases below the value obtained by SB2. The total running times are much below, namely 28:30 minutes, 16:27 minutes, 43:19 minutes, and 32:42 minutes, respectively.

Of particular interest is the famous 10×10–problem T1–P02. Therefore we did a lot of more runs with different parameter settings on this instance, the results are shown in Table 12. We used Brucker et al.'s branch and bound procedure as an exact method. Again the column 'Opt' shows the best values we obtained. Additionally we present the number of tree nodes in the branch and bound procedure which is performed finally after a partial selection of arcs is introduced into the complete problem. The column 'Time (3)' only gives the time needed for this optimization phase (3). The difference between 'Total Time' and 'Time' is the time needed to solve the subproblems to optimality and to perform the constraint propagation phase. The last column 'Arcs' shows the number of arcs introduced by the constraint propagation phase (2). We can see that a major reason why this problem is so difficult is the substantial difference between the optimal makespan of a subproblem and the makespan of a subproblem as it will be scheduled in the optimal complete schedule. Only six times the optimal solution with the makespan of 930 is found and the running time is almost the same as for the Brucker et al. (1992) algorithm used to solve T1–P02, see Table 12.

What is striking is that obviously there is only one "difficult edge" within the 10×10 problem. If we number the operations from 1 to 100 (as presented in Brucker et al.) then there are two operations labeled 22 and 57 which have to be processed on the same machine. A sequence "57 before 22" is in the branch and bound procedure for a long time apparently a bad branch, so the opposite direction "22 before 57" is chosen first. However the latter cannot lead to an optimal solution even if we fix only the order of these two operations. The best value that can be obtained, including that operation 22 is processed before 57, is 937. However, if we only fix the order of these two operations such that 57 comes first then the algorithm of Brucker et al. (1992) can find the optimum within 7:28 minutes on a PC–386. This is substantially faster than running the algorithm from the scratch. So, these disjunctive arcs connecting operations 22 and 57 provide to be a very good candidate for branching. Starting with such candidates first certainly could accelerate the existing branch and bound algorithm.

Additional detailed description including the number of introduced arcs during the constraint propagation phase, the number of nodes in the branch and bound tree in the overall optimization phase and the time needed for the latter is depicted in Table 13 for four out of the 50 10×10 problems. It turns out if the upper bound on the subproblems is really tight, for instance equals their optimal makespan, then almost no additional branching is necessary in order to find a feasible schedule.

We include Table 14 in order to provide a base for comparing the algorithm running times on different computers for 8 benchmark problems.

| | BJS | | 10_5 | | 10_5.5 | | 10_5.10 | | 20_5 | | 20_5.5 | | 20_5.10 | | 50_5 | |
|---|---|---|---|---|---|---|---|---|---|---|---|---|---|---|---|---|
| Problem | Time | Opt | Time | Opt | Time | Opt | Time | Opt | Time | Opt | Time | Opt | Time | Opt | Time | Opt |
| 1010-01 | 1:52 | 781 | 0:31 | 790 | 1:56 | 788 | 0:50 | 781* | 0:23 | 884 | 0:30 | 823 | 0:54 | 785 | 0:13 | 884 |
| 1010-02 | 0:39 | 862 | 0:18 | 879 | 0:17 | 862* | 0:27 | 862* | 0:23 | 980 | 0:24 | 862* | 0:30 | 862* | 0:14 | 980 |
| 1010-03 | 0:55 | 849 | 0:14 | 935 | 0:26 | 849* | 0:26 | 849* | 0:24 | 939 | 0:20 | 883 | 0:32 | 849* | 0:51 | 974 |
| 1010-04 | 1:47 | 932 | 0:14 | 977 | 0:19 | 943 | 1:57 | 943 | 0:27 | 1109 | 0:19 | 1003 | 0:23 | 942 | 1:06 | 1003 |
| 1010-05 | 4:39 | 900 | 0:12 | 1149 | 0:19 | 923 | 1:08 | 900* | 0:28 | 966 | 0:26 | 927 | 1:01 | 909 | 0:58 | 1136 |
| 1010-06 | 1:55 | 841 | 0:12 | 932 | 0:15 | 882 | 0:44 | 851 | 0:24 | 938 | 0:23 | 869 | 0:36 | 843 | 0:53 | 965 |
| 1010-07 | 2:56 | 770 | 0:14 | 812 | 0:53 | 779 | 1:49 | 770* | 0:27 | 923 | 0:30 | 812 | 1:14 | 778 | 1:00 | 934 |
| 1010-08 | 1:22 | 823 | 0:13 | 1015 | 0:50 | 851 | 2:54 | 836 | 0:26 | 965 | 0:24 | 920 | 0:58 | 834 | 1:02 | 930 |
| 1010-09 | 0:40 | 804 | 0:10 | 884 | 0:14 | 821 | 0:16 | 804* | 0:21 | 932 | 0:17 | 862 | 0:20 | 835 | 0:49 | 850 |
| 1010-10 | 19:48 | 891 | 0:19 | 995 | 2:03 | 912 | 7:11 | 892 | 0:33 | 1151 | 0:52 | 926 | 2:08 | 892 | 1:04 | 1057 |
| 1010-11 | 3:00 | 899 | 0:14 | 1003 | 0:25 | 929 | 0:56 | 924 | 0:23 | 1194 | 0:34 | 943 | 0:47 | 929 | 0:58 | 996 |
| 1010-12 | 0:54 | 795 | 0:12 | 858 | 0:23 | 798 | 0:52 | 795* | 0:22 | 896 | 0:29 | 795* | 0:41 | 795* | 0:53 | 878 |
| 1010-13 | 0:12 | 861 | 0:12 | 951 | 0:14 | 896 | 0:14 | 861* | 0:27 | 908 | 0:20 | 899 | 0:23 | 861* | 1:01 | 881 |
| 1010-14 | 5:53 | 901 | 0:17 | 1018 | 6:56 | 932 | 3:35 | 903 | 0:25 | 1072 | 0:40 | 960 | 1:22 | 910 | 1:07 | 1004 |
| 1010-15 | 4:18 | 843 | 0:15 | 896 | 0:29 | 849 | 1:46 | 843* | 0:26 | 1005 | 0:23 | 863 | 0:38 | 846 | 1:04 | 885 |
| 1010-16 | 2:04 | 797 | 0:13 | 859 | 1:43 | 811 | 1:31 | 810* | 0:22 | 923 | 0:53 | 829 | 1:57 | 829 | 0:54 | 864 |
| 1010-17 | 5:16 | 791 | 0:13 | 893 | 1:02 | 791* | 1:38 | 791* | 0:26 | 871 | 0:32 | 821 | 1:28 | 792 | 1:01 | 843 |
| 1010-18 | 2:08 | 828 | 0:11 | 920 | 0:14 | 871 | 1:03 | 830 | 0:25 | 1049 | 0:19 | 895 | 0:23 | 830 | 0:18 | 1049 |
| 1010-19 | 1:41 | 824 | 0:10 | 978 | 0:14 | 868 | 0:52 | 824* | 0:24 | 967 | 0:31 | 836 | 0:38 | 828 | 0:53 | 897 |
| 1010-20 | 19:34 | 971 | 0:14 | 1130 | 0:41 | 1013 | 3:52 | 971* | 0:29 | 1085 | 1:19 | 982 | 5:53 | 971* | 1:11 | 1150 |
| 1010-21 | 1:47 | 820 | 0:12 | 880 | 0:13 | 828 | 0:34 | 821 | 0:24 | 887 | 0:19 | 854 | 0:22 | 821 | 0:53 | 884 |
| 1010-22 | 1:49 | 776 | 0:11 | 884 | 0:36 | 793 | 0:58 | 793 | 0:25 | 883 | 0:22 | 812 | 0:53 | 796 | 0:53 | 859 |
| 1010-23 | 5:34 | 810 | 0:13 | 910 | 0:18 | 810* | 1:52 | 810* | 0:23 | 845 | 0:25 | 834 | 0:47 | 810* | 0:58 | 912 |
| 1010-24 | 2:31 | 800 | 0:14 | 909 | 0:32 | 808 | 2:16 | 808 | 0:24 | 999 | 0:20 | 932 | 0:45 | 813 | 0:56 | 895 |
| 1010-25 | 0:51 | 857 | 0:12 | 917 | 0:12 | 881 | 0:10 | 857* | 0:23 | 917 | 0:21 | 857* | 0:20 | 857* | 0:57 | 956 |
| 1010-26 | 6:56 | 826 | 0:19 | 914 | 4:21 | 852 | 5:32 | 826* | 0:29 | 950 | 0:25 | 879 | 9:08 | 857 | 1:03 | 918 |
| 1010-27 | 2:36 | 738 | 0:12 | 791 | 0:33 | 741 | 1:07 | 738* | 0:22 | 897 | 0:32 | 746 | 0:46 | 738* | 0:49 | 829 |
| 1010-28 | 1:25 | 869 | 0:13 | 936 | 0:15 | 905 | 0:33 | 870 | 0:22 | 1003 | 0:19 | 899 | 0:25 | 878 | 0:54 | 979 |
| 1010-29 | 15:04 | 936 | 0:16 | 1056 | 4:20 | 960 | 4:09 | 948 | 0:27 | 1086 | 0:42 | 967 | 2:22 | 948 | 1:01 | 984 |
| 1010-30 | 6:41 | 896 | 0:14 | 971 | 1:08 | 908 | 2:00 | 899 | 0:25 | 1033 | 3:36 | 897 | 4:19 | 897 | 1:03 | 1001 |
| 1010-31 | 1:39 | 859 | 0:12 | 933 | 0:11 | 866 | 0:23 | 864 | 0:26 | 912 | 0:18 | 914 | 0:28 | 864 | 0:59 | 912 |
| 1010-32 | 0:43 | 816 | 0:14 | 994 | 0:12 | 822 | 1:42 | 816* | 0:25 | 907 | 0:23 | 830 | 0:34 | 822 | 0:56 | 944 |
| 1010-33 | 0:53 | 902 | 0:11 | 1024 | 0:56 | 906 | 1:38 | 906 | 0:29 | 1013 | 0:34 | 944 | 0:38 | 907 | 1:01 | 1024 |
| 1010-34 | 2:16 | 825 | 0:12 | 916 | 0:18 | 828 | 3:23 | 828 | 0:24 | 904 | 0:25 | 831 | 1:27 | 828 | 0:56 | 880 |
| 1010-35 | 1:48 | 892 | 0:12 | 938 | 0:22 | 898 | 0:40 | 895 | 0:23 | 1004 | 0:23 | 896 | 0:39 | 896 | 0:53 | 957 |
| 1010-36 | 1:46 | 875 | 0:11 | 1109 | 0:11 | 894 | 0:19 | 884 | 0:25 | 1134 | 0:20 | 885 | 0:53 | 884 | 0:55 | 1027 |
| 1010-37 | 3:56 | 912 | 0:30 | 913 | 1:53 | 912* | 2:34 | 912* | 0:25 | 988 | 0:22 | 940 | 0:46 | 912* | 0:56 | 950 |
| 1010-38 | 0:32 | 804 | 0:10 | 865 | 0:12 | 824 | 0:13 | 824 | 0:25 | 860 | 0:25 | 825 | 0:42 | 813 | 1:02 | 830 |
| 1010-39 | 6:17 | 901 | 0:13 | 991 | 0:40 | 922 | 6:49 | 918 | 0:24 | 1026 | 1:00 | 916 | 5:48 | 916 | 1:00 | 956 |
| 1010-40 | 1:04 | 869 | 0:13 | 950 | 0:31 | 874 | 1:05 | 869* | 0:23 | 1005 | 0:46 | 900 | 0:59 | 874 | 0:54 | 1074 |
| 1010-41 | 4:19 | 910 | 0:17 | 958 | 5:50 | 910* | 3:22 | 910* | 0:21 | 1041 | 2:23 | 920 | 3:23 | 910* | 0:52 | 980 |
| 1010-42 | 0:30 | 884 | 0:16 | 951 | 0:23 | 889 | 0:35 | 889 | 0:22 | 953 | 0:27 | 937 | 0:35 | 890 | 0:58 | 997 |
| 1010-43 | 2:25 | 903 | 0:16 | 984 | 1:12 | 920 | 1:33 | 903* | 0:24 | 1024 | 0:22 | 976 | 0:55 | 904 | 1:02 | 937 |
| 1010-44 | 0:56 | 811 | 0:16 | 855 | 0:37 | 816 | 0:42 | 816 | 0:23 | 865 | 0:36 | 816 | 0:48 | 816 | 0:54 | 869 |
| 1010-45 | 3:16 | 893 | 0:14 | 977 | 1:45 | 894 | 1:51 | 893* | 0:28 | 956 | 0:26 | 922 | 1:05 | 893* | 1:07 | 980 |
| 1010-46 | 1:05 | 810 | 0:17 | 876 | 1:08 | 812 | 1:11 | 812 | 0:24 | 1071 | 1:59 | 836 | 0:48 | 810* | 0:56 | 944 |
| 1010-47 | 3:17 | 835 | 0:15 | 903 | 0:51 | 856 | 0:38 | 837 | 0:25 | 1033 | 0:22 | 883 | 0:29 | 841 | 1:00 | 929 |
| 1010-48 | 16:49 | 955 | 0:17 | 1036 | 9:04 | 975 | 14:25 | 967 | 0:26 | 1097 | 4:38 | 978 | 6:47 | 967 | 0:54 | 1056 |
| 1010-49 | 4:34 | 950 | 0:13 | 1037 | 0:34 | 950* | 0:51 | 950* | 0:27 | 1017 | 0:21 | 960 | 1:30 | 950* | 1:01 | 1001 |
| 1010-50 | 0:39 | 882 | 0:15 | 920 | 0:32 | 905 | 0:28 | 889 | 0:26 | 963 | 0:24 | 916 | 0:32 | 889 | 0:59 | 974 |
| Σ Time | 3:05:31 | | 11:58 | | 59:43 | | 1:37:34 | | 20:44 | | 34:40 | | 1:12:39 | | 46:12 | |
| φ Time | 3:42 | | 0:14 | | 1:11 | | 1:57 | | 0:24 | | 0:41 | | 1:27 | | 0:55 | |
| % of BJS | 100.00 | | 6.45 | | 32.19 | | 52.59 | | 11.18 | | 18.69 | | 39.16 | | 24.90 | |
| Σ Opt | | 42779 | | 47172 | | 43527 | | 42992 | | 49059 | | 44412 | | 43121 | | 47598 |
| % above BJS | | 0.0 | | 10.27 | | 1.75 | | 0.50 | | 14.68 | | 3.82 | | 0.80 | | 11.27 |

(1|3)

Table 7.   Constraint propagation results on 50 random 10×10 problems, part 1.

| Problem | BJS Time Opt | 10_6 Time Opt | 10_6.5 Time Opt | 10_6.10 Time Opt | 20_6 Time Opt | 20_6.5 Time Opt | 20_6.10 Time Opt | 50_6 Time Opt |
|---|---|---|---|---|---|---|---|---|
| 1010-01 | 1:52 781 | 1:03 790 | 1:43 785 | 1:16 781* | 0:50 824 | 1:30 790 | 3:24 785 | 0:23 824 |
| 1010-02 | 0:39 862 | 0:29 862* | 0:29 862* | 0:35 862* | 0:41 937 | 0:46 862* | 1:05 862* | 0:15 937 |
| 1010-03 | 0:55 849 | 0:36 883 | 1:00 849* | 0:51 849* | 0:37 877 | 0:49 858 | 1:05 858 | 1:33 905 |
| 1010-04 | 1:47 932 | 0:19 1047 | 0:17 942 | 1:10 932* | 0:39 1022 | 0:44 943 | 0:50 932* | 1:37 1003 |
| 1010-05 | 4:39 900 | 0:19 942 | 0:30 909 | 1:50 900* | 1:08 962 | 1:31 909 | 1:59 900* | 2:11 1048 |
| 1010-06 | 1:55 843 | 0:21 911 | 0:39 855 | 1:11 841* | 0:40 1009 | 1:06 851 | 1:31 841* | 1:41 875 |
| 1010-07 | 2:56 770 | 0:40 781 | 1:10 779 | 2:19 770* | 0:47 959 | 1:49 778 | 2:24 778 | 2:08 887 |
| 1010-08 | 1:22 823 | 0:21 982 | 1:31 839 | 3:18 823* | 0:41 1013 | 0:45 823* | 1:17 823* | 1:52 973 |
| 1010-09 | 0:40 804 | 0:14 836 | 0:16 804* | 0:22 804* | 0:33 925 | 0:26 810 | 0:35 804* | 1:20 922 |
| 1010-10 | 19:48 891 | 0:46 921 | 4:59 897 | 12:12 891* | 1:22 947 | 4:47 897 | 8:35 892 | 3:17 1043 |
| 1010-11 | 3:00 899 | 0:21 953 | 1:08 922 | 0:56 899* | 0:41 1004 | 0:49 935 | 2:01 929 | 1:43 1064 |
| 1010-12 | 0:54 795 | 0:17 881 | 0:41 795* | 1:26 795* | 0:37 865 | 0:54 795* | 0:50 795* | 1:38 850 |
| 1010-13 | 0:12 861 | 0:19 959 | 0:17 887 | 0:16 861* | 0:42 901 | 0:31 866 | 0:32 861* | 1:49 874 |
| 1010-14 | 5:53 901 | 0:49 948 | 3:46 923 | 7:02 901* | 0:55 1013 | 2:46 942 | 2:23 903 | 2:06 1096 |
| 1010-15 | 4:18 843 | 0:24 866 | 2:15 843* | 3:31 843* | 0:46 866 | 1:33 843* | 3:43 843* | 1:50 980 |
| 1010-16 | 2:04 797 | 0:20 870 | 3:04 811 | 1:26 810* | 0:40 835 | 1:35 811 | 2:09 797* | 1:42 922 |
| 1010-17 | 5:16 791 | 0:24 861 | 1:41 793 | 3:17 791* | 0:51 845 | 1:09 793 | 3:12 791* | 2:02 824 |
| 1010-18 | 2:08 828 | 0:21 954 | 0:34 830 | 1:15 830 | 0:36 877 | 0:50 830 | 0:53 830 | 0:20 877 |
| 1010-19 | 1:41 824 | 0:18 891 | 0:26 824* | 0:44 824* | 0:36 869 | 0:53 824* | 0:58 824* | 1:45 869 |
| 1010-20 | 19:34 971 | 0:39 1054 | 1:20 971* | 4:16 971* | 1:06 1132 | 3:49 988 | 6:48 971* | 2:42 1033 |
| 1010-21 | 1:47 820 | 0:19 851 | 0:19 837 | 1:40 820* | 0:37 946 | 0:32 840 | 1:35 820* | 1:19 913 |
| 1010-22 | 1:49 776 | 0:21 809 | 0:55 776* | 1:25 776* | 0:40 826 | 0:49 793 | 1:30 776* | 1:48 807 |
| 1010-23 | 5:34 810 | 0:20 925 | 0:36 810* | 1:36 810* | 0:42 880 | 0:57 820 | 2:00 810* | 1:37 861 |
| 1010-24 | 2:31 800 | 0:22 910 | 0:34 808 | 2:12 808 | 0:37 911 | 0:32 877 | 1:42 808 | 1:41 913 |
| 1010-25 | 0:51 857 | 0:17 921 | 0:16 857* | 0:12 857* | 0:35 900 | 0:32 857* | 0:44 857* | 1:28 905 |
| 1010-26 | 6:56 826 | 1:18 859 | 13:02 859 | 5:09 826* | 1:00 948 | 4:47 859 | 5:28 826* | 2:14 1006 |
| 1010-27 | 2:36 738 | 0:20 783 | 0:49 741 | 1:50 738* | 0:38 769 | 1:17 741 | 2:11 738* | 1:36 868 |
| 1010-28 | 1:25 869 | 0:19 936 | 0:19 909 | 0:50 870 | 0:44 954 | 0:45 890 | 1:01 870 | 1:50 965 |
| 1010-29 | 15:04 936 | 0:25 967 | 3:11 950 | 12:11 945 | 0:43 1031 | 2:52 952 | 5:38 945 | 2:02 1007 |
| 1010-30 | 6:41 896 | 0:39 928 | 2:24 900 | 2:33 896* | 1:05 926 | 5:45 897 | 9:28 897 | 2:29 964 |
| 1010-31 | 1:39 859 | 0:18 934 | 0:18 866 | 0:45 864 | 0:37 907 | 0:27 887 | 0:39 864 | 1:35 907 |
| 1010-32 | 0:43 816 | 0:22 1008 | 1:21 816* | 0:59 816* | 0:37 997 | 0:40 823 | 0:51 816* | 1:38 902 |
| 1010-33 | 0:53 902 | 0:22 1018 | 2:12 906 | 0:44 906 | 0:48 970 | 0:57 914 | 2:49 904 | 1:47 1002 |
| 1010-34 | 2:16 825 | 0:19 975 | 0:49 828 | 0:55 825* | 0:36 920 | 0:45 834 | 1:15 828 | 1:36 866 |
| 1010-35 | 1:48 892 | 0:23 898 | 0:51 898 | 1:32 892* | 0:36 984 | 0:41 896 | 0:55 896 | 1:30 1000 |
| 1010-36 | 1:46 875 | 0:18 966 | 0:17 875* | 1:18 875* | 0:40 901 | 0:31 889 | 0:41 884 | 1:37 918 |
| 1010-37 | 3:56 912 | 0:28 914 | 2:35 912* | 4:00 912* | 0:42 985 | 1:05 912* | 6:12 912* | 1:44 1030 |
| 1010-38 | 0:32 804 | 0:19 865 | 0:24 825 | 0:32 812 | 0:43 881 | 0:40 813 | 0:39 804* | 1:36 860 |
| 1010-39 | 6:17 901 | 0:35 916 | 3:48 904 | 5:17 901* | 0:51 979 | 3:03 916 | 5:56 901* | 1:59 936 |
| 1010-40 | 1:04 869 | 0:17 1020 | 0:45 869* | 0:53 869* | 0:42 874 | 0:54 874 | 1:14 869* | 1:33 950 |
| 1010-41 | 4:19 910 | 1:03 916 | 6:40 910* | 3:33 910* | 1:07 918 | 3:25 910* | 3:45 910* | 2:03 985 |
| 1010-42 | 0:30 884 | 0:24 937 | 0:46 889 | 1:22 889 | 0:52 935 | 0:59 889 | 1:02 884* | 1:32 968 |
| 1010-43 | 2:25 903 | 0:27 984 | 2:01 904 | 1:51 903* | 0:47 977 | 1:32 904 | 1:42 903* | 1:48 982 |
| 1010-44 | 0:56 811 | 0:21 890 | 0:32 816 | 0:57 816 | 0:38 866 | 0:46 816 | 1:32 816 | 1:35 921 |
| 1010-45 | 3:16 893 | 0:49 894 | 2:33 893* | 4:14 893* | 0:57 970 | 1:40 894 | 2:16 893* | 2:18 938 |
| 1010-46 | 1:05 810 | 0:38 834 | 0:46 812 | 1:26 810* | 0:50 904 | 0:52 841 | 1:28 812 | 2:03 1022 |
| 1010-47 | 3:17 835 | 0:25 906 | 1:43 841 | 3:59 835* | 0:43 936 | 0:54 841 | 4:04 837 | 1:45 868 |
| 1010-48 | 16:49 955 | 0:55 978 | 10:08 967 | 13:16 964 | 0:50 1069 | 5:48 967 | 14:08 957 | 2:43 1028 |
| 1010-49 | 4:34 950 | 0:22 1087 | 0:25 950* | 1:36 950* | 0:48 1058 | 0:45 960 | 1:06 950* | 2:14 1093 |
| 1010-50 | 0:39 882 | 0:25 926 | 0:41 905 | 0:49 889 | 0:50 1005 | 0:56 911 | 2:10 889 | 1:54 989 |
| | | | | | | | | |
| Σ Time | 3:05:31 | 23:30 | 1:29:46 | 2:08:49 | 38:03 | 1:15:10 | 2:11:55 | 1:28:28 |
| φ Time | 3:42 | 0:28 | 1:47 | 2:34 | 0:45 | 1:30 | 2:38 | 1:46 |
| % of BJS | 100.00 | 12.67 | 48.39 | 69.44 | 20.51 | 40.52 | 71.11 | 47.69 |
| | | | | | | | | |
| Σ Opt | 42779 | 45947 | 43153 | 42855 | 46809 | 43365 | 42895 | 47181 |
| % above BJS | 0.0 | 7.41 | 0.87 | 0.18 | 9.42 | 1.37 | 0.27 | 10.29 |

Table 7.   Constraint propagation results on 50 random 10×10 problems, part 2.

| Problem | BJS Time | BJS Opt | 10_7 Time | 10_7 Opt | 10_7.10 Time | 10_7.10 Opt | 20_7 Time | 20_7 Opt | 50_7 Time | 50_7 Opt | 10_8 Time | 10_8 Opt | 50_8 Time | 50_8 Opt |
|---|---|---|---|---|---|---|---|---|---|---|---|---|---|---|
| 1010-01 | 1:52 | 781 | 2:05 | 790 | 2:29 | 781* | 1:31 | 790 | 0:42 | 790 | 2:40 | 781* | 1:09 | 781* |
| 1010-02 | 0:39 | 862 | 0:40 | 879 | 0:47 | 862* | 1:25 | 890 | 0:13 | 890 | 1:52 | 862* | 0:19 | 862* |
| 1010-03 | 0:55 | 849 | 1:06 | 849* | 1:32 | 849* | 1:32 | 858 | 4:14 | 883 | 2:28 | 849* | 9:18 | 870 |
| 1010-04 | 1:47 | 932 | 0:34 | ·968 | 2:01 | 932* | 0:47 | 968 | 3:45 | 1065 | 1:30 | 943 | 7:07 | 943 |
| 1010-05 | 4:39 | 900 | 0:49 | 915 | 3:03 | 900* | 1:56 | 915 | 5:33 | 957 | 4:52 | 900* | 17:11 | 900* |
| 1010-06 | 1:55 | 841 | 0:39 | 869 | 1:27 | 841* | 1:07 | 904 | 4:01 | 1020 | 2:58 | 851 | 6:59 | 851 |
| 1010-07 | 2:56 | 770 | 1:25 | 779 | 4:00 | 770* | 2:13 | 779 | 5:45 | 862 | 4:57 | 778 | 20:25 | 778 |
| 1010-08 | 1:22 | 823 | 0:43 | 890 | 1:28 | 823* | 1:15 | 837 | 4:48 | 1021 | 2:24 | 823* | 9:29 | 845 |
| 1010-09 | 0:40 | 804 | 0:34 | 810 | 0:51 | 804* | 0:46 | 836 | 2:46 | 836 | 1:01 | 804* | 4:22 | 804* |
| 1010-10 | 19:48 | 891 | 4:48 | 897 | 12:45 | 891* | 4:43 | 896 | 11:13 | 903 | 17:25 | 891* | 38:11 | 891* |
| 1010-11 | 3:00 | 899 | 1:28 | 899* | 3:32 | 899* | 2:03 | 935 | 5:01 | 944 | 3:54 | 899* | 14:09 | 919 |
| 1010-12 | 0:54 | 795 | 0:29 | 865 | 1:30 | 795* | 1:02 | 875 | 3:04 | 819 | 1:03 | 814 | 4:29 | 821 |
| 1010-13 | 0:12 | 861 | 0:34 | 950 | 0:31 | 861* | 0:56 | 881 | 3:27 | 903 | 1:12 | 938 | 5:01 | 928 |
| 1010-14 | 5:53 | 901 | 3:46 | 937 | 11:10 | 901* | 2:48 | 901* | 7:43 | 975 | 6:55 | 932 | 23:12 | 944 |
| 1010-15 | 4:18 | 843 | 2:14 | 843* | 4:30 | 843* | 2:49 | 843* | 5:42 | 896 | 5:56 | 843* | 15:28 | 843* |
| 1010-16 | 2:04 | 797 | 1:08 | 834 | 2:04 | 797* | 1:09 | 874 | 3:46 | 863 | 2:01 | 835 | 8:56 | 835 |
| 1010-17 | 5:16 | 791 | 0:54 | 880 | 5:27 | 791* | 1:17 | 829 | 5:00 | 876 | 3:43 | 798 | 11:40 | 799 |
| 1010-18 | 2:08 | 828 | 0:37 | 969 | 1:56 | 828* | 1:06 | 1029 | 0:18 | 1029 | 1:19 | 831 | 0:11 | 831 |
| 1010-19 | 1:41 | 824 | 0:37 | 870 | 1:35 | 824* | 1:14 | 857 | 3:58 | 872 | 1:44 | 834 | 7:03 | 857 |
| 1010-20 | 19:34 | 971 | 10:31 | 988 | 22:16 | 971* | 6:16 | 971* | 12:37 | 1004 | 15:10 | 971* | 48:53 | 971* |
| 1010-21 | 1:47 | 820 | 0:34 | 861 | 1:47 | 820* | 0:39 | 861 | 2:56 | 889 | 1:20 | 840 | 4:27 | 841 |
| 1010-22 | 1:49 | 776 | 0:46 | 795 | 1:38 | 776* | 1:20 | 793 | 4:15 | 873 | 1:31 | 817 | 8:28 | 821 |
| 1010-23 | 5:34 | 810 | 0:39 | 885 | 2:00 | 810* | 1:13 | 875 | 3:53 | 877 | 2:50 | 833 | 10:00 | 839 |
| 1010-24 | 2:31 | 800 | 0:56 | 848 | 3:14 | 800* | 1:11 | 859 | 3:31 | 887 | 3:20 | 808 | 7:17 | 808 |
| 1010-25 | 0:51 | 857 | 0:23 | 926 | 0:20 | 857* | 0:39 | 881 | 2:35 | 881 | 0:42 | 881 | 3:17 | 857* |
| 1010-26 | 6:56 | 826 | 4:36 | 859 | 7:32 | 826* | 4:33 | 859 | 9:47 | 859 | 8:25 | 826* | 29:47 | 857 |
| 1010-27 | 2:36 | 738 | 0:40 | 773 | 2:10 | 738* | 1:25 | 738* | 4:16 | 768 | 3:11 | 738* | 11:02 | 738* |
| 1010-28 | 1:25 | 869 | 0:45 | 927 | 2:05 | 869* | 1:33 | 895 | 4:18 | 935 | 2:03 | 870 | 8:26 | 870 |
| 1010-29 | 15:04 | 936 | 1:06 | 973 | 8:14 | 936* | 1:11 | 1036 | 6:18 | 1039 | 3:24 | 1044 | 19:18 | 1044 |
| 1010-30 | 6:41 | 896 | 2:08 | 908 | 6:02 | 896* | 9:03 | 897 | 10:44 | 897 | 18:21 | 896* | 33:20 | 897 |
| 1010-31 | 1:39 | 859 | 0:33 | 949 | 1:14 | 859* | 1:03 | 890 | 3:20 | 887 | 1:08 | 865 | 5:20 | 907 |
| 1010-32 | 0:43 | 816 | 0:45 | 830 | 1:10 | 816* | 0:58 | 832 | 3:39 | 926 | 1:15 | 823 | 6:03 | 823 |
| 1010-33 | 0:53 | 902 | 0:56 | 914 | 1:18 | 902* | 1:40 | 906 | 4:38 | 958 | 4:16 | 904 | 12:52 | 906 |
| 1010-34 | 2:16 | 825 | 0:32 | 880 | 1:27 | 825* | 0:54 | 852 | 3:23 | 893 | 1:19 | 846 | 5:58 | 866 |
| 1010-35 | 1:48 | 892 | 0:51 | 898 | 2:20 | 892* | 1:19 | 898 | 3:54 | 934 | 2:33 | 898 | 8:29 | 898 |
| 1010-36 | 1:46 | 875 | 0:34 | 905 | 1:38 | 875* | 1:05 | 899 | 3:45 | 899 | 1:32 | 875* | 6:51 | 899 |
| 1010-37 | 3:56 | 912 | 2:46 | 912* | 4:51 | 912* | 1:13 | 955 | 4:30 | 1088 | 5:27 | 912* | 13:04 | 912* |
| 1010-38 | 0:32 | 804 | 0:35 | 881 | 0:50 | 804* | 1:02 | 877 | 3:27 | 857 | 1:48 | 813 | 6:20 | 825 |
| 1010-39 | 6:17 | 901 | 2:22 | 904 | ·6:11 | 901* | 3:16 | 904 | 5:54 | 933 | 8:31 | 901* | 23:04 | 904 |
| 1010-40 | 1:04 | 869 | 0:33 | 920 | 1:15 | 869* | 1:22 | 874 | 3:18 | 898 | 2:21 | 869* | 7:09 | 869* |
| 1010-41 | 4:19 | 910 | 2:19 | 910* | 4:04 | 910* | 2:52 | 918 | 5:24 | 918 | 8:21 | 910* | 14:24 | 910* |
| 1010-42 | 0:30 | 884 | 0:45 | 937 | 1:44 | 884* | 1:41 | 905 | 3:42 | 947 | 1:26 | 889 | 6:32 | 894 |
| 1010-43 | 2:25 | 903 | 1:34 | 904 | 3:09 | 903* | 1:48 | 922 | 4:03 | 966 | 4:24 | 903* | 10:17 | 904 |
| 1010-44 | 0:56 | 811 | 1:07 | 816 | 1:38 | 811* | 1:24 | 816 | 4:11 | 816 | 2:42 | 816 | 9:52 | 816 |
| 1010-45 | 3:16 | 893 | 1:30 | 893* | 3:44 | 893* | 2:04 | 915 | 6:09 | 946 | 5:54 | 893* | 16:39 | 893* |
| 1010-46 | 1:05 | 810 | 1:39 | 812 | 2:15 | 810* | 2:21 | 826 | 5:22 | 834 | 3:52 | 810* | 13:55 | 810* |
| 1010-47 | 3:17 | 835 | 1:02 | 908 | 6:59 | 835* | 2:02 | 1037 | 5:28 | 964 | 4:18 | 898 | 17:39 | 927 |
| 1010-48 | 16:49 | 955 | 11:42 | 967 | 18:41 | 957 | 4:36 | 967 | 10:55 | 967 | 18:53 | 967 | 30:22 | 967 |
| 1010-49 | 4:34 | 950 | 0:42 | 974 | 2:40 | 950* | 1:21 | 962 | 6:23 | 1014 | 4:23 | 950* | 13:32 | 963 |
| 1010-50 | 0:39 | 882 | 1:05 | 916 | 3:46 | 882* | 1:14 | 936 | 5:24 | 974 | 3:13 | 889 | 9:33 | 905 |
| Σ Time | 3:05:31 | | 1:22:06 | | 3:10:50 | | 1:35:57 | | 4:02:58 | | 3:37:47 | | 10:26:49 | |
| φ Time | 3:42 | | 1:38 | | 3:49 | | 1:55 | | 4:51 | | 4:21 | | 12:32 | |
| % of BJS | 100.00 | | 44.26 | | 102.87 | | 51.72 | | 130.97 | | 117.39 | | 337.88 | |
| Σ Opt | | 42779 | | 44466 | | 42781 | | 44453 | | 45932 | | 43361 | | 43643 |
| % above BJS | | 0.0 | | 3.94 | | 0.005 | | 3.91 | | 7.37 | | 1.36 | | 2.02 |

(3|3)

Table 7.    Constraint propagation results on 50 random 10×10 problems, part 3.

10 Jobs, 10 Machines

| Problem | T1-P04 | | T1-P05 | | T2-P16 | | T2-P17 | | T2-P18 | | T2-P19 | | T2-P20 | | Σ Time | % of BJS | Σ Opt | % above BJS-Opt |
|---|---|---|---|---|---|---|---|---|---|---|---|---|---|---|---|---|---|---|
| | Time | Opt | Time | Opt | Time | Opt | Time | Opt | Time | Opt | Time | Opt | Time | Opt | | | | |
| BJS | 29:44 | 1234 | 1:49 | 943 | 3:09 | 945 | 0:47 | 784 | 4:06 | 848 | 20:52 | 842 | 20:35 | 902 | 1:21:02 | 100 | 6498 | 0.00 |
| 10_5 | 0:12 | 1354 | 0:12 | 1019 | 0:15 | 996 | 0:11 | 944 | 0:13 | 934 | 0:13 | 875 | 0:19 | 940 | 1:35 | 1.95 | 7062 | 8.68 |
| 10_5.5 | 0:19 | 1275 | 0:25 | 948 | 0:13 | 982 | 0:15 | 800 | 0:12 | 921 | 3:46 | 842* | 6:02 | 907 | 11:12 | 13.82 | 6675 | 2.72 |
| 10_5.10 | 2:28 | 1236 | 1:08 | 947 | 1:08 | 956 | 0:12 | 784* | 0:40 | 857 | 7:44 | 842* | 15:10 | 902* | 28:30 | 35.17 | 6524 | 0.40 |
| 10_6 | 0:18 | 1332 | 0:19 | 1054 | 0:27 | 988 | 0:14 | 823 | 0:25 | 923 | 2:09 | 843 | 0:25 | 917 | 4:17 | 5.29 | 6880 | 5.88 |
| 10_6.5 | 2:00 | 1242 | 0:35 | 956 | 1:11 | 982 | 0:13 | 797 | 0:35 | 848* | 7:56 | 842* | 4:28 | 907 | 16:58 | 20.94 | 6574 | 1.17 |
| 10_6.10 | 3:59 | 1238 | 2:02 | 947 | 1:59 | 945* | 0:25 | 784* | 1:17 | 848* | 13:11 | 842* | 20:26 | 902* | 43:19 | 53.46 | 6506 | 0.12 |
| 10_7 | 2:44 | 1246 | 0:37 | 1018 | 1:05 | 982 | 0:28 | 828 | 0:55 | 926 | 4:37 | 842* | 1:24 | 908 | 11:50 | 14.60 | 6750 | 3.88 |
| 10_7.10 | 17:48 | 1234* | 2:03 | 946 | 1:13 | 945* | 0:32 | 784* | 4:15 | 848* | 13:57 | 842* | 19:06 | 902* | 58:54 | 72.69 | 6501 | 0.05 |
| 10_8 | 6:37 | 1242 | 3:03 | 947 | 3:26 | 956 | 0:56 | 801* | 2:15 | 883 | 11:43 | 842* | 4:01 | 912 | 32:01 | 39.51 | 6583 | 1.31 |
| 20_5 | 0:25 | 1384 | 0:27 | 1037 | 0:29 | 1150 | 0:21 | 829 | 0:25 | 967 | 0:29 | 918 | 0:30 | 970 | 3:06 | 3.83 | 7255 | 11.65 |
| 20_5.5 | 0:21 | 1275 | 0:21 | 958 | 0:28 | 982 | 0:20 | 801 | 0:20 | 916 | 4:07 | 846 | 0:31 | 912 | 6:29 | 8.00 | 6690 | 2.96 |
| 20_5.10 | 4:43 | 1244 | 2:06 | 946 | 0:57 | 982 | 0:18 | 784* | 0:33 | 861 | 4:56 | 842* | 2:54 | 907 | 16:27 | 20.30 | 6566 | 1.05 |
| 20_5A | 0:23 | 1354 | 0:22 | 1064 | 0:27 | 996 | 0:21 | 852 | 0:22 | 947 | 0:23 | 918 | 0:26 | 955 | 2:44 | 3.37 | 7086 | 9.05 |
| 20_5A.10 | 1:08 | 1246 | 0:42 | 948 | 1:10 | 956 | 0:23 | 787* | 0:19 | 861 | 6:27 | 842* | 3:48 | 907 | 13:37 | 16.80 | 6547 | 0.75 |
| 20_6 | 0:54 | 1339 | 0:40 | 1031 | 0:58 | 982 | 0:35 | 888 | 0:39 | 985 | 0:54 | 871 | 0:44 | 1040 | 5:26 | 6.71 | 7136 | 9.82 |
| 20_6.5 | 2:33 | 1253 | 1:10 | 958 | 2:12 | 982 | 0:30 | 797 | 0:39 | 862 | 4:54 | 847 | 1:12 | 914 | 11:23 | 14.05 | 6613 | 1.77 |
| 20_6.10 | 6:54 | 1236 | 0:37 | 946 | 0:47 | 956 | 0:34 | 787 | 3:01 | 848* | 9:49 | 842* | 9:02 | 902* | 32:42 | 40.35 | 6517 | 0.29 |
| 20_6A | 0:38 | 1332 | 1:48 | 1018 | 1:00 | 988 | 0:32 | 901 | 0:41 | 952 | 1:02 | 869 | 0:46 | 940 | 5:03 | 6.23 | 7000 | 7.73 |
| 20_6A.10 | 3:36 | 1238 | 1:01 | 947 | 1:50 | 945* | 0:30 | 787 | 0:57 | 848* | 12:09 | 842* | 10:09 | 902* | 30:09 | 37.21 | 6509 | 0.17 |
| 20_7 | 3:17 | 1246 | | 996 | | 982 | 0:47 | 801 | 1:22 | 882 | 8:14 | 842* | 1:57 | 1012 | 18:28 | 22.79 | 6761 | 4.05 |
| 50_5 | 0:59 | 1395 | 1:01 | 1006 | 1:03 | 1070 | 0:51 | 846 | 0:55 | 909 | 1:05 | 918 | 0:57 | 1073 | 6:51 | 8.45 | 7217 | 11.07 |
| 50_6 | 2:19 | 1451 | 1:45 | 1004 | 2:11 | 987 | 1:30 | 897 | 1:29 | 924 | 2:09 | 898 | 2:03 | 1064 | 13:26 | 16.58 | 7225 | 11.19 |
| 50_7 | 8:13 | 1258 | 4:06 | 1004 | 6:50 | 982 | 2:54 | 801 | 4:24 | 987 | 7:58 | 850* | 5:31 | 953 | 39:56 | 49.28 | 6835 | 5.19 |
| 50_8 | 20:59 | 1242 | 7:54 | 996 | 17:41 | 982 | 4:19 | 803 | 10:31 | 897 | 27:02 | 842* | 18:17 | 912 | 1:46:43 | 131.7 | 6674 | 2.71 |

Table 8. Constraint propagation results on 7 10×10 benchmark problems.

| 15 Jobs, 10 Machines | | | | | | | |
|---|---|---|---|---|---|---|---|
| **Problem** | **T2-P22** | | **T1-P23** | | **T2-P24** | |
| | Time | Opt | Time | Opt | Time | Opt |
| BJS | 2:51:22 | 927 | 2:20:29 | 1032 | --- | 935 |
| 15_5 | 0:21 | 1038 | 1:20 | 1070 | 0:47 | 1016 |
| 15_5.10 | 9:28 | 932 | 14:14 | 1032* | 6:37:04 | 936 |
| 15_5.20 | 1:28:43 | 827* | 26:04 | 1032* | --- | --- |
| 30_5 | 0:37 | 1105 | 0:40 | 1098 | 0:40 | 1112 |
| 15_7 | 1:19 | 959 | 7:30 | 1032* | 0:55 | 1030 |
| 15_7.10 | 1:24:20 | 936 | 26:10 | 1032* | --- | --- |
| 15_7.20 | 1:59:26 | 927* | 4:18:28 | 1032* | --- | --- |
| 30_7 | 2:06 | 1002 | 2:23 | 1088 | 2:02 | 1082 |
| 15_8 | 3:37 | 972 | 13:18 | 1032* | 1:51 | 1031 |
| 15_8.10 | 1:54:01 | 931 | 42:22 | 1032* | --- | --- |
| 15_9 | 7:18 | 954 | 14:50 | 1032* | 6:36 | 981 |
| 15_9.10 | 1:48:25 | 931 | 5:37:57 | 1032* | --- | --- |
| 15_10 | 3:27:27 | 939 | 29:08 | 1032* | 12:00:00 | 938 |

Table 9.    Constraint propagation on 3 15×10 benchmark problems.

| Problem | 1010-04 Time | 1010-04 Opt | 1010-05 Time | 1010-05 Opt | 1010-09 Time | 1010-09 Opt | 1010-12 Time | 1010-12 Opt | 1010-13 Time | 1010-13 Opt | 1010-24 Time | 1010-24 Opt | 1010-25 Time | 1010-25 Opt | 1010-26 Time | 1010-26 Opt | 1010-30 Time | 1010-30 Opt | Σ Time | % of CP | Σ Opt | % above CP-Opt |
|---|---|---|---|---|---|---|---|---|---|---|---|---|---|---|---|---|---|---|---|---|---|---|
| CP | 1:19:46 | 932 | 2:20:20 | 900 | 10:31 | 804 | 15:22 | 795 | 5:38 | 861 | 5:02:05 | 800 | 3:53 | 857 | 1:26:10 | 826 | 5:10:33 | 896 | 15:54:18 | 100 | 7671 | 0.00 |
| 10_5.0 | 1:20 | 977 | 1:41 | 1149 | 1:44 | 884 | 1:58 | 858 | 1:33 | 951 | 2:06 | 909 | 1:32 | 917 | 2:14 | 914 | 2:19 | 971 | 16:27 | 1.72 | 8530 | 11.20 |
| 10_5.5 | 2:23 | 943 | 7:03 | 923 | 4:29 | 821 | 6:25 | 798 | 2:47 | 896 | 31:21 | 808 | 2:31 | 881 | 1:10:09 | 852 | 36:47 | 908 | 2:43:55 | 17.18 | 7830 | 2.07 |
| 10_5.10 | 3:01:22 | 943 | 43:38 | 900* | 4:57 | 804* | 38:21 | 795* | 10:13 | 861* | 9:07 | 808 | 3:21 | 857* | 34:12 | 826* | 45:37 | 899 | 6:10:48 | 38.86 | 7693 | 0.29 |
| 10_5.15 | 2:53:20 | 932* | 54:55 | 900* | 5:25 | 804* | 10:51 | 795* | 42:32 | 861* | 6:04 | 800* | 3:51 | 857* | 40:19 | 826* | 54:44 | 896* | 6:32:01 | 41.08 | 7671 | 0.00 |
| 20_5.0 | 3:06 | 1109 | 3:06 | 966 | 2:57 | 932 | 3:04 | 896 | 3:28 | 908 | 2:37 | 999 | 3:22 | 917 | 3:20 | 950 | 3:09 | 1033 | 28:09 | 2.95 | 8710 | 13.55 |
| 20_5.5 | 3:34 | 1003 | 19:54 | 927 | 3:41 | 862 | 5:32 | 795* | 24:18 | 899 | 3:17 | 932 | 4:28 | 857* | 15:35 | 879 | 37:31 | 897 | 1:57:50 | 12.35 | 8051 | 4.95 |
| 20_5.10 | 1:18:01 | 942 | 14:53 | 909 | 4:36 | 835 | 10:16 | 795* | 9:37 | 861* | 11:49 | 813 | 5:56 | 857* | 1:14:39 | 857 | 1:56:17 | 897 | 5:26:04 | 34.17 | 7766 | 1.24 |
| 20_5.15 | 5:11 | 942 | 49:03 | 900* | 8:55 | 804* | 10:32 | 795* | 8:57 | 861* | 11:26 | 808 | 5:08 | 857* | 1:21:08 | 857 | 2:49:46 | 896* | 5:50:06 | 36.69 | 7720 | 0.64 |
| 10_6.0 | 3:15 | 1047 | 4:12 | 942 | 4:35 | 836 | 4:58 | 881 | 2:54 | 959 | 3:19 | 910 | 3:02 | 921 | 28:48 | 859 | 7:52 | 928 | 1:02:55 | 6.59 | 8283 | 7.98 |
| 10_6.5 | 4:24 | 942 | 44:34 | 909 | 6:39 | 804* | 8:05 | 795* | 12:09 | 887 | 40:01 | 806 | 6:04 | 859 | 1:56:34 | 859 | 1:17:27 | 900 | 5:15:57 | 33.11 | 7761 | 1.17 |
| 10_6.10 | 5:13:06 | 932* | 1:01:51 | 900* | 10:43 | 804* | 12:56 | 795* | 7:51 | 861* | 15:36 | 808 | 4:06 | 857* | 38:40 | 826* | 46:02 | 896* | 8:30:51 | 53.53 | 7679 | 0.10 |
| 10_6.15 | 3:03:56 | 932* | 1:40:01 | 900* | 15:16 | 804* | 14:14 | 795* | 7:47 | 861* | 10:51 | 800* | 5:27 | 857* | 43:11 | 826* | 2:40:14 | 896* | 9:00:57 | 56.69 | 7671 | 0.00 |
| 20_6.0 | 8:40 | 1022 | 8:12 | 962 | 6:39 | 925 | 8:09 | 865 | 7:05 | 901 | 5:33 | 911 | 6:12 | 900 | 9:15 | 948 | 19:37 | 926 | 1:19:22 | 8.32 | 8360 | 8.98 |
| 20_6.5 | 11:42 | 943 | 28:43 | 909 | 8:25 | 810 | 13:00 | 795* | 8:06 | 866 | 7:00 | 877 | 7:20 | 857* | 40:41 | 859 | 1:08:11 | 897 | 3:13:08 | 20.24 | 7813 | 1.85 |
| 20_6.10 | 10:43 | 932* | 57:07 | 900* | 8:18 | 804* | 15:47 | 795* | 13:11 | 861* | 28:51 | 808 | 7:46 | 857* | 25:37 | 826* | 2:30:13 | 897 | 5:17:33 | 33.28 | 7680 | 0.12 |
| 20_6.15 | 10:58 | 932* | 1:39:23 | 900* | 13:12 | 804* | 16:56 | 795* | 12:01 | 861* | 12:39 | 800* | 8:18 | 857* | 34:00 | 826* | 3:51:18 | 896* | 7:18:45 | 45.98 | 7671 | 0.00 |
| 10_7.0 | 8:48 | 968 | 26:55 | 915 | 10:49 | 810 | 12:29 | 865 | 6:19 | 950 | 10:30 | 848 | 7:04 | 926 | 1:38:20 | 908 | 1:09:22 | 908 | 4:10:36 | 26.26 | 8049 | 4.93 |
| 10_7.5 | 9:34 | 932* | 1:13:04 | 900* | 15:39 | 804* | 17:33 | 795* | 9:52 | 861* | 19:32 | 808 | 8:19 | 857* | 37:41 | 826* | 3:25:46 | 896* | 6:37:00 | 41.60 | 7679 | 0.10 |
| 10_7.10 | 3:03:18 | 932* | 1:25:40 | 900* | 21:58 | 804* | 20:46 | 795* | 11:08 | 861* | 14:00 | 800* | 9:16 | 857* | 51:01 | 826* | 3:08:23 | 896* | 9:45:30 | 61.35 | 7671 | 0.00 |
| 10_7.15 | 3:08:17 | 932* | 1:58:31 | 900* | 22:23 | 804* | 21:31 | 795* | 11:40 | 861* | 14:02 | 800* | 9:05 | 857* | 51:01 | 826* | 4:21:41 | 896* | 11:38:11 | 73.16 | 7671 | 0.00 |
| 20_7.0 | 14:51 | 968 | 44:26 | 915 | 10:48 | 836 | 22:26 | 875 | 13:40 | 881 | 15:37 | 859 | 11:24 | 881 | 1:27:01 | 859 | 1:53:09 | 897 | 5:33:22 | 34.93 | 7971 | 3.91 |
| 20_7.5 | 3:41:20 | 943 | 1:17:56 | 900* | 13:30 | 804* | 27:27 | 795* | 15:45 | 861* | 59:49 | 808 | 13:06 | 857* | 46:22 | 826* | 3:08:56 | 896* | 11:04:11 | 69.60 | 7690 | 0.25 |
| 20_7.10 | 2:55:18 | 932* | 1:41:19 | 900* | 16:24 | 804* | 30:50 | 795* | 18:26 | 861* | 22:11 | 800* | 12:49 | 857* | 50:46 | 826* | 4:31:09 | 896* | 11:39:12 | 73.27 | 7671 | 0.00 |
| 20_7.15 | 3:10:52 | 932* | 2:14:01 | 900* | 25:04 | 804* | 31:28 | 795* | 18:39 | 861* | 19:47 | 800* | 13:15 | 857* | 1:03:11 | 826* | 4:43:21 | 896* | 12:59:38 | 81.70 | 7671 | 0.00 |

Table 10. Constraint propagation compared to Carlier / Pinson (1989).

| Problem | 1010-04 Time | 1010-04 Opt | 1010-05 Time | 1010-05 Opt | 1010-09 Time | 1010-09 Opt | 1010-12 Time | 1010-12 Opt | 1010-13 Time | 1010-13 Opt | 1010-24 Time | 1010-24 Opt | 1010-25 Time | 1010-25 Opt | 1010-26 Time | 1010-26 Opt | 1010-30 Time | 1010-30 Opt | Σ Time | % of BJS | Σ Opt | % above BJS-Opt |
|---|---|---|---|---|---|---|---|---|---|---|---|---|---|---|---|---|---|---|---|---|---|---|
| BJS | 1:47 | 932 | 4:39 | 900 | 0:40 | 804 | 0:54 | 795 | 0:12 | 861 | 2:31 | 800 | 0:51 | 857 | 6:56 | 826 | 6:41 | 896 | 25:11 | 100 | 7671 | 0.00 |
| 10_5_0 | 0:14 | 977 | 0:12 | 1149 | 0:10 | 884 | 0:12 | 858 | 0:12 | 951 | 0:14 | 909 | 0:12 | 917 | 0:19 | 914 | 0:14 | 971 | 1:59 | 7.88 | 8530 | 11.20 |
| 10_5_5 | 0:19 | 943 | 0:19 | 923 | 0:14 | 821 | 0:23 | 798 | 0:14 | 896 | 0:32 | 808 | 0:12 | 881 | 4:21 | 852 | 1:08 | 908 | 7:42 | 30.58 | 7830 | 2.07 |
| 10_5_10 | 1:57 | 943 | 1:08 | 900* | 0:16 | 804* | 0:52 | 795* | 0:14 | 861* | 2:16 | 808 | 0:10 | 857* | 5:32 | 826* | 2:00 | 899 | 14:25 | 57.25 | 7693 | 0.29 |
| 20_5_0 | 0:27 | 1109 | 0:28 | 966 | 0:21 | 932 | 0:22 | 896 | 0:27 | 908 | 0:24 | 999 | 0:23 | 917 | 0:29 | 950 | 0:25 | 1033 | 3:46 | 14.96 | 8710 | 13.55 |
| 20_5_5 | 0:19 | 1003 | 0:26 | 927 | 0:17 | 862 | 0:29 | 795* | 0:20 | 899 | 0:20 | 932 | 0:21 | 857* | 0:25 | 879 | 3:36 | 897 | 6:33 | 26.01 | 8051 | 4.95 |
| 20_5_10 | 0:23 | 942 | 1:01 | 909 | 0:20 | 835 | 0:41 | 795* | 0:23 | 861* | 0:45 | 813 | 0:20 | 857* | 9:08 | 857 | 4:19 | 897 | 17:20 | 68.83 | 7766 | 1.24 |
| 10_6_0 | 0:19 | 1047 | 0:19 | 942 | 0:14 | 836 | 0:17 | 881 | 0:19 | 959 | 0:22 | 910 | 0:17 | 921 | 1:18 | 859 | 0:39 | 928 | 4:04 | 16.15 | 8283 | 7.98 |
| 10_6_5 | 0:17 | 942 | 0:30 | 909 | 0:16 | 804* | 0:41 | 795* | 0:17 | 887 | 0:34 | 808 | 0:16 | 857* | 13:02 | 859 | 2:24 | 900 | 18:17 | 72.60 | 7761 | 1.17 |
| 10_6_10 | 1:10 | 932* | 1:50 | 900* | 0:22 | 804* | 1:26 | 795* | 0:16 | 861* | 2:12 | 808 | 0:12 | 857* | 5:09 | 826* | 2:33 | 896* | 15:10 | 60.23 | 7679 | 0.10 |
| 20_6_0 | 0:39 | 1022 | 1:08 | 962 | 0:33 | 925 | 0:37 | 865 | 0:42 | 901 | 0:37 | 911 | 0:35 | 900 | 1:00 | 948 | 1:05 | 926 | 6:56 | 27.53 | 8360 | 8.98 |
| 20_6_5 | 0:44 | 943 | 1:31 | 909 | 0:26 | 810 | 0:54 | 795* | 0:31 | 866 | 0:32 | 877 | 0:32 | 857* | 4:47 | 859 | 5:45 | 897 | 15:42 | 62.34 | 7813 | 1.85 |
| 20_6_10 | 0:50 | 932* | 1:59 | 900* | 0:35 | 804* | 0:50 | 795* | 0:32 | 861* | 1:42 | 808 | 0:44 | 857* | 5:28 | 826* | 9:28 | 897 | 22:08 | 87.89 | 7680 | 0.12 |
| 10_7_0 | 0:34 | 968 | 0:49 | 915 | 0:34 | 810 | 0:29 | 865 | 0:34 | 950 | 0:56 | 848 | 0:23 | 926 | 4:36 | 859 | 2:08 | 908 | 11:03 | 43.88 | 8049 | 4.93 |
| 10_7_10 | 2:01 | 932* | 3:03 | 900* | 0:51 | 804* | 1:30 | 795* | 0:31 | 861* | 3:14 | 800* | 0:20 | 857* | 7:32 | 826* | 6:02 | 896* | 25:04 | 99.54 | 7671 | 0.00 |
| 20_7_0 | 0:47 | 968 | 1:56 | 915 | 0:46 | 836 | 1:02 | 875 | 0:56 | 881 | 1:11 | 859 | 0:39 | 881 | 4:33 | 859 | 9:03 | 897 | 20:53 | 82.93 | 7971 | 3.91 |

Table 11.   Constraint propagation compared to Brucker / Jurisch / Sievers (1991, 1992).

| Subproblem | Opt | #Nodes | Time (3) | Total Time | #Arcs |
|------------|-----|--------|----------|------------|-------|
| 10_5       | 951  | 1269 | 17:04   | 17:27   | 121 |
| 10⁻5.5     | 942  | 799  | 11:10   | 11:33   | 86  |
| 10⁻5.10    | 942  | 1340 | 18:16   | 18:38   | 65  |
| 10⁻5.15    | 937  | 4173 | 57:46   | 58:09   | 42  |
| 10⁻5.16    | 930* | 4461 | 1:02:00 | 1:02:23 | 40  |
| 10_6       | 951  | 1396 | 18:45   | 20:33   | 108 |
| 10⁻6.5     | 942  | 1240 | 17:10   | 18:57   | 76  |
| 10⁻6.10    | 937  | 3342 | 46:50   | 48:36   | 53  |
| 10⁻6.13    | 937  | 3986 | 55:00   | 56:46   | 42  |
| 10⁻6.14    | 930* | 4353 | 1:00:27 | 1:02:13 | 39  |
| 10_7       | 942  | 1290 | 17:35   | 27:49   | 83  |
| 10⁻7.10    | 937  | 4168 | 58:37   | 1:08:50 | 39  |
| 10⁻7.12    | 930* | 4912 | 1:08:30 | 1:18:35 | 32  |
| 10⁻8       | 939  | 2982 | 41:44   | 1:21:49 | 66  |
| 20⁻5       | 980  | 36   | 0:28    | 1:10    | 235 |
| 20_5.5     | 951  | 1147 | 14:48   | 15:28   | 128 |
| 20⁻5.10    | 942  | 980  | 13:49   | 14:28   | 91  |
| 20⁻5.25    | 937  | 4573 | 1:03:48 | 1:04:27 | 30  |
| 20⁻5.28    | 930* | 4592 | 1:04:46 | 1:05:25 | 18  |
| 20⁻5A      | 969  | 165  | 2:04    | 2:45    | 242 |
| 20_5A.10   | 951  | 732  | 10:02   | 10:42   | 116 |
| 20⁻6       | 951  | 115  | 1:26    | 4:15    | 191 |
| 20⁻6.5     | 951  | 780  | 10:38   | 13:26   | 106 |
| 20⁻6.10    | 939  | 2796 | 38:55   | 41:43   | 72  |
| 20⁻6.16    | 937  | 3561 | 49:03   | 51:48   | 51  |
| 20_6.18    | 930* | 3866 | 53:25   | 56:10   | 46  |
| 20⁻6A      | 951  | 924  | 12:18   | 15:41   | 133 |
| 20⁻6A.10   | 942  | 1336 | 19:17   | 22:39   | 70  |
| 20⁻7       | 942  | 595  | 8:05    | 18:40   | 98  |
| 20⁻7.17    | 937  | 4722 | 1:06:22 | 1:16:48 | 21  |
| 20_7.18    | 930* | 5083 | 1:11:11 | 1:21:36 | 18  |
| 50⁻5       | 1080 | 12   | 0:09    | 1:36    | 264 |
| 50⁻6       | 964  | 260  | 3:35    | 10:03   | 198 |
| 50⁻7       | 942  | 1114 | 14:25   | 50:19   | 116 |
| 50⁻8       | 939  | 2914 | 40:35   | 3:28:03 | 73  |
| 10x10      | 930  | 4598 |         | 1:08:36 |     |
| 22->57     | 937  | 4325 |         | 1:01:08 |     |
| 57->22     | 930* | 588  |         | 7:28    |     |

Table 12.   Constraint propagation on the standard 10×10 problem.

| Problem 1010-05 | | | | | |
| Subproblem | Opt | #Nodes | Time (3) | Total Time | #Arcs |
|---|---|---|---|---|---|
| BJS | 900 | 341 | | 4:39 | |
| 10_5 | 1149 | 1 | 0:00 | 0:12 | 364 |
| 10_5.5 | 923 | 15 | 0:10 | 0:19 | 246 |
| 10_5.10 | 900* | 78 | 1:00 | 1:08 | 195 |
| 10_6 | 942 | 3 | 0:01 | 0:19 | 342 |
| 10_6.5 | 909 | 20 | 0:14 | 0:30 | 224 |
| 10_6.10 | 900* | 119 | 1:35 | 1:50 | 188 |
| 10_7 | 915 | 14 | 0:07 | 0:49 | 298 |
| 10_7.10 | 900* | 166 | 2:24 | 3:03 | 149 |
| 10_8 | 900* | 170 | 2:17 | 4:52 | 184 |
| 20_5 | 966 | 1 | 0:00 | 0:28 | 371 |
| 20_5.5 | 927 | 6 | 0:04 | 0:26 | 323 |
| 20_5.10 | 909 | 60 | 0:40 | 1:01 | 248 |
| 20_6 | 962 | 1 | 0:00 | 1:08 | 384 |
| 20_6.5 | 909 | 42 | 0:27 | 1:31 | 238 |
| 20_6.10 | 900* | 80 | 0:56 | 1:59 | 203 |
| 20_7 | 915 | 10 | 0:05 | 1:56 | 316 |
| 50_5 | 1136 | 1 | 0:00 | 0:58 | 390 |
| 50_6 | 1048 | 1 | 0:00 | 2:11 | 405 |
| 50_7 | 957 | 3 | 0:01 | 5:33 | 346 |
| 50_8 | 900* | 70 | 0:47 | 17:11 | 200 |

| Problem 1010-10 | | | | | |
| Subproblem | Opt | #Nodes | Time (3) | Total Time | #Arcs |
|---|---|---|---|---|---|
| BJS | 891 | 1393 | | 19:48 | |
| 10_5 | 995 | 5 | 0:03 | 0:19 | 329 |
| 10_5.5 | 912 | 163 | 1:51 | 2:03 | 229 |
| 10_5.10 | 892 | 534 | 6:59 | 7:11 | 178 |
| 10_6 | 921 | 14 | 0:08 | 0:46 | 297 |
| 10_6.5 | 897 | 334 | 4:22 | 4:59 | 206 |
| 10_6.10 | 891* | 897 | 11:36 | 12:12 | 154 |
| 10_7 | 897 | 234 | 3:00 | 4:48 | 235 |
| 10_7.10 | 891* | 818 | 10:59 | 12:45 | 146 |
| 10_8 | 891* | 844 | 10:40 | 17:25 | 169 |
| 20_5 | 1151 | 1 | 0:00 | 0:33 | 401 |
| 20_5.5 | 926 | 44 | 0:27 | 0:52 | 280 |
| 20_5.10 | 892 | 118 | 1:43 | 2:08 | 210 |
| 20_6 | 947 | 17 | 0:08 | 1:22 | 330 |
| 20_6.5 | 897 | 276 | 3:38 | 4:47 | 219 |
| 20_6.10 | 892 | 551 | 7:25 | 8:35 | 173 |
| 20_7 | 896 | 144 | 1:53 | 4:43 | 246 |
| 50_5 | 1057 | 1 | 0:00 | 1:04 | 390 |
| 50_6 | 1043 | 3 | 0:01 | 3:17 | 373 |
| 50_7 | 903 | 75 | 0:48 | 11:13 | 278 |
| 50_8 | 891* | 731 | 9:31 | 38:11 | 177 |

Table 13.   Constraint propagation on 4 random 10×10 problems in detail, parts 1,2.

| Problem 1010-15 | | | | | |
|---|---|---|---|---|---|
| Subproblem | Opt | #Nodes | Time (3) | Total Time | #Arcs |
| BJS | 843 | 350 | | 4:18 | |
| 10_5 | 896 | 6 | 0:02 | 0:15 | 344 |
| 10‾5.5 | 849 | 33 | 0:20 | 0:29 | 240 |
| 10‾5.10 | 843* | 135 | 1:38 | 1:46 | 177 |
| 10‾6 | 866 | 5 | 0:02 | 0:24 | 338 |
| 10‾6.5 | 843* | 146 | 1:57 | 2:15 | 195 |
| 10_6.10 | 843* | 268 | 3:14 | 3:31 | 150 |
| 10‾7 | 843* | 90 | 1:05 | 2:14 | 230 |
| 10‾7.10 | 843* | 282 | 3:23 | 4:30 | 130 |
| 10‾8 | 843* | 212 | 2:36 | 5:56 | 182 |
| 20‾5 | 1005 | 1 | 0:00 | 0:26 | 418 |
| 20_5.5 | 863 | 10 | 0:04 | 0:23 | 320 |
| 20‾5.10 | 846 | 30 | 0:21 | 0:38 | 218 |
| 20‾6 | 866 | 9 | 0:03 | 0:46 | 357 |
| 20‾6.5 | 843* | 70 | 0:55 | 1:33 | 227 |
| 20‾6.10 | 843* | 266 | 3:06 | 3:43 | 168 |
| 20_7 | 843* | 57 | 0:39 | 2:49 | 245 |
| 50‾5 | 885 | 1 | 0:00 | 1:04 | 375 |
| 50‾6 | 980 | 1 | 0:00 | 1:50 | 388 |
| 50‾7 | 896 | 1 | 0:00 | 5:42 | 390 |
| 50‾8 | 843* | 221 | 2:50 | 15:28 | 208 |

| Problem 1010-20 | | | | | |
|---|---|---|---|---|---|
| Subproblem | Opt | #Nodes | Time (3) | Total Time | #Arcs |
| BJS | 971 | 1639 | | 19:34 | |
| 10_5 | 1130 | 1 | 0:00 | 0:14 | 367 |
| 10‾5.5 | 1013 | 36 | 0:32 | 0:41 | 241 |
| 10‾5.10 | 971* | 291 | 3:42 | 3:52 | 200 |
| 10‾6 | 1054 | 6 | 0:04 | 0:39 | 350 |
| 10‾6.5 | 971* | 60 | 0:48 | 1:20 | 218 |
| 10_6.10 | 971* | 287 | 3:45 | 4:16 | 174 |
| 10‾7 | 988 | 602 | 8:22 | 10:31 | 206 |
| 10‾7.10 | 971* | 1770 | 20:08 | 22:16 | 123 |
| 10‾8 | 971* | 498 | 5:56 | 15:10 | 164 |
| 20‾5 | 1085 | 2 | 0:01 | 0:29 | 373 |
| 20_5.5 | 982 | 77 | 0:56 | 1:19 | 225 |
| 20‾5.10 | 971* | 440 | 5:30 | 5:53 | 182 |
| 20‾6 | 1132 | 1 | 0:00 | 1:06 | 366 |
| 20‾6.5 | 988 | 213 | 2:50 | 3:49 | 221 |
| 20‾6.10 | 971* | 478 | 5:47 | 6:48 | 173 |
| 20_7 | 971* | 194 | 2:13 | 6:16 | 206 |
| 50‾5 | 1150 | 1 | 0:00 | 1:11 | 375 |
| 50‾6 | 1033 | 11 | 0:07 | 2:42 | 335 |
| 50‾7 | 1004 | 49 | 0:38 | 12:37 | 242 |
| 50‾8 | 971* | 317 | 4:03 | 48:53 | 172 |

Table 13.    Constraint propagation on 4 random 10×10 problems in detail, parts 3,4.

| | Brucker/Jurisch/Sievers | | | Adams/Balas/Zawack | | | |
| | PC (386, 25MHz) | | (SUN 4/20) | (VAX 780/11) | | | |
| | Time | Opt | Time | Time | SB1 | Time | SB2 |
|---|---|---|---|---|---|---|---|
| T1-P02 | 1:08:36 | 930 | 16:15 | 0:10 | 1015 | 14:11 | 930* |
| T1-P04 | 29:44 | 1234 | 7:44 | 0:06 | 1306 | 25:03 | 1239 |
| T1-P05 | 1:49 | 943 | 0:28 | 0:13 | 962 | 18:21 | 943* |
| T2-P16 | 3:09 | 945 | 0:52 | 0:07 | 1021 | #4:00 | 978 |
| T2-P17 | 0:47 | 784 | 0:14 | 0:05 | 796 | #3:12 | 787 |
| T2-P18 | 4:06 | 848 | 0:58 | 0:10 | 891 | #3:45 | 859 |
| T2-P19 | 20:52 | 842 | 5:08 | 0:07 | 875 | #4:00 | 860 |
| T2-P20 | 20:35 | 902 | 5:16 | 0:10 | 924 | #4:49 | 914 |

Table 14.    Performance of three different computers.

## 3.5  Conclusions

Backtracking can easily be accompanied by a constraint propagation phase
before as well as during the search. A guaranteed 3–consistency can substantially
reduce the search tree for some problems and correspondingly can reduce the
variable domains. For other problems it may be difficult to draw conclusions
from a local consistency phase such as arc– or path–consistency. Even tight
upper bounds on the objective value cannot overcome these difficulties. However
using upper bounds on the objective value of subproblems might be a reasonable
approach and also an optimal approach if these subproblem upper bounds are
not below the real subproblem objective value if the subproblem is extracted
from an optimal solution of the overall problem. The upper bound can be easily
computed by some fast heuristics such as priority rules. Only in case of an
optimal solution of the subproblems a certain percentage should be added to the
objective value in order to avoid to many fault decisions. The constraint
propagation phase can really extract a lot of knowledge about the structure of
the subproblems in order to use this knowledge in a subsequent optimization
phase. This optimization phase need not be performed by an exact algorithm a
good heuristic would suffice, in particular because a lot of problem specific

knowledge is already existing. Moreover the knowledge about the performance of the constraint propagation phase applied to the subproblems can yield extremely useful criteria for a modified exact algorithm, respectively a prospective backtrack search.

The latter however, a backtrack search including the propagation of constraints in order to reach arc–consistency is implicitly realized in recent constraint based logic programming languages such as CHIP, CHARME, PROLOG III, or cc(FD). First results under the use of CHARME with respect to job shop and resource constraint project scheduling are very encouraging [234].

# 4. Decomposition Based Learning

## 4.1 Opportunistic Scheduling Heuristics

**Priority rules** are probably the most frequently applied heuristics for solving (job shop) scheduling problems in practice because of their ease of implementation and their low time complexity. In Section 2 we applied 12 different rules to decide which of the operations or jobs in the conflict set (with respect to the application of the algorithm of Giffler and Thompson (1960)) gets highest priority.

The quality of the schedules obtained by the **shifting bottleneck heuristic** suggested by Adams, Balas, and Zawack (1988) respectively its extension Balas et al. (1992) heavily depends on the sequence in which the one machine problems are solved and, furthermore, their arc directions are included into the whole (partial) schedule. Sequence changes, i.e. disregarding the bottleneck machine as to be scheduled next, may yield substantial improvements.

In analogy to the shifting bottleneck heuristic where the best machine sequence is to be determined we can also aim to find the best sequence of all job pair subproblems and successively solve them. Consider a job shop scheduling problem consisting of the set J of jobs. Then a disjunctive arc belongs to exactly one of the $(|J| \cdot (|J| - 1))/2$ job pairs. A certain amount of arc directions belonging to the optimal job shop schedule coincides with optimal solutions of these job pair problems. Hence, at step k of the **job pair heuristic** a pair of jobs

---

[234]    Drexl / Sprecher (1993), Pesch / Drexl / Salewski (1993)

is chosen which is still connected by disjunctive arcs. Only those disjunctive arcs became directed which result from transitivity via certain other jobs. Respecting these transitivities the new job pair problem is solved to optimality [235] and the obtained arc directions are included into the whole schedule. To reduce cycling arcs connecting operations of some other jobs are also included if they are resulting from transitivities. The final schedule heavily depends on the sequence how the job pair problems are solved and their arc directions included into the partial schedule. Sequence changes may yield substantial improvements, however, a complete enumeration of the search tree is even less acceptable as in the case of the shifting bottleneck procedure.


## 4.2    Constraint Propagation, Local Consistency, and Genetic Based Learning

In the previous section we have seen that the job shop scheduling problem is a typical representative of a binary constraint satisfaction problem (CSP), i.e., generally speaking, there is a set of variables each of which has its own domain of values. Find an assignment of values to variables such that a set of constraints on variable pairs is satisfied [236].    Assume that there is an upper bound on the makespan of an optimal schedule of the underlying job shop scheduling problem. Then computing heads and tails assigns to each operation an interval of possible start times. Considering variable domains as possible operation start times where the variables define the operations in a schedule then the disjunctive graph illustrates the job shop scheduling constraint satisfaction problem, hence it corresponds to the constraint graph [237]. Consistency checks, or roughly speaking propagation of constraints will make implicitly defined constraints more visible and will prune the search tree in a branch and bound algorithm [238].    Obviously, n–consistency, where n is the number of operations, immediately implies that a feasible schedule can be generated easily, however, to achieve n–consistency is in general not practicable. Moreover, worse upper bounds on the makespan of an optimal schedule will hardly reduce variable domains, i.e. only a few arc directions are fixed during the constraint propagation process. The better the bounds the more arc

---

[235]    cf. Brucker (1988)

[236]    see Dechter / Pearl (1988) and Meseguer (1989)

[237]    see Montanari (1974)

[238]    cf. Van der Wal / Kolen (1992)

directions can be fixed [239].

In Section 3 we deterministically fixed upper bounds for certain subproblem schedules. A succeeding propagation phase implied fixing of a number of operation sequences. Our goal now is to find genetically best upper bounds on subproblem schedules and then perform a propagation phase respecting these bounds.[240]

The strategy in Section 2 controls a sequence of priority rules or the machine inclusion sequence for the SB–heuristic and learns to find best combinations in both cases.

Recall, each individual of the **priority rule based genetic algorithm** (for short: P–GA) is a string of $n-1$ entries $(pr_1, pr_2, ..., pr_{n-1})$ where $n-1$ is the number of non–trivial operations in the underlying problem instance. An entry $pr_i$ represents one rule of the set of twelve priority rules described earlier. The entry in the i–th position says that a conflict in the i–th iteration of the Giffler / Thompson algorithm should be resolved using priority rule $pr_i$. Ties are broken by a random choice. Within a genetic framework a best sequence of priority rules has to be determined. The crossover operator is straightforward. Obviously, the simple crossover, where the substrings of two cut strings are exchanged, applies and always yields feasible offspring. Heuristic information already occurs in the encoding scheme and a particular improvement step is dropped. The mutation operator simply switches a string position to another one, i.e. the priority rule of a randomly chosen string entry is replaced by a new rule randomly chosen.

In the first implementation from Section 2 the genetic algorithm serves as a meta–strategy to optimally control the use of priority rules, whereas the genetic algorithm controls the selection of nodes in the enumeration tree of the shifting bottleneck heuristic in a second implementation, the **shifting bottleneck based genetic algorithm** (for short: SB–GA). The SB–heuristic tries to determine the best single machine sequence which can also be achieved by a genetic strategy, even in a more effective way.

---

239  see the previous section

240  Only little work is done on genetic algorithms with respect to (job shop) scheduling, see Dorndorf / Pesch (1993c), Husbands et al. (1991), Starkweather et al. (1992), and Kanet / Sridharan (1991). There are several possibilities to apply an improvement heuristic. During the recombination phase an improvement step, like tabu search or simulated annealing can be applied to all or several of the solutions in a population (Chapter II) and Johnson (1990a) in case of the traveling salesman problem, Dorndorf / Pesch (1993a,b) and Aarts et al. (1991) for the job shop scheduling problem, and Thiel / Voß (1992) for the multiconstraint knapsack problem).

Hence, an individual is encoded over the alphabet from 1 to the number of machines and a (partial) string just describes the sequence in which the single machine solutions are considered for inclusion. As a crossover operator one can use any traveling salesman crossover. The difference between the shifting bottleneck heuristic and a genetic approach is that the bottleneck is no longer a decision criterion for the choice of the next machine.

Here we are continuing in this line in a third implementation, the **job pair based genetic algorithm** (for short: 2J–GA). It also optimally controls the sequence of job pairs. Hence, the length of an individual corresponds to the number $(|J| \cdot (|J| - 1))$ of pairs of jobs where J is the set of jobs in the underlying problem. Each entry of an individual represents a job pair and a (partial) string just describes the sequence in which the job pair solutions are considered for inclusion. Again, we can apply all kinds of traveling salesman crossover; in our case we took the Mühlenbein–Gorges–Schleuter crossover [241] as described in Chapter I.

Constraint propagation in the sense of arc– or path–consistency generally fixes only a few most often trivial arc directions even if tight upper bounds (in the limit the optimal value) of the makespan in an optimal schedule are applied. Remedy yields to consider subproblems (e.g. one–machine or two–job problems). Hence, two new genetic based approaches were applied [242]. In the first one, the **one machine constraint propagation based genetic algorithm** (1MCP–GA), each entry of an individual of the SB–GA is replaced by an upper bound on the makespan of the corresponding one–machine problem. In the second approach, the **two job constraint propagation based genetic algorithm** (2JCP–GA), each entry of an individual of the 2J–GA is replaced by an upper bound on the makespan of the corresponding two job problem. In fact, the entry contains an integer value $\alpha$ between 0 and 31 providing an upper bound which is $\alpha$ % above the optimal makespan of the considered subproblem. The entries are encoded as binary numbers in order to apply the simple one point crossover as well as a simple mutation (switching 0 and 1) as in the case of P–GA. In both cases the upper bound percentages were modified. Whenever a new population is generated a local decision rule in the sense of constraint propagation in order to achieve arc– and path–consistency is applied simultaneously to each subproblem (corresponding to an entry of an individual) with respect to its upper bound which is $\alpha$ % above the optimal makespan of the subproblem. The number of

---

[241]  see Gorges–Schleuter (1989), Mühlenbein et al. (1987, 1988)

[242]  Pesch (1993b)

newly fixed arc directions, divided by the number of arcs which were included into a cycle during the constraint propagation process on the subproblems, defines the fitness of an individual. Regard, that an individual need not represent one schedule, because there are still some arc directions not fixed. An individual of the population corresponds to a partial schedule. However each population is transformed to a population of feasible solutions, where each individual of a population is assessed in order to judge its contribution to a schedule. Therefore Giffler / Thompson's algorithm is applied with respect to the partial schedule respresenting indidvidual. Ties are broken with respect to the complete schedules that are attached to the parents (partial schedules) of the considered offspring (partial schedule). Hence, the next operation is chosen as in one of the parents corresponding complete schedules with the same probability [243]. In the first population of complete schedules ties were broken randomly.

### 4.3 Computational Results

Our implementation uses the algorithm of Baker (1987) for selection and an elitist strategy, i.e. the best individual in each population always survived. It is written in PASCAL and all problems were either solved on a DECstation 3100 under the operating system Ultrix (P–GA and SB–GA, see Section 2) or on a VAX 8650 under VMS (2J–GA, 2JCP–GA, 1MCP–GA).

Details of implementation and computational comparisons of P–GA, SB–GA, the shifting bottlenck heuristic, some priority rules as well as a random choice of priority rules to detect the next operation if during the execution of the Giffler / Thompson algorithm a choice point is reached, are extensively presented in Section 2. Table 15 contains the results for the 3 best known test problems for job shop scheduling (Fisher / Thompson (1963): the 6 × 6, the 5 × 20, and the 10 × 10 problem the optima of which are 55, 1165, and 930, respectively. The makespan of the best solution found by each of the heuristics is shown in column "f", column "t sec" refers to the computation time needed for the correponding algorithm. Besides column "SB–GA" which presents the shifting bottleneck based genetic algorithm results of Section 2 (on a DECstation 3100), the columns "2J–GA", "2JCP–GA", and "1MCP–GA" contain the results obtained by the three genetic learning strategies. Population sizes where always set to 80,

---

[243]   cf. GT–crossover in Yamada / Nakano (1992)

the crossover rate was fixed to 1, the mutation rate was kept 0 in case of 2J–GA
and 0.01 for 2JCP–GA and 1MCP–GA. There was no inversion and no scaling of
fitness values. An elitest strategy took place in all implementations. These
parameters are within a reasonable amount of test runs found to be empirically
the best ones. The initial populations are always generated randomly and the
makespan reported provides the best value obtained within three runs. Thus the
times are average running times on these three runs. A run was stopped if there
was no improvement within 50 successive generations.

As expected, SB–GA dominates all other genetic algorithm implementations if
both, the quality of the obtained solutions and the running time is considered.
The best results, however, were obtained by the 1MCP–GA, an optimal solution
could be found for all three standard problems. The job pair based genetic
algorithm approaches performed well, however, considering the 10 x 10 problem
they seem to run into a dead–end, fixing some worse arc directions. If the
number of jobs compared to the number of machines is quite large then 2J–GA
and 2JCP–GA seem to be more effective. Columns "GLS–ALU" and "SA–ALU"
contain the best of two results obtained by two different implementations for
each, a genetic local search strategy and simulated annealing, respectively [244].
These four algorithms got the same time restrictions and the part after the
decimal point is truncated. Their values are average results over 5 runs on a
VAX 8650. The neighbourhood structure for the simulated annealing and the
genetic local search versions are the same: (i) Reversing an edge on a longest
path in the graph, and (ii) reversing an edge on a longest path in the graph such
that this edge is incident to an edge of the arc set A. For details we refer to
Aarts et al. (1991). In the genetic local search algorithms a local search
procedure is performed after the recombination step of the genetic algorithm
such that each member of the population is made locally optimal with respect to
the two neighbourhood structures. The last two columns contain the results
(there are no times) obtained by Nakano and Yamada (1991, 1992). Their
Giffler / Thompson based crossover operator solves conflicts (in the offspring)
equally distributed with respect to parents' solutions. Column "TS–DT" presents
the excellent tabu search results of Dell'Amico / Trubian (1993) on a 80386PC,
33 MHz, using an extension of the above mentioned neighbourhood structure, cf.
the simulated annealing approach of Van Laarhoven et al. (1992).

---

[244]   see Aarts et al. (1991)

| problem | 2J-GA f | 2J-GA t sec | 2JCP-GA f | 2JCP-GA t sec | 1MCP-GA f | 1MCP-GA t sec | SB-GA f | SB-GA t sec | GLS-ALU f | GLS-ALU t sec | SA-ALU f | SA-ALU t sec | TS-DT f | TS-DT t sec | GA-NY f | GA-YN f | OPT |
|---|---|---|---|---|---|---|---|---|---|---|---|---|---|---|---|---|---|
| 6 x 6 | 55 | 8.1 | 55 | 11.4 | 55 | 8.5 | 55 | 19.7 | 55 | 9.4 | 55 | 9.4 | 55 | 2.4 | 55 | 55 | 55 |
| 5 x 20 | 1193 | 95.3 | 1175 | 121.1 | 1165 | 101.0 | 1178 | 95.7 | 1294 | 88.4 | 1216 | 88.4 | 1165 | 160.1 | 1215 | 1184 | 1165 |
| 10 x 10 | 937 | 100.0 | 937 | 146.9 | 930 | 104.4 | 938 | 106.7 | 978 | 99.4 | 969 | 99.4 | 935 | 155.8 | 965 | 930 | 930 |

Table 15.  The 3 Fisher / Thompson standard problems.

P-GA is less powerful than the other genetic based approaches (we dropped the results therefore and refer to Section 2, but it has its advantages: Ease of implementation, also in a more general framework, as well as robustness to problem changes (SB-GA and 2J-GA seem to be much more sensitive) can give P-GA the user's preference.

Besides the 3 well known benchmark problems we considered 35 benchmark problems listed in Adams et al. (1988) [245]. Problems 31 to 35 are neglected because the first part of the shifting bottleneck algorithm [246] solves these set of problems easily to optimality. Hence, Table 16 contains only the results for the remaining ones.

Fine tuning of the control parameters is of some influence on the quality of the solutions [247] but changing the instance most often also changes the best parameter configuration. So, the configuration of parameters was kept as in the Fisher / Thompson instances in case of 2J-GA, 2JCP-GA, and 1MCP-GA. The SB-GA parameters are those from Section 2. The initial population was always randomly generated. Column "SB-ABZ" contains the makespans that Adams et al. (1988) obtained from their shifting bottleneck implementation. Columns "f-GLS" and "f-SA" contain the best of two results obtained by two different implementations for each, a genetic local search strategy and simulated annealing, respectively. These four algorithms got the same time restrictions from column "t sec". These three columns are headed by the name "ALU" indicating that their values are taken from Aarts et al. (1991); the part after the decimal point is truncated. The "TS-DT" contains the makespans and running times of the excellent tabu search implementation by Dell'Amico and Trubian (1993). Finally the last column "opt" presents the smallest makespan published in Carlier / Pinson (1990), Brucker et al. (1991), Applegate and Cook (1991), or Dell'Amico and Trubian (1993), for the particular problem instance. An asterisk (*) says that the value is proven to be optimal.

The values in the "ALU" column are average results over 5 runs on a VAX 8650. The neighbourhood structure for the simulated annealing and the genetic local search versions are those already described above. Obviously, the first 15 problems are easy and all implementations can easily reach the optimum or find a near-optimal solution. Only TS-DT and 1MCP-GA found the optimum in all

---

[245]   cf. Taillard (1993)

[246]   see Adams et al. (1988)

[247]   See Schaffer (1989), De Jong / Spears (1989), Grefenstette (1986) who used a genetic algorithm for the control parameter setting.

cases. The set of 10 x 10 problems is still solved to optimality by TS–DT and 1MCP–GA. Slightly worse solutions yield the implementation 2JCP–GA, and the ALU simulated annealing and genetic local search implementation, but their results are still close to the minimum makespan. The 2J–GA cannot compete any longer seriously, also its results are not to worse. Obviously, wrongly fixed arcs, leading to promising solutions in the job pair subproblem can hardly be turned into the right direction. For the larger problems, the last 15 ones, TS–DT and 1MCP–clearly dominate. The 2JCP–GA behaves comparable to the shifting bottleneck algorithms and still dominates the simulated annealing version although the latter needed a substantial amount of running time. The tabu search implementation is undoubtly the winner in both, the results as well as the running time. It is immediately followed by the one machine constraint propagation based genetic algorithm. For instance, 1MCP–GA was able to find the second best result ever published for problem instance 38. The 2–job constraint propagation based genetic algorithm and the shifting bottleneck algorithm are quite comparable. The latter, however, is dominating on smaller problem instances as well as quadratic instances while the first one is more suitable for rectangle problems where the number of jobs is significantly higher than the number of machines. Simulated annealing and the genetic local search approach of Aarts et al. (1991) is slightly worse than 2JCP–GA but most often better than the job pair based genetic algorithm.

The genetic algorithms relying on job pairs (2J–GA and 2JCP–GA) in all cases need a substantial amount of computation time if the number of jobs becomes large. This is not a surprising effect because an increasing number of jobs heavily influences the string lengths which corresponds to the number of job pairs. Although achieving path consistency is fast, a repeated application drives the overall running time upward.

In general we can conclude that the genetic based approaches are much more robust on different problem types than tabu search. There are no drastic time differences and also for the easy problems it takes some time to realize that they are easy.

| problem | 2J-GA | | 2JCP-GA | | 1MCP-GA | | SB-ABZ | ALU | | | TS-DT | | opt |
|---|---|---|---|---|---|---|---|---|---|---|---|---|---|
| | f | t sec | f | t sec | f | t sec | f | f-GLS | f-SA | t sec | f | t sec | |
| 5 machines, 10 jobs | | | | | | | | | | | | | |
| 01 | 666 | 18.4 | 666 | 19.6 | 666 | 18.4 | 666 | 666 | 666 | 17 | 665 | 0.1 | 666*BJS |
| 02 | 671 | 17.9 | 658 | 19.2 | 655 | 18.2 | 669 | 659 | 658 | 18 | 655 | 18.8 | 655*CP |
| 03 | 601 | 14.8 | 597 | 15.0 | 597 | 16.1 | 605 | 609 | 616 | 21 | 597 | 21.6 | 597*BJS |
| 04 | 610 | 14.3 | 599 | 14.9 | 590 | 14.2 | 593 | 594 | 590 | 16 | 590 | 32.2 | 590*BJS |
| 05 | 593 | 16.2 | 593 | 18.4 | 593 | 13.9 | 593 | 593 | 593 | 13 | 593 | 0.3 | 593*BJS |
| 5 machines, 15 jobs | | | | | | | | | | | | | |
| 06 | 928 | 23.6 | 926 | 25.9 | 926 | 25.2 | 926 | 926 | 926 | 25 | 926 | 0.3 | 926*CP |
| 07 | 894 | 28.5 | 890 | 33.7 | 890 | 27.8 | 890 | 890 | 890 | 42 | 890 | 0.6 | 890*BJS |
| 08 | 863 | 29.7 | 863 | 31.2 | 863 | 26.0 | 863 | 863 | 863 | 39 | 863 | 0.3 | 863*BJS |
| 09 | 951 | 27.4 | 951 | 30.9 | 951 | 28.5 | 951 | 951 | 951 | 35 | 951 | 0.2 | 951*BJS |
| 10 | 958 | 19.1 | 958 | 22.5 | 958 | 20.2 | 959 | 958 | 958 | 17 | 958 | 0.2 | 958*BJS |
| 5 machines, 20 jobs | | | | | | | | | | | | | |
| 11 | 1222 | 41.1 | 1222 | 44.3 | 1222 | 39.5 | 1222 | 1222 | 1222 | 59 | 1222 | 0.4 | 1222*BJS |
| 12 | 1039 | 40.5 | 1039 | 45.4 | 1039 | 41.5 | 1039 | 1039 | 1039 | 50 | 1039 | 0.2 | 1039*BJS |
| 13 | 1150 | 33.3 | 1150 | 39.7 | 1150 | 37.2 | 1150 | 1150 | 1150 | 50 | 1150 | 0.4 | 1150*BJS |
| 14 | 1292 | 28.7 | 1292 | 29.9 | 1292 | 30.2 | 1292 | 1292 | 1292 | 21 | 1292 | 0.4 | 1292*BJS |
| 15 | 1214 | 48.5 | 1217 | 50.1 | 1207 | 42.8 | 1207 | 1207 | 1207 | 76 | 1207 | 1.2 | 1207*BJS |
| 10 machines, 10 jobs | | | | | | | | | | | | | |
| 16 | 989 | 103.9 | 959 | 127.6 | 945 | 100.0 | 978 | 976 | 969 | 88 | 945 | 97.4 | 945*CP |
| 17 | 807 | 108.5 | 788 | 117.8 | 784 | 89.9 | 787 | 791 | 785 | 91 | 784 | 21.7 | 784*CP |
| 18 | 891 | 115.4 | 852 | 121.5 | 848 | 88.9 | 859 | 856 | 856 | 96 | 848 | 63.1 | 848*BJS |
| 19 | 881 | 140.1 | 851 | 139.3 | 842 | 123.1 | 860 | 859 | 854 | 93 | 842 | 103.8 | 842*BJS |
| 20 | 935 | 140.0 | 925 | 141.8 | 902 | 94.5 | 914 | 913 | 911 | 107 | 902 | 71.7 | 902*BJS |
| 10 machines, 15 jobs | | | | | | | | | | | | | |
| 21 | 1102 | 186.7 | 1074 | 212.8 | 1070 | 199.9 | 1084 | 1084 | 1078 | 243 | 1048 | 198.8 | 1048 DT |
| 22 | 974 | 189.8 | 935 | 201.2 | 933 | 189.0 | 944 | 944 | 950 | 254 | 933 | 191.4 | 927*BJS |
| 23 | 1034 | 182.1 | 1032 | 189.4 | 1032 | 176.2 | 1032 | 1032 | 1032 | 242 | 1032 | 1.8 | 1032*CP |
| 24 | 998 | 190.0 | 948 | 197.2 | 940 | 178.9 | 976 | 970 | 960 | 234 | 941 | 181.8 | 935*AC |
| 25 | 1021 | 189.0 | 1002 | 184.6 | 989 | 181.8 | 1017 | 1010 | 1003 | 254 | 979 | 191.7 | 977*AC |
| 10 machines, 20 jobs | | | | | | | | | | | | | |
| 26 | 1245 | 189.2 | 1218 | 227.2 | 1218 | 221.5 | 1224 | 1236 | 1221 | 469 | 1218 | 22.1 | 1218*BJS |
| 27 | 1271 | 201.2 | 1268 | 293.6 | 1268 | 201.5 | 1291 | 1300 | 1275 | 492 | 1242 | 254.2 | 1242 DT |
| 28 | 1271 | 195.6 | 1240 | 266.6 | 1234 | 218.5 | 1250 | 1264 | 1242 | 455 | 1216 | 186.4 | 1216*DT |
| 29 | 1226 | 187.8 | 1209 | 244.0 | 1204 | 237.8 | 1239 | 1260 | 1225 | 471 | 1182 | 281.3 | 1182 DT |
| 30 | 1359 | 194.5 | 1355 | 214.5 | 1355 | 189.6 | 1355 | 1386 | 1355 | 441 | 1355 | 10.4 | 1355*CP |
| 15 machines, 15 jobs | | | | | | | | | | | | | |
| 36 | 1289 | 290.7 | 1311 | 295.3 | 1273 | 278.4 | 1305 | 1310 | 1307 | 602 | 1278 | 238.4 | 1268*CP |
| 37 | 1507 | 287.7 | 1454 | 329.7 | 1414 | 291.6 | 1423 | 1449 | 1440 | 636 | 1409 | 242.2 | 1402 CP |
| 38 | 1325 | 318.8 | 1292 | 347.3 | 1204 | 308.1 | 1255 | 1283 | 1235 | 635 | 1203 | 256.6 | 1203 DT |
| 39 | 1287 | 304.1 | 1275 | 338.1 | 1233 | 301.5 | 1273 | 1279 | 1258 | 592 | 1242 | 237.8 | 1233*BJS |
| 40 | 1301 | 301.2 | 1291 | 317.0 | 1242 | 291.1 | 1269 | 1260 | 1254 | 596 | 1233 | 236.6 | 1222*AC |

Table 16.

Computational results on a VAX 8650 under VMS for 35 well known problem instances. Comparison to results of the shifting bottleneck procedure from Adams et al. (1988) ("SB-ABZ") on a VAX 780/11; to the results from Aarts et al. (1991) ("ALU") on a VAX 8650 for two genetic local search ("f-GLS") and two simulated annealing ("f-SA") approaches; and to the results from Dell'Amico / Trubian (1993) ("TS-DT") on a 386, 33 MHz processor. Column "opt" contains the value of an optimal (*) or a best known solution, which results from the exact algorithms of Brucker et al. (1991), (BJS), Applegate / Cook (1991) (AC), Carlier / Pinson (1990) (CP), or from tabu search by Dell'Amico / Trubian (1993) (DT).

Figure 9 finally shows the behaviour of all described methods from Chapter III and recent successful heuristics from the literature for the 5×20 and 10×10 Fisher / Thompson benchmark problems. Performance is compared in terms of percentages above the optimal solution. Regard, only three methods were capable to solve the 10×10 problem optimally, namely GA–YN (Yamada / Nakano (1992)), SB–ABZ (Adams et al. (1988)), and 1MCP–GA (Pesch (1993a)). Two methods could solve the 5×20 problem to optimality, TS–DT (Dell'Amico / Trubian (1993)) and 1MCP–GA (Pesch (1993a)).[248]    The results of 1MCP–GA are not included in the figure because they are optimal in both cases.

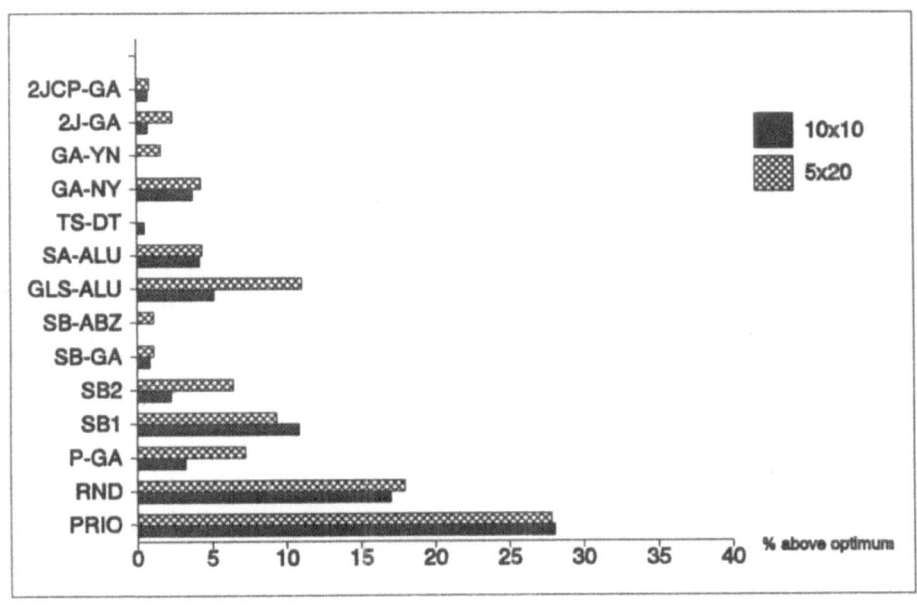

Figure 9.    Performance of all heuristics on the Fisher / Thompson problems.

---

[248]    Not included in the figure are the results of Nowicki and Smutnicki (1993). They were able to solve the standard problems 5×20 and 10×10 to optimality in only a fraction of time other methods needed. The author received their paper while this book was already in press.

## 4.4   Conclusions

We presented a special type of probabilistic machine learning by population
genetics. Learning to find a best sequence of priority rules for solving job shop
scheduling problems did not work best because of the amount of computation
time needed. This is not a big surprise because priority rules only make little
use of problem specific knowledge. In contrast, the solution of a one machine
problem or a job pair subproblem (with respect to some former decisions)
involves a lot of problem specific knowledge, the use of which may be the main
reason for the success of the genetically guided shifting bottleneck procedure.
The constraint propagation based approaches even try to avoid the use of
problem specific knowledge that is not relevant for the complete problem.
Introducing problem specific knowledge makes a genetic strategy a promising
alternative. This knowledge might be incorporated by the application of a
heuristic performing well on the particular problem, such as the shifting
bottleneck procedure for the job shop problem and the Lin–Kernighan algorithm
for the traveling salesman problem. We may also think of local search
algorithms such as tabu search or simulated annealing as "special purpose"
heuristics. So a simple genetic algorithm can do its work very well if an
individual is considered to be a sequence of local decisions on the use of the
knowledge provided by special purpose heuristics or parts of them. Problem
specific knowledge can also be introduced by decomposition of the whole problem
into easily to handle subproblems. The latter should be solved efficiently by an
exact or an approximation algorithm. Additional gains can be achieved if the
subproblems help to structure the complete problem in consideration. Those
transferring structures may be recognized by some consistency checks and their
propagation.

# IV. FLEXIBLE MANUFACTURING SYSTEMS

During the last two decades important changes in manufacturing technology have taken place. Automation in manufacturing has become an important factor in order to increase the traditionally conflicting objectives flexibility and productivity. Flexibility addresses the problem of producing a number of distinct orders in a job shop environment with respect to different production alternatives and in order to meet some objectives. Productivity, on the other hand, refers high speed production, high machine utilization, and small storage requirements, e.g. similar to an assembly line or flow shops. Therefore increased job shop productivity while maintaining production flexibility has become a desired goal. As a result, the introduction of concepts related to automated manufacturing, e.g. computer aided manufacturing (CAM), robotics, or computer integrated manufacturing (CIM) as well as an increased progress in information technology became an important issue to control the flow of orders in a factory. The introduction of numerically controlled (NC) or computer numerically controlled (CNC) machines has become a major step in manufacturing automation, in order to enable automated machines to produce complex orders (products) quickly and accurately. A CNC machine is capable to handle a number of tools provided by some tool magazine in order to perform completely different production requirements through particular tool configurations. Micro–computers are used to control the machine tool requirements.

A flexible manufacturing system (FMS) is an integrated computer controlled complex of automated material handling devices and numerically controlled machines that can simultaneously process medium sized volumes of a variety of part types (Stecke (1983)). Every material handling system has an automated part transportation system and possibly an automated tool transportation system. The latter can transfer between tool magazines as well as to a central tool storage.[249]

The design and use of flexible manufacturing systems involves considering several intricate problems such as design, planning, scheduling, and control

---

[249] The number of FMS publications is immense and among others we refer to recent books of Tempelmeier / Kuhn (1993), the dissertations of Kuhn (1990) and Tetzlaff (1990), Zeestraten (1989) and the excellent survey papers of Gunasekaran et al. (1993) and Tempelmeier / Kuhn (1992), or Oerlemans (1992a,b), Suri (1985), Buzacott / Yao (1986), Kusiak (1986), Van Looveren et al. (1986), Zijm (1988).

problems[250]. Design problems involve issues such as selection of equipment, for instance, determining the appropriate number of machine tools of each type, material handling, layout, labour issues, and the size of buffers [251]. Planning problems involved in the operation of FMSs are aggregate planning (over a planning horizon from several days to months), part type selection (choose a subset of part types for immediate and simultaneous manufacture), resource (or machine or tool) grouping, production ratio determination (with respect to the selected part types, resource allocation and loading subject to technological and capacity constraints. FMS scheduling problems are concerned with running the FMS during real time once it has been set up during the planning stage which is in advance of actual production [252]. This phase involves to determine the optimal sequence at which the parts of the selected part types are to be input into the system. Appropriate scheduling methods (exact and heuristic algorithm) have to be developed. In Chapters II and III we have focused on this field. FMS control problems are associated with the continuous monitoring of the system, in order to meet production requirements and order demands. Maintainance and inspection policies as well as a policy to handle machine breakdowns belong to this field.

In this final Chapter IV we are going to discuss design and planning problems in FMSs. A problem that frequently arises in flexible manufacturing planning is the process of dividing a manufacturing system into machine cells by grouping machines and parts, e.g. based on the similarity of part manufacturing characteristics. Parts that have to undergo similar operations require the same set of machines and are grouped together to part families each of which is assigned to the corresponding machines forming a machine cell. This is the philosophy of group technology [253] in order to maintain a high efficiency if correlated parts of the manufacturing system are organized into cells [254]. Tooling, loading, and scheduling decisions within each cell should preferably be made as far as possible independently of the other cells. This requires that machine groups and part families are as homogeneous as possible and that a minimum interference occurs between cells. A major goal is to reduce transportation costs and setup costs, both are often neglible as long as they are

---

[250]    Stecke (1985)

[251]    Tetzlaff (1990)

[252]    Stecke (1985)

[253]    the first time applied by Mitrofanov (1959), see also Burbidge (1963, 1975)

[254]    Schönberger (1982), Hyer / Wemmerlöv (1989)

restricted to a particular group technology cell. Moreover size reduction and decomposition are the main issues in order to find a way to handle the FMS production planning and scheduling problems. Group technology became a very attractive field of research in recent years and a vast number of publications in the FMS area is dedicated to that kind of planning problems.[255] Let us have a closer look to the model presented by Venugopal and Narendan (1992). Consider processing of n parts on at most m machines and define

$p_{ij}$      the processing time (hour/part) of part type j on machine i;

$T_i$      available time on machine i in a given period of time;

$N_j$      number of parts from part type j required to be processed in a given period of time;

$(d_{ij})$      an m×n machine–part type incidence matrix where

$d_{ij}$      is the workload on machine i induced by part type j, i.e. $d_{ij} = N_j \cdot p_{ij} / T_i$ determines the percentage of available time of machine i requested by part type j;

$(x_{i\ell})$      m×m (machine) cell membership matrix, where at most m nonempty manufacturing cells can be created, and

$x_{i\ell}$      equals 1 if machine i is assigned to cell $\ell$, 0 otherwise.

Let $(m_{\ell j})$ be a m×n–matrix of average cell load by part type j, i.e.

$$m_{\ell j} = \frac{\sum\limits_{i=1}^{m} x_{i\ell} \cdot d_{ij}}{\max\left\{ \sum\limits_{i=1}^{m} x_{i\ell} \, , \, b_\ell \right\}} \quad \text{if cell } \ell \text{ is assigned at least one machine, 0}$$

otherwise.

where $\sum\limits_{i=1}^{m} x_{i\ell} \cdot d_{ij}$ is the total load of cell $\ell$ induced by part type j which then is divided by the number of machines in cell $\ell$. The number of machines in cell $\ell$ is desired to be at least $b_\ell$

The mathematical programming formulation of the grouping problem is now:

---

[255] We refer to the surveys by Kuhn (1990), Oerlemans (1992a), Crama / Oosten (1992), Ruben et al. (1993), and Shafer / Rogers (1993a,b), Wemmerlöv / Hyer (1987), King / Nakornchai (1982), Miltenburg / Zhang (1991).

$$\text{Min} \sum_{i=1}^{m} \sum_{\ell=1}^{m} x_{i\ell} \cdot \sum_{j=1}^{n} (d_{ij} - m_{\ell j})^2$$

subject to

$$\sum_{\ell=1}^{m} x_{i\ell} = 1 \qquad \text{for all } 1 \leq i \leq m;$$

$$x_{i\ell} = 0 \text{ or } 1 \qquad \text{for all } 1 \leq i,\ell \leq n.$$

The constraints ensure that each machine is assigned to only one cell. The expression $x_{i\ell} \cdot \sum_{j=1}^{n} (d_{ij} - m_{\ell j})^2$ in the objective function describes the load variance between the machine i request of part type j and the average cell $\ell$ load of part type j if machine i should belong to cell $\ell$. Hence the objective is to balance the workload of a machine and the cell to which it belongs. Let us illustrate the model by the problem indicated in Figure 1. Figure 1a shows the matrix $(d_{ij})$ for 14 machines and 20 part types. Its solution into 5 nonempty manufacturing cells, respectively. part type families to machine cell assignments, is exhibited in Figure 1b. The column sums of Figure 1b divided by the number of machines assigned to the cell are just the values $m_{\ell j}$, i.e. this value describes the average cell load by part type j. Register, if production is required over only one period then for each row rounding up the sum over all row entries to the next integral value determines the number of machines of the same type necessary to process all parts. Hence there is need for 3 copies of machine 11. The bounds $b_{\ell}$ are assumed to be 2 for all cells. Figure 1b shows an ideal situation where the machine–part incidence matrix represented by the machine workload matrix could be transformed into a block diagonal matrix without any exceptional elements. That means, there is a part and machine to cell assignment such that all parts in a cell are manufactured only in this cell. Costly transportations to some other manufacturing cells which probably require additional setups are not necessary. However, in most practical cases only a near–block diagonal matrix can be found which includes still some exceptional elements. Furthermore, it may occur that not all machines included into a cell are supposed to process each part type assigned to that cell. FMS scheduling problems, such as job shop or flow shop, now arise in each particular cell. For instance, in our case the period of time a machine is available is not always

| | 1 | 2 | 3 | 4 | 5 | 6 | 7 | 8 | 9 | 10 | 11 | 12 | 13 | 14 | 15 | 16 | 17 | 18 | 19 | 20 |
|---|---|---|---|---|---|---|---|---|---|---|---|---|---|---|---|---|---|---|---|---|
| 1 | 0.8 | | | 0.9 | | 0.5 | | | 0.3 | 0.2 | 0.3 | | | | | | | | | |
| 2 | | | | | 0.9 | | | | | | | 0.7 | | | | | | 0.8 | 0.4 | |
| 3 | | | 0.5 | | | | | | | | | | | 0.3 | | | 0.2 | | | |
| 4 | | | | | | | 0.1 | | | | | | | | 0.2 | | | | | 0.3 |
| 5 | 0.4 | | | 0.5 | | 0.2 | | | 0.5 | 0.6 | 0.7 | | | | | | | | | |
| 6 | | | 0.2 | | | | | | | | | | | 0.3 | | | 0.4 | | | |
| 7 | | | | | | | 0.6 | | | | | | | | 0.7 | | | | | 0.8 |
| 8 | | 0.2 | | | | | | 0.3 | | | | | 0.4 | | | 0.5 | | | | |
| 9 | | 0.6 | | | | | | 0.7 | | | | | 0.8 | | | 0.9 | | | | |
| 10 | | | | | | | 0.2 | | | | | | | | 0.3 | | | | | 0.4 |
| 11 | | | | | 0.5 | | | | | | | 0.6 | | | | | | 0.7 | 0.8 | |
| 12 | | | 0.5 | | | | | | | | | | | 0.6 | | | 0.7 | | | |
| 13 | | | | | 0.9 | | | | | | | 0.8 | | | | | | 0.9 | 0.5 | |
| 14 | | 0.5 | | | | | | 0.3 | | | | | 0.6 | | | 0.8 | | | | |

| | 3 | 14 | 17 | 7 | 15 | 20 | 2 | 8 | 13 | 16 | 1 | 4 | 6 | 9 | 10 | 11 | 5 | 12 | 18 | 19 |
|---|---|---|---|---|---|---|---|---|---|---|---|---|---|---|---|---|---|---|---|---|
| 3 | 0.5 | 0.3 | 0.2 | | | | | | | | | | | | | | | | | |
| 6 | 0.2 | 0.3 | 0.4 | | | | | | | | | | | | | | | | | |
| 12 | 0.5 | 0.6 | 0.7 | | | | | | | | | | | | | | | | | |
| 4 | | | | 0.1 | 0.2 | 0.3 | | | | | | | | | | | | | | |
| 7 | | | | 0.6 | 0.7 | 0.8 | | | | | | | | | | | | | | |
| 10 | | | | 0.2 | 0.3 | 0.4 | | | | | | | | | | | | | | |
| 8 | | | | | | | 0.2 | 0.3 | 0.4 | 0.5 | | | | | | | | | | |
| 9 | | | | | | | 0.6 | 0.7 | 0.8 | 0.9 | | | | | | | | | | |
| 14 | | | | | | | 0.5 | 0.3 | 0.6 | 0.8 | | | | | | | | | | |
| 1 | | | | | | | | | | | 0.8 | 0.9 | 0.5 | 0.3 | 0.2 | 0.3 | | | | |
| 5 | | | | | | | | | | | 0.4 | 0.5 | 0.2 | 0.5 | 0.6 | 0.7 | | | | |
| 11 | | | | | | | | | | | | | | | | | 0.9 | 0.7 | 0.8 | 0.4 |
| 2 | | | | | | | | | | | | | | | | | 0.5 | 0.6 | 0.7 | 0.8 |
| 13 | | | | | | | | | | | | | | | | | 0.9 | 0.8 | 0.9 | 0.5 |

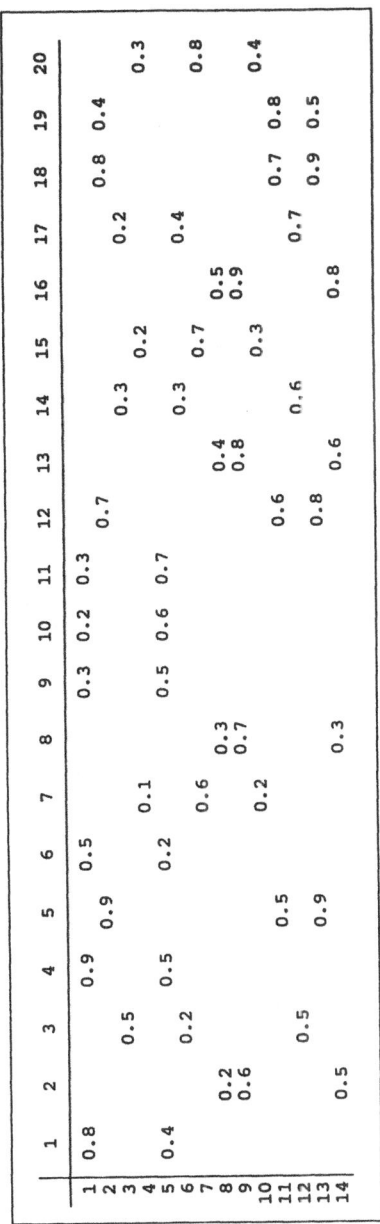

**Figure 1a.** Matrix of machine work loads per part type.

**Figure 1b.** A solution into 5 manufacturing cells.

sufficient in order to manufacture all parts, a problem that has to be resolved possibly by use of some priority rules.

Contrary to our example, many methods described in the literature have not considered key factors such as machine capacity or work load balance among machines. Parts grouping has been performed at the level of long horizon planning and interpreted as a way of parts aggregation. Based on only the part–machine incidence matrix a commonly applied objective is to minimze the number of exceptional elements or in other words to create mutually distinct clusters of parts usually under additional cell size criteria (lower and upper bounds on the machine number in a cell). Others considered certain types of cost functions most often however depending on the number of intercell part movements. Solution methods are discribed in McCormick / Schweitzer / White (1972) – the so-called bond energy algorithm. This method is most widely spread and received a number of extensions by Arabie / Hubert (1990), Askin et al. (1991), Hilger et al. (1991), or Ng (1991). Lexicographic ordering of rows or columns of the part–machine incidence matrix is applied by King (1980); Chan / Milner (1982), King / Nakornchai (1982), and Chandrasekharan / Rajagopalan (1986) introduced some variants. Methods clustering machines based on similarity measures between machines or parts are described by Chandrasekharan / Rajagopalan (1986), Kusiak / Heragu (1987), Chu (1989), Boe / Cheng (1991), Kusiak / Cho (1992), or Miltenburg / Zhang (1991). Local search based approaches have been used by McCormick et al. (1972), Follonier (1992), Venugopal / Narendran (1992).[256] The next section [257] presents a generally applicable new clustering approach in order to create manufacturing cells in an FMS. The method can be applied to numerous other problems arising in completely different fields such as biology, electrical engineering, medicine, etc. By this reason we are going to describe the ideas in a fairly general setting.

---

[256] There are other methods which do not perfectly fit into the above classification, cf. Askin / Subramanian (1987), Kusiak / Cheng (1990), Kusiak / Chow (1987), Venugopal / Narendran (1993), Song / Hitomi (1992), Vannelli / Hall (1993), Wu / Salvendy (1993), Balasubramanian / Panneerselvam (1993), and Liang / Dutta (1993), Lee / Garzia–Dias (1993).

[257] which basically consists of the contents of Dorndorf / Pesch (1992b)

# 1. Clustering in Cellular Manufacturing

## 1.1 Introduction and Background

A task that frequently arises in qualitative data analysis is to uncover natural groupings, or types, of objects each of which is characterized by several attributes. One can think of these objects as vertices of an edge–weighted graph G; each positive or negative weight represents some measure of similarity or dissimilarity, respectively, of an edge defining object pair. A clustering of the objects into groups is a partition of the graph, which means a partition of the vertex set of G into non–overlapping subsets. The set of edges connecting vertices of different subsets from some partition of G is called a cut. In order to find groups as homogeneous as possible positive edges should appear within groups and negative edges in the cut. Hence, a best clustering is one with a minimal cut weight.

Cut minimization subject to some additional constraints arises in many applications, and the literature straddles all quantitative, scientific disciplines, as demonstrated by the remarkable variety in the reference section of Dubes / Jain (1980).

In VLSI (Very Large Scale Integration) components of an electronic circuit have to be placed on a silicon wafer. The components communicate with each other through wires. Hence, grouping highly connected components will meet two basic objectives, minimization of the total wire length and minimization of the area required for placing the components of the chip [258].

Let us consider one particular problem in more detail. Advanced technologies in the production of integrated circuits enable the designer to place more and more transistor like devices, gates say, on a chip. Gates are arranged linearly in a row where the connections in between are the nets. Nets may share the same row, called track, if they have no gate in common. An assignment of nets to tracks preserving this property is called a feasible track assignment. The area of a gate matrix layout, i.e. the area of an arrangement of the electronic devices on a chip, is proportional to the product of the number of gates and the number of tracks. The objective is to determine a permutation of the gates such that the number of assigned tracks to realize the layout is minimum. An instance of such

---

[258]    see Kernighan / Lin (1970), Dunlop / Kernighan (1985), Lengauer (1990), Feo / Khellaf (1990), and Barahona et al. (1988)

a linear layout problem consists of an m×n matrix $(d_{ij})$, the net–gate matrix, whose rows correspond to the nets $N_i$, $i = 1, ..., m$, and whose columns represent the gates $G_j$, $j = 1, ..., n$[259]. The matrix contains only 0 and 1 entries where all gates $G_j$ with $d_{ij} = 1$ must be connected by one net $N_i$. Connections are realized only in rows such that for any permutation of the gates a net is defined by the part of the row (all columns) from the leftmost to the rightmost gate (column) containing a 1–entry in the considered row. Hence connections of a net are established over this area even if there are some entries of 0 in between. Consider Figure 2 [260] the net–gate matrix (part (a) of Figure 2) of which consists of 8 gates G0, A, B, C, D, E, F, and G8 and 11 nets. A layout of 7 tracks is shown in (b), and the optimal layout in (c) consists of only 6 tracks, three of which (no. 1, 3, 4) contain more than one net.

The problem can easily be transformed into some type of a clustering problem. Identify each net of the net–gate matrix with an object whose attributes are the 0–1 entries of gate connections. Any two vertices in the corresponding edge–weighted graph G are adjacent, the edge is weighted by $-\infty$ if the vertex representing nets share a common gate. Otherwise, the weight is some positive value. A partitioning of the vertices may correspond to a layout such that the clusters correspond to the tracks. But in some cases the number of clusters only gives a lower bound on the number of necessary tracks. This is easy to see. Consider for instance nets 5 and 6 of the net–gate matrix. Then the edge connecting the vertices 5 and 6 is weighted positively. Hence they might occure in the same cluster however there is no feasible layout with respect to the considered permutation of gates such that nets 5 and 6 may share the same track. Thus, additionally, the problem arises to find the "best" permutation of gates such that the number of tracks is minimal (over all possible gate permutations). The permutation problem on the other hand can be translated into a traveling salesman problem; but we are not going to clarify this correspondence. There is a somewhat different interpretation of the gate matrix layout problem in terms of clustering. Consider a permutation of gates, as in Figure 2, then a net of the net–gate matrix consists not only of those gates having 1–entries in the corresponding net position, the net also straddles all gates in between. For instance, net 5 from the net–gate matrix in Figure 2 connects gate G0 to E straddling gates A, B, C, D. This can be expressed more formally by considering for any permutation $\pi : \{1, ..., n\} \to \{1, ..., n\}$ of gates

---

[259]   Möhring (1990)

[260]   Möhring (1990)

the augmented net–gate matrix $(ad_{ij})$ where $ad_{ij} = 1$ if there are gates $G_r$ and $G_s$ with $\pi(r) \leq j \leq \pi(s)$ and $d_{ir} = d_{is} = 1$. Now tracks and clusters correspond to each other if the clustering problem is extended to the augmented net–gate matrix.

| | GO | A | B | C | D | E | F | G8 |
|---|---|---|---|---|---|---|---|---|
| 1 | 1 | 1 | 1 | | | | | |
| 2 | 1 | 1 | | 1 | | | | |
| 3 | 1 | | 1 | 1 | | | | |
| 4 | 1 | | | 1 | | | | |
| 5 | 1 | | | | | 1 | | |
| 6 | | | 1 | | 1 | | | |
| 7 | | | | 1 | 1 | | | |
| 8 | | | | | 1 | 1 | | |
| 9 | | 1 | | | | 1 | | |
| 10 | | | | | 1 | 1 | | |
| 11 | | | | | | 1 | 1 | |

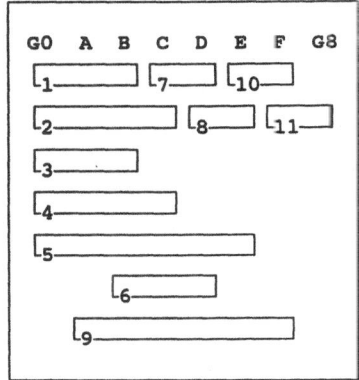

(a)   Net–gate matrix                                   (b)   Layout

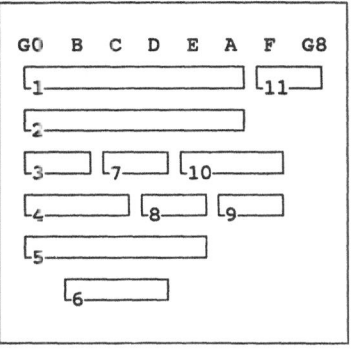

(c)   Optimal layout

Figure 2.   Gate matrix layout.

Consider another motivation: the problem of program paging of computer memory. A computer program may be thought of as a set of blocks, for instance subroutines or procedures. Connections, like procedure calls, between these blocks occur via data or control flows. In paged memory organizations these blocks have to be assigned to pages of a fixed size such that the number of calls between blocks of different pages is minimized [261].

There are other applications from economics like grouping in flexible manufacturing systems [262], or workload balancing in service networks [263], from biology, from political and social science etc. which can all be translated into some type of a graph partitioning problem [264]. Let us have briefly a closer look to some grouping problems arising in FMS design and planning problems. A flexible machine is capable to manufacture completely different jobs via the change of tools. However, tool changes are costly as they require a machine powerdown sometimes even for hours. So a major goal is to distribute jobs among groups in such a way that each job group needs only one machine setup and no additional tool changes. Thus the number of tools required by all jobs in one group may not exceed the number of tools that can be stored into the flexible machine's tool magazine simultaneously in order to perform all job operations of that group. The problem as well as some solution methods are described in Tang and Denardo (1988a,b), Oerlemans (1992a), or Crama / Oerlemans (1991a,b, 1992), Crama et al. (1991), Spieksma (1992), Kuhn (1990), Follonier (1992), O'Grady / Menon (1987). Besides, lower bounds can be obtained by traveling salesman solutions where the jobs correspond to the cities, a distance between any two jobs is the minimal number of tool switches necessary in order to process these jobs consecutively.

Grouping in FMS design (machine as well as part grouping) has already been introduced and a number of publications have handled this problem as a clustering problem based on different measures of machine (respectively part)

---

[261]   see Kral (1965), Kernighan (1971), and the remarks on compiler construction in Johnson et al. (1991)

[262]   cf. King / Nakornchai (1982), Kumar / Kusiak / Vannelli (1986), or Ahmadi / Tang (1991)

[263]   cf. Bodin / Levy (1991) and Van Driessche / Piessens (1992)

[264]   see for instance Benders (1962), Barthélemy / Monjardet (1981), or Garfinkel / Nemhauser (1970)

similarity or dissimilarity [265]. [266]    Many    of these approaches can be translated into partitioning the vertex set of a complete graph into non–overlapping subsets. Let us illustrate that on our introductory group technology problem in order to create manufacturing cells. By the way, grouping in cellular manufacturing actually encompasses two clustering steps which, in our case, can be seen as mirror images of each other. Grouping of parts from the set of n part types to part families and grouping of machines to machine cells as well are these two steps followed by a part family machine cell assignment. These steps are neither completely independent nor complementary to each other although many authors determine a part (respectively machine) grouping leaving open the question of determining the best corresponding machine (respectively part) grouping. Consider any two machines i and h and the workload $d_{ij}$ and $d_{hj}$, respectively, with respect to part type j. The Minkowski metric may serve as some measure of similarity of machines h and i [267],    i.e. we say machines h and i are similar  with respect to part type j if

$$|d_{ij} - d_{hj}| \leq \left[ |d_{ij} - m_{\ell j}|^p + |d_{hj} - m_{\ell j}|^p \right]^{1/p}.$$

Obviously, the result heavily depends on the parameter p and the values $m_{\ell j}$. The latter can be calculated only if the number of machines that belong to a manufacturing cell $\ell$ is already determined. Therefore, an initial random partition of the machines is considered as a starting solution. Part type j belongs to a family of part types which are supposed to be manufactured mainly in cell $\ell$, i.e. this family of part types is assigned to machine cell $\ell$. A measure of similarity is the value $w_{hi}$ defined as twice the number of part types satisfying the forementioned formula subtracted by the number of part types. The obtained similarity measure serves as edge weight in the graph representation of the machine grouping problem where the machines are identified with the vertices of the graph. Based on the currently calculated edge weights a new vertex partition is obtained. The corresponding machine groups yield new values $m_{\ell j}$ in order to

[265]    see Miltenburg / Zhang (1991)

[266]    Kusiak / Cho (1992) also include alternative process plans in their incidence matrix, Kusiak (1985, 1988) took technological and geometrical attributes into account, see Shafer / Rogers (1993a,b), Ribeiro / Pradin (1993).

[267]    cf. Kusiak / Cho (1992), Kusiak (1985)

recompute, respectively to learn after a number steps, improved edge weights [268]. Preliminary computational results for a fixed value p = 2 are very encouraging. A sequence of partitions is generated as long as the value of the objective function is decreasing.[269]

Figure 3 summarizes several graph partitioning problems; for each problem besides its name there is a brief description as well as some references and a remark on its complexity. In all of these problems the vertex set V of the underlying graph is partitioned into k subsets of a possibly predefined size. Hence the number of vertices $|V_i|$ in subset i is bounded by some lower bound $\ell_i$ and some upper bound $u_i$ . Edge weights may be positive or negative. The objective function is to minimize – to maximize in case of the max–cut problem – the external costs, i.e. the sum of the edge weights of all edges connecting vertices of different subsets. Other objectives closely related to optimal facility locations are listed in Pinter / Pesti (1991) or Dubes / Jain (1980).

---

[268] For an analog type of parameter learning concerning the determination of batches in acrylic–glass production we refer to Friedrich et al. (1991).

[269] In the sequel we restrict our attention to the partitioning part, see Dorndorf / Pesch (1992b).

| min–cut | $k = 2$; easy; Ford / Fulkerson (1974) |
| k–way–partitioning | $k > 2$; NP–complete; Chopra / Rao (1991, 1993), Minoux / Pinson (1987) |
| k–partitioning | $k \geq 2$; $\ell_i \leq u_i$ for $i = 1$ ,..., k; NP–complete; Garey / Johnson / Stockmeyer (1976), Garey / Johnson (1979) |
| clique–partitioning | $k$ not fixed; NP–complete for positive and negative edge weights; Wakabayashi (1986), Grötschel / Wakabayashi (1989, 1990a,b) |
| m–dimensional matching | $k \geq 2$; $\ell_i = u_i = m$ for all $i = 1$ ,..., k; NP–complete; easy for $m = 2$; Feo / Khellaf (1990), Papadimitriou / Steiglitz (1982) |
| max–cut | $k = 2$; NP–complete; Garey / Johnson / Stockmeyer (1976) |
| hypergraph partitioning | Sanchis (1989), Kahng (1989) |

Figure 3.    Graph partitioning problems.[270]

## 1.2   The Clique–Partitioning Problem

The objective of cluster analysis is to uncover natural groupings, or types, of n objects each of which is characterized by m attributes. Describe in some sense the best partitioning of all objects into disjoint subsets or classes (also called

---

[270] Linear and non–linear binary problem formulations (see Lukes (1975), Barnes (1982), Barnes et al. (1988), Kumar et al. (1986), Roucairol / Hansen (1987)) as well as set partitioning problem formulations (see Minoux / Pinson (1987), Anily / Federgruen (1991)) led to the development of sophisticated exact and approximation algorithms. We refer the reader to the branch and bound or branch and cut algorithms from Koontz et al. (1975), Christofides / Brooker (1976), Roucairol / Hansen (1987), Faigle et al. (1987), Grötschel / Wakabayashi (1989, 1990a,b), Chopra / Rao (1991), Deza et al. (1992), the column generation method of E.L. Johnson et al. (1991), or a dynamic programming approach from Lukes (1975). See also Jensen (1969), Donath / Hoffman (1973), Kumar et al. (1986), Feo / Khellaf (1990), and Feo et al. (1992). The most prominent representative among the approximation algorithms is the algorithm of Kernighan and Lin (1970), the predecessor of the famous paper of Lin and Kernighan (1973) on the traveling salesman problem. Modifications of the Kernighan–Lin algorithm are used in Fiduccia / Mattheyses (1982), Krishnamurty (1984), Bui et al. (1987). For other local search approaches based on simulated annealing (Johnson et al. (1989) and Amorim et al. (1992)), tabu search (Amorim et al. (1992)), neural networks (Koenig et al. (1991)), genetic algorithm hybrids (Mühlenbein et al. (1988), Cohoon et al. (1991), Bhuyan et al. (1991), Von Laszewski (1991), Van Driessche / Piessens (1992), and Kamidoi et al. (1991, 1992a,b)) see the references. More about data analysis and clustering can be found in the books of Hand (1981), Hartigan (1975), Anderberg (1973), and Späth (1985).

clusters) such that each cluster is as homogeneous as possible with respect to all attributes. We can think of all kinds of objects such as animals, countries, computers, markets, companies, etc., as we will see later. The set of n objects may be described by some $n \times m$ matrix $D = (d_{ij})$ where each entry $d_{ij}$, $i = 1 ,..., n$ and $j = 1 ,..., m$, represents the property of object i with respect to attribute j . In general there are two steps to be performed during the clustering process. Firstly, there must be some measure of similarity between distinct objects and secondly according to these similarities the objects must be clustered into groups. Only the last step corresponds to the clique–partitioning problem. To find some measure of similarity for qualitative data we can proceed in the following way [271]. For two objects i and j let $\bar{m}$ be the number of attributes k of the same value (property), i.e. where $d_{ik} = d_{jk}$ . Then some measure of similarity is the value $w_{ij} := 2 \cdot \bar{m} - m$ , where $w_{ij} = -m$ says that both objects are completely different while increasing values $w_{ij}$ correspond to increasing similarity of the objects. The objects may be considered as vertices of an undirected complete graph with edge weights $w_{ij}$ where

$$w_{ij} = 2 \cdot |\{ k \mid d_{ik} = d_{jk} , k = 1 ,..., m \}| - m , \text{ for all } 1 \leq i < j \leq n . \quad (1)$$

Hence, clustering of the objects corresponds to partitioning the complete graph of n vertices into complete subgraphs, or cliques, such that the sum of all edge weights in all cliques is maximum, or equivalently

$$\min \sum_{1 \leq k < \ell \leq n} \sum_{i \in V_k, j \in V_\ell} w_{ij}$$

where $V_k$ and $V_\ell$ are the vertex sets of cliques k and $\ell$, respectively.[272]
Obviously, if all edge weights are nonnegative or nonpositive the problem can easily be solved. If the graph has negative as well as positive edge weights the clique–partitioning problem is NP–complete [273].
If we assign to each edge (i,j) a binary variable $x_{ij}$ where

---

[271]    see for instance Hartigan (1975), Späth (1985), or Grötschel / Wakabayashi (1989)

[272]    More about the analysis of graph theoretic clustering methods can be found in Augustson / Minker (1970), Corneil / Woodward (1978), Day (1977, 1981), Hubert (1974), Matula (1977), and Oglivie / Olson (1972).

[273]    cf. Wakabayashi (1986), Dyer / Frieze (1985)

$$x_{ij} = \begin{cases} 1 & \text{if vertices i and j are in the same clique} \\ \\ 0 & \text{otherwise} \end{cases}$$

then the following model gives a complete description of the clique–partitioning problem [274].

$$\max \sum_{1 \leq i < j \leq n} w_{ij} \cdot x_{ij}$$

s. t.

$$x_{ij} + x_{jk} - x_{ik} \leq 1 \qquad \text{for } 1 \leq i < j < k \leq n$$

$$x_{ij} - x_{jk} + x_{ik} \leq 1 \qquad \text{for } 1 \leq i < j < k \leq n$$

$$-x_{ij} + x_{jk} + x_{ik} \leq 1 \qquad \text{for } 1 \leq i < j < k \leq n$$

$$x_{ij} \in \{0,1\} \qquad \text{for } 1 \leq i < j \leq n.$$

The constraints assure that for each triangle (a clique of only three vertices) in the graph if two edges belong to the same clique then the whole triangle belongs to this clique. For qualitative data analysis the definition of the edge weights as mentioned before can easily be derived if the clusters are considered as equivalence classes of an equivalence relation whose m projections to each of the m attributes (or equivalence classes) lead to a maximum of coincidence [275]. However, if quantitative attributes have to be considered or in case of weighted or missing attributes other edge weight computation formulas apply. For instance, in case of the micro problem which is mentioned below the criterion $d_{ik} = d_{jk}$ is replaced by the more general similarity criterion

$$\frac{|d_{ik} - d_{jk}|}{\max \{d_{ik}, d_{jk}, 1\}} \leq \alpha$$

where $\alpha = 0.3$. In case of the classification of cetacea (dolphines, porpoises, and whales), which is also mentioned below, some information is missing (entry "*" in the data matrix). Therefore the weights $w_{ij}$ are calculated such that only

---

[274]   see Grötschel / Wakabayashi (1989, 1990a), Marcotorchino / Michaud (1980a,b, 1981), Opitz / Schader (1984a,b), Schader /Tüshaus (1985), and Tüshaus (1983), see also Rao (1971)

[275]   see Grötschel / Wakabayashi (1989) or Zahn (1964)

comparable attributes (no entry "*") of the objects i and j are considered. Hence,

$$w_{ij} = 2 \cdot |\{ k \mid d_{ik} = d_{jk} , d_{ik} \neq * , k = 1 ,..., m \}| -$$

$$|\{ k \mid d_{ik} \neq * , d_{jk} \neq * , k = 1 ,..., m \}|, \quad 1 \leq i < j \leq n . \qquad (2)$$

The second term corresponds to the number of comparable attributes for both objects. If no information is missing then we get the usual definition of the edge weights.

The classification of 137 companies [276] with respect to their need of different groups of employees (secretaries, programers, engineers etc.) seems to give a less interesting result. For 25 groups of employees each company indicated whether the group is needed (entry "1") or not (entry "0"). The optimal solution (see Table 5) has two clusters, one containing only two companies (no. 10 and no. 107). Both companies mentioned to require all kinds of employee groups. Hence, for a useful clustering weighted attributes seem to be indispensible, for instance, where the weighting factor for each attribute property is inversely proportional to the number of the attribute's property occurences over all objects [277].

The latter weight function (2) is also used in our additional problem instance of classification of lecturers. The Association of Business Administration Lecturers from Germany, Austria, and Switzerland has 13 commissions, each of which covers a special subject such as marketing, banking and finance, operations research, production management, etc. Each member of the association is supposed to be a member of some commissions as well as to express interest in their subjects. To each lecturer we assigned a 0,1–tuple where an entry "1" indicates the membership in the corresponding commission, or the interest in its subject, respectively. Commission membership always implies interest in its subject but not vice versa. Therefore a membership entry "1" yields also an interest entry "1". If we use the first weight function (1) we get an optimal solution resulting in one cluster only. However, this is not a big surprise. For instance, if two lecturers have no common interest or are not members of a particular commission then the corresponding "0" entries do not provide some characteristic information that these two lecturers have in common. Entries of "1" give more information. Therefore, we used the second type of weight

---

[276]   Späth (1977)

[277]   see Anderberg (1973)

function (2) where a "0" entry in our data set was considered as "∗". All relevant data as well as the best solution we could find, consisting of 114 clusters, is provided in Dorndorf / Pesch (1992b).

We are leaving aside the question whether a partition into disjoint clusters will be an appropriate setting for the companies and lecturers problem instances in order to yield meaningful interpretations.[278]

## 1.3   An Ejection Chain Heuristic

We now present a powerful heuristic based on the ejection chain idea of Glover (1991a, 1992a,c) and the variable depth local search approach of Kernighan and Lin (1970) for the graph partitioning problem.

Consider a feasible solution X of the clique–partitioning problem, i.e. a partition of the vertices into a number of cliques. A solution X' is defined to be a neighbour of X if X' is obtained by the following operation (called a move): Move a vertex u from its current cluster in X to another perhaps non–existing and by vertex u newly defined cluster. Define the neighbourhood of X to consist of all solutions X' obtained from X by one move [279].   Obviously, the right sequence of moves will lead from any solution to an optimal solution of the clique–partitioning problem. In the simplest version of our local search algorithm we are always proceeding in a greedy way. Thus, each move to some neighbour of a feasible solution is always a most improving move, i.e. among all neighbours that yield an improved objective function value that one with the highest objective function value is chosen. Usually this process will end up in some suboptimal solution.[280]

**Simple Local Search.** In an n vertex graph the number of nonempty clusters is at most n. Starting from a random partition where each vertex randomly is assigned to one of the n clusters some empty clusters may be left over. Empty clusters are not assumed to be the same. We also tested implementations with only one additional cluster which is empty. However, these implementations did

---

278   More information about the problem of a meaningful interpretation and aggregation of related attributes can be found in Grötschel / Wakabayashi (1989), Opitz / Schader (1984), or Anderberg (1973).

279   see Règnier (1965)

280   The same neighbourhood structure was employed by Amorim et al. (1992) in their simulated annealing and tabu search approach.

not save time. Let $v(u)$ be the label of the clique containing vertex $u$. Let $V_1, ..., V_k, ..., V_n$ be the clusters or cliques of the current solution where several of the $V_i$ may be empty. Thus, vertex $u$ is contained in clique $V_{v(u)}$. For each vertex $u$ we can define internal costs $I(u)$ and external costs $E(u,k)$ where

$$I(u) = \sum_{\substack{i \in V_{v(u)} \\ i \neq u}} w_{ui} \quad \text{and} \quad E(u,k) = \sum_{\substack{i \in V_k \\ i \neq u}} w_{ui}$$

$I(u)$ is the sum of all edge weights of all edges connecting $u$ and all other vertices in the clique containing vertex $u$. $E(u,k)$ is defined with respect to clique $V_k$. It is the sum of all edge weights of edges in the complete graph connecting $u$ and all vertices in clique $V_k$. Obviously, $E(u,v(u)) = I(u)$. Each move of a vertex $u$ from clique $V_{v(u)}$ to clique $V_k$ yields the gain $g(u,k) = E(u,k) - I(u)$. Hence the best move is that one for which $g(u,k)$ is maximum (and greater zero) for all vertices $u$ and all cliques $V_k$. Let

$$g(u^*,k^*) = \max_{u \in V} \{ \max_{\substack{1 \leq k \leq n \\ k \neq v(u)}} \{g(u,k)\} \}$$

and let $E = (E(u,k))_{u=1,...,n;k=1,...n}$ be an $n \times n$ matrix containing all external costs. If vertex $u^*$ is moved from clique $V_{v(u^*)}$ to clique $V_{k^*}$ then the new entries $E(u,v(u^*))$ are $E(u,v(u^*)) - w_{u^*u}$ for all $u \in V \setminus \{u^*\}$. The entries $E(u,k^*)$ are changed to $E(u,k^*) + w_{u^*u}$ for all $u \in V \setminus \{u^*\}$. Hence the number of elements changed is of order $O(n)$. To find the best move at most $n^2$ entries in the matrix $E$ have to be considered. However, if we organize the entries of each row in $E$ as a heap, the best move can be found within $O(n)$ steps, and updating all heaps requires again $O(n \cdot \log n)$ steps [281]. Hence, the neighbourhood can be explored with time complexity $O(n \cdot \log n)$ instead of $O(n^2)$ required by Amorim et al. (1992). Both implementations are based on the same memory space complexity. Thus, for each row $u$, $1 \leq u \leq n$, an array $h_u$ contains the entries of $E$ corresponding to row $u$. Array $h_u$ is organized as a heap, i.e.

$$h_u[k] \geq h_u[2 \cdot k] \quad \text{and} \quad h_u[k] \geq h_u[2 \cdot k+1] \quad \text{for } 1 \leq k \leq \lfloor n/2 \rfloor .$$

For each vertex $u \in V$ the best move can be made in constant time because the

---

[281]   cf. Aho / Hopcroft / Ullmann (1974)

highest external costs can be found in at most the first three entries of array $h_u$ .

**Ejection Chain Heuristic.** The local search algorithm as described above might stop after a few iterations in some local optimum far away from the global optimum. To avoid this type of premature convergence, the idea of Kernighan and Lin (1970) may be adapted to the clique–partitioning problem. The algorithm is a special case of a more general approach introduced first by Glover (1991a, 1992a,c). The main idea is similar to the one presented in tabu search [283]

(i)     Greedily look for the best move which is not necessarily improving.

(ii)    Avoid cycling via a dynamically organized list of forbidden (tabu) moves.

(iii)   Perform an improving sequence of moves.

In case of the clique–partitioning problem we get the algorithm of Figure 4.

Initalization:      Randomly generate a feasible solution, compute all external and
            internal costs and construct the heaps.
*repeat*
        Duplicate the current solution and mark all vertices as non–tabu.
        *for* i := 1 *to* n *do*
        *begin* {Find a sequence of n best moves}
                Among all non–tabu vertices determine that one which maximizes the gain
                g(u,k) for all u, k $\in$ V. {Intensification part}
                Let g(u*,k*) be maximum (not necessarily positive).
                Perform the corresponding move and mark vertex u* tabu.
                Update all heaps and matrix E.
        *end;*
        {Improve the solution}
        Among all subsequences of the first 0 $\leq$ r $\leq$ n moves of the sequence of n moves let
        r* be a number of moves of a subsequence such that these first r* moves improve
        the solution at most.
        If r* > 0 replace the current solution by its duplicate after performing these r*
        moves.
*until* r* = 0;

Figure 4.    Ejection–Chain Heuristic.

Compressing a sequence of moves into a single compound move is a main idea of an ejection chain based local search algorithm. Hence the neighbourhood for a simple type of moves becomes embedded, level by level, in successively larger

---

[282]    see Glover (1986, 1989c, and 1990a)

neighbourhoods that represent more complex moves [283].    Contrary to our case, the component moves carried forward level by level need not in general yield feasible solutions. In order to avoid substantial infeasibility each component move is usually forced by a component move on the preceding level, hence the name ejection chain. In case of the m–dimensional matching problem we would get a series of ejecting vertices, one by one, over the set of clusters, leading finally to a new feasible solution. For the clique partitioning problem we kept the ejection process fairly rudimentary because each component move never violates feasibility. Nevertheless, an implementation of Glover's idea of ejection chains can easily be achieved. The neighbourhood of a component move in the chain will be much smaller than in our implementation, i.e. each move of a vertex to a new cluster ejects a new vertex from that cluster, and so on, until all n vertices are considered or the chain stops with a move into an empty cluster or a cluster consisting of only tabu vertices. So, the ejection chain idea leads to a larger diversification of the search while we focused on search intensification resulting from the larger neighbourhood of the component moves and the chain length of n which is the maximum possible chain length where each vertex is considered at most once. The quality of the obtained results as well as the running times let us drop the diversification part of an ejection chain based search (this, however, is completely in contrast to experiments on the traveling salesman problem).

The time complexity of the algorithm described in Figure 4 in its initalization step is $O(n^2 \cdot \log n)$ and is determined by the time needed to construct n heaps. This is also the overall time complexity for a single *repeat ... until* iteration where the *for ... do* loop contributes with a complexity of $O(n^2 \cdot \log n)$.[284]

## 1.4   Computational Results of the Heuristics

We tested our algorithms on real world problem instances which are taken from Grötschel and Wakabayashi (1989); see also Vescia (1985) for the cetacea instance, Opitz / Schader (1984) for the workers instance, Hartigan (1975) for the cars instance, and Späth (1977) for the companies instance. Sources of the

---

[283]   Glover (1991a)

[284]   More information about the complexity of (variable depth) local search algorithms is presented in Johnson / Papadimitriou / Yannakakis (1988) and in Papadimitriou / Steiglitz (1982).

remaining instances can also be found in Amorim et al. (1992). They include the largest ones whose optimal solutions have been published in the literature. Moreover, we used 3 random data sets of 300 vertices each and a new real world problem 'lecturers' of 797 vertices of which we do not know the optimal solution. The simple local search algorithm (henceforth called 1–opt) as well as the ejection chain algorithm (for short: EC) were both run 100 times each on all real world problems. Initial feasible solutions were randomly generated. All runs were performed on a Silicon Graphics Personal Iris 4D/30 computer (MIPS R3000 processor, 30 Mhz) under the operating system UNIX. The algorithms are implemented in C. In Table 1 we can see that all practical problems from the literature – even the larger ones – can be solved to optimality in less than 1 second on average (column $\bar{t}$ sec'). The third column shows the number of times the optimum was found within 100 runs.

| problem | n | times | $\bar{t}$ sec |
|---|---|---|---|
| wild cats | 30 | 100 | 0.026 |
| cars | 33 | 100 | 0.037 |
| workers | 34 | 100 | 0.038 |
| cetacea | 36 | 100 | 0.032 |
| micro | 40 | 96 | 0.064 |
| UNO | 54 | 100 | 0.080 |
| UNO 1a | 158 | 100 | 0.807 |
| UNO 1b | 139 | 100 | 0.641 |
| UNO 2a | 158 | 100 | 0.930 |
| UNO 2b | 145 | 100 | 0.798 |
| companies | 137 | 100 | 0.704 |

Table 1.    EC-local search

| problem | n | times | $\bar{i}$ | $\bar{t}$ sec |
|---|---|---|---|---|
| wild cats | 30 | 79 | 24.9 | 0.009 |
| cars | 33 | 85 | 28.7 | 0.013 |
| workers | 34 | 89 | 29.5 | 0.013 |
| cetacea | 36 | 100 | 28.2 | 0.013 |
| micro | 40 | 42 | 35.6 | 0.019 |
| UNO | 54 | 90 | 46.9 | 0.038 |
| UNO 1a | 158 | 100 | 149.0 | 0.481 |
| UNO 1b | 139 | 100 | 132.8 | 0.380 |
| UNO 2a | 158 | 100 | 152.0 | 0.557 |
| UNO 2b | 145 | 100 | 140.5 | 0.466 |
| companies | 137 | 100 | 132.7 | 0.419 |

Table 3.    1-opt local search

The micro data set seems to be the most difficult one. It is the only problem instance where the optimum was not found in all runs. The 4 suboptimal results were only 0.6 % below the optimum. Amorim et al. (1992) did their runs on a comparable computer, a Sun Sparc workstation, and the average computation times in both implementations, simulated annealing (parameter: number of iterations in a series is $n^2/2$, factor by which the temperature is reduced after each series of iterations equals 0.85, and the initial temperature equals 1) and tabu search (parameters: number of iterations in a series is $n/2$ and the tabu list length equals 5) were more or less the same. Their average running times for the real world problems are about 10 to 20 times higher and there is no recognizable advantage of their simulated annealing or tabu search implementation compared to our 1–opt local search heuristic. If we sum up the running times for all real world problems then we get 2.408 seconds for the 1–opt local search heuristic, 4.157 seconds for the EC–local search heuristic, 113.16 seconds for the simulated annealing, and 89.24 seconds for the tabu search algorithm of Amorim et al. (1992).

If we consider an iteration of the EC–algorithm as one run in the *repeat ... until* loop or in other words as a sequence of n best moves to determine the best subsequence then each of the above problems needs between 2 and at most 5 iterations to be solved. The last iteration is always needed to check that no further improvement of the current solution is possible. Table 2a shows the average length of the most improving subsequence (the average of the r*) in iterations i , i = 1, 2, 3, 4, over 100 runs. It also shows that the maximum number of improving iteration is 2 or 3, and 4 only for the micro problem. We can conclude that many of the 100 runs already found the optimum within 1 iteration, i.e. the current solution was improved only once. In particular 'UNO 1b', 'UNO 2b', and 'companies' never used a second improving iteration for all 100 runs. Column ī of Table 2a presents the average number of iterations for each problem over 100 runs. It always includes a final non–improving iteration.

| problem | n | iteration | | | | ī | iteration | | | |
|---------|---|-----------|---|---|---|---|-----------|---|---|---|
| | | 1 | 2 | 3 | 4 | | 1 | 2 | 3 | 4 |
| wild cats | 30 | 23.7 | 1.02 | 0.11 | – | 2.43 | 99.55 | 0.40 | 0.05 | – |
| cars | 33 | 27.0 | 1.27 | 0.20 | – | 2.88 | 99.73 | 0.26 | 0.01 | – |
| workers | 34 | 27.9 | 0.95 | 0.06 | – | 2.72 | 99.32 | 0.67 | 0.01 | – |
| cetacea | 36 | 27.2 | 0.01 | – | – | 2.01 | 99.92 | 0.08 | – | – |
| micro | 40 | 32.4 | 13.72 | 3.86 | 0.05 | 3.57 | 88.27 | 7.62 | 4.03 | 0.08 |
| UNO | 54 | 45.4 | 3.26 | 0.21 | – | 2.51 | 98.11 | 1.78 | 0.11 | – |
| UNO 1a | 158 | 147.9 | 0.09 | – | – | 2.06 | 99.98 | 0.02 | – | – |
| UNO 1b | 139 | 131.8 | – | – | – | 2.00 | 100.00 | – | – | – |
| UNO 2a | 158 | 151.1 | 0.11 | – | – | 2.07 | 99.98 | 0.02 | – | – |
| UNO 2b | 145 | 139.5 | – | – | – | 2.00 | 100.00 | – | – | – |
| companies | 137 | 131.7 | – | – | – | 2.00 | 100.00 | – | – | – |

Table 2a.  Average number of moves in each        Table 2b.  Average improvement in
iteration, i. e. average r*.                                  for each iteration.

| lecturers | iteration | | | | | | |
|---|---|---|---|---|---|---|---|
| | 1 | 2 | 3 | 4 | 5 | 6 | 7 |
| $\overline{r^*}$ | 680.4 | 106.87 | 45.25 | 11.93 | 2.18 | 0.75 | 0.04 |
| $\overline{\%}$ | 97.34 | 2.54 | 0.10 | 0.02 | 0.00 | 0.00 | 0.00 |

Table 2c.    EC-local search on the lecturers problem instance.

Row $\overline{r^*}$ contains the average number of moves in each iteration, i. e. average $r^*$.

Row $\overline{\%}$ contains the average improvement in % for each iteration.

Table 2b shows the average improvement in %. The percent values are evaluated with respect to the overall improvement. Almost the whole improvement of a random initial solution is achieved during the first iteration.

We also compared the results obtained by the EC–algorithm to those that can be obtained by the 1–opt local search algorithm which runs in $\mathcal{O}(n \cdot \log n)$ time for a single iteration. The results are shown in Table 3. For each problem we used the same 100 random initial solutions as for the EC–algorithm. The third column of Table 3 shows how often the optimum was found within 100 runs.

Column $\bar{i}$ gives the average number of iterations needed to be locally optimal over 100 runs. In this case an iteration simply means a move. The last column $\bar{t}$ exhibits the average time for 100 runs.

The worst solution found for the 'micro' data set was 12.3% below the optimum value while the EC–algorithm reached 0.6% below the optimum value in the worst case. The number of moves in the first iteration of the EC–algorithm empirically is of the same order as the number of moves to reach a suboptimum with respect to 1–opt local search. In both cases it is $\mathcal{O}(n)$. Hence in both cases empirically the time complexity of both algorithms is of the same order. This also holds for the new problem instance 'lecturers' to reach suboptimal solutions. Within 100 runs all suboptimal solutions from the EC–heuristic ranged from 12845 (worst solution) to 12983 (best solution found). The average of all 100 local optima is 12913 and the average running time was 45.86 seconds. The average running time over 100 runs of the 1–opt heuristic was 14.71 seconds. The results are 12951 for the best solution, 12867 on average, and 12697 for the worst solution. Both algorithms found the best solution only once. The 1–opt heuristic needed over 100 runs on average 707.7 moves to be suboptimal. The number of iterations of the EC–heuristic ranged from 3 to 8 (including the non–improving suboptimality check), and 4.5 on average. Table 2c describes in its first row the average number of moves of the most improving subsequences for all iterations while the second row contains the percentage improvement on average for each iteration (rounded to two positions behind the decimal point).

Besides the real world problems we tested both heuristics on three randomly generated data sets R1, R2, and R3 (see Table 4). The entries of these data sets are randomly chosen from the sets $\{0,1\}$, $\{0,1,2\}$, and $\{0,1,2,3\}$, respectively, where each element of the sets has the same probability. To get smaller random problems too, we only took the first p rows of each of these matrices. For instance, R1.p corresponds to the problem described by the first p rows of matrix R1. Table 4 exhibits the results obtained by both algorithms for 15 random problems of size 25 up to 300 objects (rows). The 'best' column contains the best result and enclosed in brackets the number of times this result is obtained within 100 runs. The $\bar{x}$ column gives the average of all 100 local optima and column 'worst' presents the worst results. Both algorithms got the same set of 100 randomly generated initial solutions for each of the 15 problems. The results confirm the superiority of the EC–heuristic to the 1–opt local search heuristic. Although the average running time over 100 runs $\bar{t}$ for the EC–heuristic is between 3 and 5 times the average running time of 1–opt local search the empirical time complexity of both algorithms is about $n^2 \cdot \log n$ .

Compared to the real world problems the random problems of Table 4 seem to be more difficult. The reason may be that the random problems are completely unstructured. The edge lengths connecting the vertices of the random problems are almost uniformly distributed, hence almost equal. There are no clusters inherited when the object–attribute structure is translated into a weighted graph. Thus, it is not a surprise that the algorithms presented above behave completely different because they try to find clusters where in fact there are none in the underlying data. Tests based on these type of random problems are less useful to draw conclusions about the behavior of heuristics on practical data. They demonstrate, however, the robustness of both heuristics. The EC–heuristic seems to be more robust with respect to changes of data than the 1–opt local search heuristic.

| problem | EC-heuristic | | | | 1-opt-heuristic | | | |
|---------|--------------|---|---|---|-----------------|---|---|---|
|         | best(times) | $\bar{x}$ | worst | $\bar{t}$ sec | best(times) | $\bar{x}$ | worst | $\bar{t}$ sec |
| R1.25  | 255(45)   | 252.9   | 251     | 0.02 | 255(43)  | 250.7   | 185   | 0.006 |
| R1.50  | 834(12)   | 822.1   | 788     | 0.09 | 834(1)   | 803.7   | 710   | 0.02 |
| R1.100 | 3173(1)   | 3089.5  | 2949    | 0.57 | 3124(1)  | 3022.2  | 2873  | 0.13 |
| R1.200 | 11525(6)  | 11177.1 | 10259   | 2.85 | 11455(3) | 10846.3 | 9750  | 0.62 |
| R1.300 | 24896(27) | 24503.0 | 23035   | 6.92 | 24896(3) | 23861.1 | 22556 | 1.52 |
| R2.25  | 39(60)    | 38.6    | 38      | 0.02 | 39(28)   | 38.0    | 35    | 0.005 |
| R2.50  | 147(79)   | 146.8   | 145     | 0.08 | 147(60)  | 145.8   | 138   | 0.02 |
| R2.100 | 326(6)    | 321.9   | 313     | 0.46 | 323(3)   | 315.5   | 296   | 0.14 |
| R2.200 | 1030(1)   | 1009.9  | 976     | 2.58 | 1020(1)  | 967.5   | 923   | 0.73 |
| R2.300 | 1959(1)   | 1901.6  | 1848(1) | 7.24 | 1878(1)  | 1804.8  | 1735  | 1.88 |
| R3.25  | 5(100)    | -       | -       | 0.02 | 5(100)   | -       | -     | 0.007 |
| R3.50  | 17(100)   | -       | -       | 0.10 | 17(68)   | 16.5    | 14    | 0.03 |
| R3.100 | 76(36)    | 75.4    | 75      | 0.61 | 75(15)   | 73.1    | 68    | 0.21 |
| R3.200 | 252(22)   | 250.0   | 243     | 3.00 | 252(1)   | 243.4   | 243   | 1.03 |
| R3.300 | 487(3)    | 480.3   | 470     | 7.49 | 480(2)   | 467.0   | 442   | 2.55 |

Table 4.   Results of the random problems (100 runs).

## 1.5   An EC–Based Branch and Bound Algorithm

The success of the heuristics has motivated the development of a branch and
bound algorithm for the clique–partitioning problem. We first outline the
algorithm and then fill in the details in the subsequent sections.

The formulation of the clique–partitioning problem in the second chapter
suggests a binary structure of a branch and bound tree. At the root all $x_{ij}$ are
still free; at each branch a free edge (i,j) is either *included* (fix $x_{ij}$ to 1) or
*excluded* (fix $x_{ij}$ to 0).

Our method for systematically exploring the tree is depth–first search, or
backtracking. Although it is not immediately apparent from the structure of the
search tree, the algorithm distinguishes between two types of branches:

*Explicit* branches correspond to 'free' decisions made according to a rule
described below.

*Implicit* branches are a logical consequence of explicitly or implicitly fixing
other edges. Suppose, for example, that after including an edge (i,j) we now
have $x_{ij} = 1$ and $x_{ik} = 0$ while $x_{jk}$ is still free. Since a solution must be
transitive $x_{jk}$ has to be 0.

Instead of explicit or implicit branches we will occasionally speak of explicit or implicit edges or variables, meaning the edge (variable) fixed at the branch.

Figure 5 shows the pseudo code for the outline of the search. The notation functionName(&argument) indicates a "call by reference": The argument may be (here it will be) changed in the subroutine.

At the beginning, the function testsBeforeBranching calculates a lower and an initial upper bound for the value of the optimal solution. It then tries to fix some edges implicitly by employing a logical test before explicit branching begins. The function chooseAndFixFirstEdge does what its name indicates: It selects and fixes an edge for the first explicit branch. Afterwards, the depth–first search loop is entered.

```
testsBeforeBranching();
chooseAndFixFirstEdge(&newFixedEdge);
done := FALSE;
while not done do
    if   (logicalTests(newFixedEdge)) then
         branch(&newFixedEdge);
    else
         done := backtrack(&newFixedEdge);
```

Figure 5.        Outline of the EC–based branch and bound search.

The routine logicalTests performs two tasks: It implicitely fixes edges as a consequence of the last explicit branching decision and it tests if the current node of the search tree is fathomed by comparing the new upper bounds – we use two independent bounds – with the best lower bound. If the node cannot be fathomed logicalTests returns TRUE and the search proceeds with branch where a new edge is explicitely fixed. If a node is fathomed or if the search arrives at a leaf of the tree, logicalTests returns the value FALSE and backtracking begins. The function backtrack goes back through the tree until it arrives at the most recently fixed explicit edge. All implicit variables encountered along the way are set free again. The explicit edge is then fixed to its opposite value and marked as *implicit*. It will be set free the next time it is visited again during backtracking. The result of backtrack is always FALSE, except when we return to the root of the tree and nothing is left to explore (i.e. the first branch is marked as *implicit*); then done is set to TRUE and the search stops. The best solution that has been found is optimal. Next, we take a closer look at the individual components of the algorithm.

**A Simple Upper Bound.** We want to maximize the sum of the edge weights within all cliques. An upper bound $\bar{z}$ for this sum follows from the relaxation of our model in the second section, obtained by disregarding the triangle inequalities. The initial bound $\bar{z}^{(0)}$ at the root of the search tree (node 0) is simply the sum of all positive edge weights. Whenever an edge with positive weight – a positive edge for short – is excluded or a negative edge is included this upper bound decreases. At a node $\lambda$, somewhere in the tree, we have

$$\bar{z}^{(\lambda)} = \bar{z}^{(0)} - \sum_{\substack{1 \leq i < j \leq n \\ x_{ij} \text{ fixed}}} \left((1 - x_{ij}) \cdot \max\{w_{ij}, 0\} - x_{ij} \cdot \min\{w_{ij}, 0\}\right).$$

This is the upper bound used in Tüshaus (1983) and Amorim et al. (1992). Our computational experience has shown that this bound alone cannot efficiently limit the search space. After describing an initial logical test based on the triangle inequalities we will develop another, triangle–based bound.

**An Initial Logical Test.** At the outset of the algorithm, the objective function value obtained by the EC–heuristic serves as a lower bound $\underline{z}$; an upper bound $\bar{z}^{(0)}$ is determined as in the preceeding section. The tests described here are then applied before branching. They try to show that certain edges must be included or excluded in a solution with a value of at least $\underline{z}$. The tests use the triangle inequalities of the model which ensure that a solution is transitive: For all triangles $\nabla(i,j,k)$ they prohibit every combination of the corresponding binary variables in which exactly 2 variables have the value 1. If an edge $(i,j)$ is included $(x_{ij} = 1)$ we can draw a conclusion about every triangle $\nabla(i,j,k)$ in which the edge appears:

$$x_{ij} = 1 \text{ yields} \quad \begin{cases} x_{ik} = 1 \text{ and } x_{jk} = 1, \\ \qquad \text{or} \\ x_{ik} = 0 \text{ and } x_{jk} = 0. \end{cases} \quad \text{for all } k \in V \setminus \{i,j\}.$$

Similarly, if $(i,j)$ is excluded then

$$x_{ij} = 0 \quad \text{yields} \quad \begin{cases} x_{ik} = 0 \text{ and } x_{jk} = 0 \text{ , or} \\ x_{ik} = 1 \text{ and } x_{jk} = 0 \text{ , or} \\ x_{ik} = 0 \text{ and } x_{jk} = 1 \text{ .} \end{cases}$$

$$\text{for all } k \in V \setminus \{i,j\} \text{ .}$$

These observations can be used to design logical tests. We only describe the test for $x_{ij} = 1$; the test for the opposite assumption $x_{ij} = 0$ works similarly.

The question to be answered is: What is the value of the best solution – or: find a bound for this value – that can be achieved if $x_{ij}$ is set to 1? If the value is less than $\underline{z}$ then the edge $(i,j)$ cannot be included in an optimal solution and we may directly fix $x_{ij}$ to 0. In other words, the test for $x_{ij} = 1$ tries to show that $x_{ij} \neq 1$ in an optimal solution.

Consider a triangle $\nabla(i,j,k)$: If $x_{ij} = 1$ we have the choice either to exclude both edges $(i,k)$ and $(j,k)$ or to include them. If we exclude a positive or include a negative edge the initial upper bound $\overline{z}^{(0)}$ decreases. We call this decrease $\delta_{ijk}(x_{ij}, x_{ik}, x_{jk})$ . For $x_{ij} = 1$ we obtain:

$$\delta_{ijk}(1;0,0) = \max \{w_{ik}, 0\} + \max \{w_{jk}, 0\} \text{ , and}$$

$$\delta_{ijk}(1;1,1) = -\min \{w_{ik}, 0\} - \min \{w_{jk}, 0\} \text{ .}$$

An edge $(i,j)$ appears in $n - 2$ triangles. If, after setting $x_{ij} = 1$, we make the best choice for the remaining 2 edges in each of these triangles then the sum of the $n - 2$ corresponding $\delta$ values is the amount $\Delta_{ij}(1) := \Delta_{ij}(x_{ij}=1)$ by which the upper bound $\overline{z}^{(0)}$ will at least decrease:

$$\Delta_{ij}(1) = \sum_{k \in V \setminus \{i,j\}} \min \{\delta_{ijk}(1;0,0), \delta_{ijk}(1;1,1)\} \text{ .}$$

If the upper bound decreases below the lower bound $\underline{z}$, then the edge $(i,j)$ cannot be included in an optimal solution, hence

$$\overline{z}^{(0)} + \min \{w_{ij}, 0\} - \Delta_{ij}(1) < \underline{z} \quad \text{yields} \quad x_{ij} \neq 1 \text{ .}$$

The term $\min \{w_{ij}, 0\}$ takes into account the decrease in the initial bound caused by including a negative edge. It is easy to devise a similar test that uses analogous values $\Delta_{ij}(0)$ and that tries to show that an edge must be included in an optimal solution.

What is the complexity of this test? Since an edge $(i,j)$ appears in $\mathcal{O}(n)$ triangles and a single $\delta$ value can be found in constant time, the overall effort for finding $\Delta_{ij}$ is $\mathcal{O}(n)$. Given $\underline{z}$, testing all $\binom{n}{2}$ edges requires effort $\mathcal{O}(n^3)$.

**A Second Upper Bound.** The algorithm uses a second, triangle–based upper bound $\hat{z}$ besides $\bar{z}$. At the root $\hat{z}^{(0)} = \bar{z}^{(0)}$ but at an arbitrary node the bounds will usually differ.

The idea behind $\hat{z}$ is the same as the one used in the initial tests. After explicitely or implicitely fixing a variable $x_{ij}$ we try to answer the question: What is the best choice for the remaining 2 edges – provided they are still free – in all triangles in which the edge $(i,j)$

appears? Again, we only describe the argument for the case that $x_{ij}$ is set to 1. Using the $\delta$ values from the preceeding section we can define an estimate (lower bound) $\Delta_{ij}^*(1) := \Delta_{ij}^*(x_{ij}=1)$ for the decrease of $\hat{z}$ caused by setting $x_{ij} = 1$:

$$\Delta_{ij}^*(1) = \sum_{\substack{k \in V \setminus \{i,j\} \\ (i,k),(j,k) \text{ free} \\ \text{and not marked}}} \min\{\delta_{ijk}(1;0,0),\ \delta_{ijk}(1;1,1)\}\ .$$

The definition is almost the same as for $\Delta_{ij}$, the difference being that only those triangles are used in which edges $(i,k)$ and $(j,k)$ are both free and not marked. If $\min\{\delta_{ijk}(1;0,0),\ \delta_{ijk}(1;1,1)\} > 0$ then the triangle $V(i,j,k)$ can be used to lower the bound $\hat{z}$; in this case the edges $(i,j)$ and $(j,k)$ are marked and stored in a list kept at the current node of the tree. The marked–flags are needed to prevent multiple use of the same edge. During backtracking the marks stored at a node are removed again.

Let $\mu$ be the predecessor of a node $\lambda$ in the search tree; if $\lambda$ is reached by (explicitely or implicitely) including $(i,j)$ then:

$$\hat{z}^{(\lambda)} = \hat{z}^{(\mu)} + \min\{w_{ij},0\} - \Delta_{ij}^*(1) \quad \text{if edge } (i,j) \text{ is not marked.}$$

A similar argument can be used to find $\hat{z}$ for a node reached by exclusion of an edge. Note that the bound $\hat{z}$ is independent of $\bar{z}$; the tighter one may serve to fathome a branch.

**Explicit Branching.** Only positive edges are used for explicit branching. Once all positive edges are fixed, the remaining free variables can be set to 0; as we always make all necessary implicit branches, this will not violate the transitivity constraints. To select a free edge for branching we have – after some tests – chosen the following heuristic rule: Fix the free positive edge for which the expression

$$\Delta_{ij}(\overline{x}_{ij}^{(h)}) + \max\{|V_{v(i)}|, |V_{v(j)}|\}$$

is maximized; $x_{ij}^{(h)}$ is the binary status of edge $(i,j)$ in the EC–heuristic solution and $\overline{x}_{ij}^{(h)}$ is the negation of this value; $v(i)$ is the number of the clique containing vertex $i$ in this solution, and $|V_{v(i)}|$ is the vertex cardinality of this clique. The $\Delta_{ij}$ are the same as in the initial logical tests. The rationale of this rule is: The first term favours edges for which the upper bound z is likely to decrease significantly when the search proceeds in the presumably wrong direction. The second term favours edges with an endpoint in a large cluster; on average this leads to a greater number of implicit branches.

**Implicit Branching.** After explicitly fixing an edge $(i,j)$ all triangles $\nabla(i,j,k)$, $k \in V \setminus \{i,j\}$, are inspected to find the necessary implicit branches. Because an implicit branch may cause further implicit branching every new implicit edge is added to a list. When all triangles containing $(i,j)$ have been checked the same process is repeated for all edges in the list.

The deeper the search has proceeded into the tree, the higher the number of implicit branches resulting from a single explicit branching decision will be on average. This is because a new implicit branch can only be made in a triangle where 2 edges are already fixed.

## 1.6   Computational Results of the EC–Based Branch and Bound Algorithm

As the local search heuristics, the branch and bound algorithm has been implemented in C on a Silicon Graphics Personal Iris 4/D 30 (MIPS R3000 processor, 30 Mhz) workstation under Unix. It has been tested using the 11 problems from Table 1.

| problem | n | $\bar{z}^{(0)}$ | $z^*$ | t sec[1] | t sec[2] |
|---------|---|------|------|--------|--------|
| wild cats | 30 | 1400 | 1304 | 0.5 | 23 |
| cars | 33 | 1748 | 1501 | 5.3 | 155 |
| workers | 34 | 1233 | 964 | 29.8 | 198 |
| cetacea | 36 | 998 | 967 | 0.2 | 15 |
| micro | 40 | 1362 | 1034 | 898.1 | 257 |
| UNO | 54 | 906 | 778 | 128.1 | 270 |
| UNO 1a | 158 | 12322 | 12197 | 22.3 | 852 |
| UNO 1b | 139 | 11859 | 11775 | 11.8 | 518 |
| UNO 2a | 158 | 73178 | 72820 | 19.1 | 587 |
| UNO 2b | 145 | 72111 | 71818 | 13.5 | 484 |
| companies | 137 | 82691 | 81874 | 11.3 | 1187 |

Table 5.    Results of the EC-based branch and bound algorithm.

[1]Computer: Silicon Graphics Personal Iris 4D/30, under Unix

[2]Computer: Siemens 7.865, under VM/370-CMS; with MPSX/370
(from Grötschel / Wakabayashi (1989)

Table 5 shows the initial upper bound $\bar{z}^{(0)}$, the value of the optimal solution $z^*$ and the solution time including the time required for the initial EC–heuristic solution. For comparison the last column cites the time required by the branch and cut method described in Grötschel / Wakabayashi (1989, 1990a). A direct comparison of the running times obviously suffers from the different hardware performance characteristics, but even then it seems safe to conclude that our method is often significantly faster.

The computational results also show that the time required by our algorithm can vary considerably. It comes as no surprise that the problems for which the ratio $\bar{z}^{(0)}/z^*$ is relatively large (cars, workers, micro, UNO) are the most difficult ones. All running times on the real world problems (except the micro instance) for our branch and bound implementation sum up to 241.9 seconds which is only about twice the time needed in the simulated annealing or tabu search approach described in Amorim et al. (1992).

Although we got the same clusters in all optimal solutions of the real world problems, our objective function values in the optimal solution differ in four cases (micro, UNO, UNO 1b, and companies) from those obtained by Grötschel / Wakabayashi (1989) or Amorim et al. (1992). These differences already might be caused by a typo in the data matrix. Our objective function values for UNO, UNO 1b, and companies are 778, 11775, and 81874, respectively, while Grötschel and Wakabayashi got 798, 11613, and 81802, respectively. In our case in the micro instance the right hand side for the edge weight computations has to be 0.262 – instead of 0.3 – in order to reach the same objective function value.

We were not able to solve the lecturers problem instance to optimality within a time limit of 2000 seconds. Thus this problem does not appear in Table 5.


## 1.7  Conclusions

We have presented a powerful local search heuristic and an effective branch and bound algorithm for the clique–partitioning problem. Both algorithms did their job very well in case of the real world problems known from literature. Randomly generated problems often can tell us about the robustness of a solution method, however, as long as there is no real world structure underlying the random data we cannot draw conclusions about the usefulness of these methods. Test beds of real world problems are in general indispensible.

A branch and cut algorithm need not be the most powerful tool to solve problems exactly. In our case it was dominated by the branch and bound procedure employing a simple propagation of constraints in order to reach arc consistency (see Chapter III).

The algorithms of Kernighan and Lin (1970, 1973) can be seen as an application of tabu search with dynamically organized and increasing tabu list. The generalization in form of the ejection chain method [285] provides an extremely powerful tool for solving combinatorial optimization problems. The ejection chain idea can easily be applied without substantial changes to other partitioning problems mentioned in the introduction, as well as to related facility location problems. In our case even the rudimentary implementation of an ejection chain solved all real world problems. The complete version could not do faster, however, for large problem instances it may be indispensible.

---

[285]   Glover (1991a, 1992c)

# 2.  Factory Layout Planning

## 2.1  Introduction and Background

Optimal design of the physical layout of a manufacturing system is one of the most important issues that must be solved in the early stages of a system design. Facilities or complex flexible machines of a production system say, are to be laid out on the plane such as a factory floor. Good solutions are basic for an effective utilization of the system. The most frequently encountered layout types in automated manufacturing systems are circular (directed and undirected) machine layout [286], linear single or double row machine layout, and cluster machines layout [287]. Here we discuss the cluster machine layout.

A real layout faces the problem of encompassing a variety of demands such as routing and scheduling, inventory, product, and communication demands, physical factory conditions or cell sizes for manufacturing cells in a flexibles system, etc. Even the inaccuracy of the input data due to large planning horizons used for layout design purposes, short product life cycles or changing processing technologies affects the layout design objective. Therefore Kouvelis et al. (1992) did not consider facility layout under some optimization criterion; their motivation was the layout robustness for a manufacturing system.

The need in modern manufacturing environments to solve increasingly facility or plant layout problems arises with the increased use of flexible machines. Such machines are capable to handle via a tool magazine different types of operations on diverse products that require to be processed. The processed material, even a single product has different processing routes through the manufacturing shop. The shop floor design havily depends on the anticipated product flow and their tool requirements. Obviously the design of the facility layout has cost implications in the maintenance and operation of the material handling system, load balancing, production delays, work–in–process, makespan, and lots of other objectives. For instance, the cost of material flow is affected not only by the quantity of material flowing between any two machines, but also by the relative location of these machines. Hence the question arises which machines or manufacturing cells to place adjacent in a layout. Desired adjacency can be

---

[286]   Kouvelis / Kim (1992), Blazewicz et al. (1992)

[287]   cf. Heragu and Kusiak (1988) or Das (1993) for a description of the different layout types.

expressed by an index, which may be the function of predicted flow between machines or cells, technological constraints, etc.[288] Apparently the facility layout problem can be treated as a graph theoretic problem [289]. Machines or manufacturing cells are represented by the vertices of a complete edge weighted graph where the weight on an edge indicates the desire to place the corresponding machine cells adjacent. Negative weights express relationships where the manufacturing cells or machines should be separated. Positive relationships tend to be related to material flows in the facility while negative relationships tend to be caused by environmental incompatabilities such as noise, vibration, pollution, etc. Physical conditions such as material flow lines, gas or air lines, lighting fixtures and other equipment requires that not all adjacencies of the complete graph can be realized in the shop floor. Crossing communication or surply lines are impossible. Moreover negative weights on edges should not be realized as physically adjacent. Then the goal is to locate machines with a high positive affinity or relationsship next to each other and machines with a negative relationship should be separated. Hence the problem is twofold, firstly its complexity can probably be reduced by applying a cluster method such as the clique partioning method described in the previous section, and secondly for the remaining cliques or clusters we have to solve the problem to determine which machine cells should be adjacent which can be modelled as that of finding a maximum planar subgraph of the given adjacency complete graph such that the weight of all edges in the subgraph is maximum. Thus in this step of the problem solving process the relative position of the manufacturing cells or machines is the considered objective ignoring their space requirements [290]. The final outcome, the maximally planar graph of maximum weight then has to be converted into a layout design of rectangular blocks which yield all rectangular shaped manufacturing cells or machine places. In a final step the block layout should be converted into a detailed layout that shows material handling aisles and interface points.

The clique partioning approach serves the separation desires while the maximal planar graph problems serve adjacency desires. For the remainder we are going to consider only the second type of problem. In fact, there are good reasons to solve first a clique partioning problem in order to decompose the problem into smaller pieces. Usually it will be a difficult problem to judge about the

---

[288]   A recent review of the facility layout literature is given in Kusiak / Heragu (1987).

[289]   see Domschke / Drexl (1990), El–Rayah / Hollier (1970)

[290]   Armour / Buffa (1963)

relationship between any pair of manufacturing cells or machines and to fix this evaluation in a real number. The complexity of a flexible manufacturing system implies a classification of factual data by a number of entity types. Each entity type has a property associated with it. Possible entities for the facilities layout problem could be 'area', 'door available', door required', 'water required', 'heat producing', 'noise producing', 'people working', 'flammables used', etc. Each such entity expresses a desired or undesired state, and for each machine or manufacturing cell there is a set of entities expressing properties. Thus, a qualitative data analysis method applies in order to have a measure between any two cell, respectively a measure for desired adjacency or separation. Following the line in Section 1 the resulting measure can be negative.[291]

The maximum planar graph is only a rough first step in the design of the facility layout, and it is expected that the design engineer will introduce a number of changes into this diagram, such as switch adjacent blocks, resize blocks what may have an influence on some other blocks, etc. Several authors addressed the problem to support these design changes interactively and filled the gap for need of a software package that provides a user friendly interface to execute such modifications or sensitivity analysis [292]. It is not unexpected that the most recent approaches tackle the problem of facility layout planning from an artificial intelligence point of view. Specifically the problem has attracted techniques from local search based learning [293].

Finally we would like to mention the impacts of the forementioned approach to chip and VLSI layout planning. Advanced technologies in the production of

---

[291]   In a number of publications the foregoing graph theoretic approach is addressed, probably the first was Krejcirik (1969), and the suggested algorithms heavily make use of the graph theoretic nature of the problem (Seppänen / Moore (1970), Foulds / Robinson (1976, 1978, 1979), Carrie et al. (1978), Foulds (1983), Foulds et al. (1985), Hammouche / Webster (1985), Giffin et al. (1986), Montreuil et al (1987), Montreuil / Ratliff (1989), Goetschalckx (1992), Leung (1992), Hassan / Osman (1993), Heragu / Alfa (1992)). For a review on these graph based methods see Hassan / Hogg (1987). Koopmans and Beckman (1957) first modelled the problem as a quadratic assignment problem and a lot of solution methods were published during the last 30 years suggesting approaches to solve this intractable problem. Among others were also tools like CRAFT (Buffa et al. (1964)) that considers the minimization of materials handling costs. CRAFT attempts to improve the objective value of an initial layout by making two- and three- machines place assignment exchanges. The system CORELAP (Lee / Moore (1967)) uses the user specified relationship weights. Others treated the layout problem as a multi–objective scenario handling via an expert system approach (Rosenblatt (1979), Kumara et al. (1988)).

[292]   cf. Kumara et al. (1988), Wäscher / Chamoni (1987), Giffin et al. (1986), Goetschalckx (1992), Montreuil et al. (1987).

[293]   cf. Shih et al. (1992), Tam (1992), and Hassan / Osman (1993)

integrated circuits enables the designer to place more and more transistor like electronic devices on a chip. A frequently considered objective is to compute a floorplan with minimum total chip area. However, an optimal placement has also to consider signal transfer times between the devices on the chip because there are lots of time critical connections between items or gates. It is desired to keep these time critical connections below a maximum distance, hence they get a high value of adjacency relationship in a corresponding complete graph where the vertices represent the items to be placed on the chip. As there may not be any intersecting electronic transmission lines the problem arises to determine a maximally planar graph of maximum weight.[294]

For the remainder [295] assume $K_n$ to denote the complete graph on n vertices, and let $w_{ij}$ denote the weight of the edge ij connecting vertex i to j. The weight $w(G)$ of a subgraph G is defined to be the sum of the weights of its edges. A graph is said to be planar if it can be drawn on a plane by distinct points for vertices and simple curves for edges in such a way that any two such curves do not intersect anywhere other than their endpoints. The representation of G on the plane is called a topological planar graph, however, we will identify a planar graph with its topological representation and simply speak only of a planar graph. A maximum planar graph is a graph G that contains a maximum number of edges subject to the condition of being planar, i.e. the addition of any edge connecting vertices i and j in G would result in a non–planar graph. We seek a subgraph $G^*$ of $K_n$ that is maximally planar and that yields $w(G^*) = $ max $\{w(G) \mid G$ is maximally planar$\}$. The maximum weight planar graph problem is NP–complete [296] except if all edge weights are equal, a case which is solvable in linear time. Foulds and Robinson (1976) and Foulds (1983) developed an exact algorithm which maintains the maximum weight graph and at each step they introduce a planarity check. The latter can be accomplished in linear time [297]. Jünger and Mutzel (1993) attacked the problem from a polyhedral side and proved that the number of edges in G is facet bounding. Edge weights are nonnegative in most applications (e.g., in layout and space planning problems) but we do not impose nonnegativity as a necessary assumption. Throughout the following we reserve the symbol G to refer to a maximally

---

[294]   For details on the underlying chip planning problem we refer the reader to Schürmann (1988).

[295]   which can also be found in Glover / Pesch (1993)

[296]   Garey / Johnson (1979)

[297]   cf. Hopcroft / Tarjan (1974)

planar graph.

A respresentation of a planar graph in the plane devides the plane into regions bounded by the edges and the vertices connecting the edges, the set of which is called boundary. If the interior of a region (not considering its boundary) does not contain any vertex or edge then the region is called a *face*. Hence there is one infinite face the outer one. Two faces are adjacent, if their boundaries have at least one edge in common (and not only common vertices).

Let us agree on the following preliminary conventions:

- A face of a maximally planar graph will be denoted by a triple $(x,y,z)$ where $x$, $y$ and $z$ are the vertices of the face and $xy$, $yz$ and $yx$ are its edges.
- Define a move $\alpha : G \rightarrow G'$ transforming a given graph $G$ into a new one $G'$ to be legitimate if $G'$ is also a maximally planar graph.
- Denote the weight of a move $\alpha$ by $w(\alpha)$ (which equals $w(G') - w(G)$ if the move $\alpha$ transforms $G$ into $G'$).
- For a given edge $ij$ of a maximally planar graph, let $Cut(i,j)$ denote the unique edge $pq$ such that $(i,p,j)$ and $(i,q,j)$ are faces of the maximally planar graph.
- Define an **edge-cut** to be a **move** that replaces a given edge $ij$ of $G$ with the edge $pq$ such that $pq = Cut(i,j)$.
- Denote the operation that produces an edge-cut move by $EdgeCut(i,j)$ hence $w(EdgeCut(i,j)) = w(p,q) - w(i,j)$ for $pq = Cut(i,j)$

## 2.2 Fundamental Results

There are only two cases of maximally planar graphs consisting of 3 or 4 vertices, both of which are unique. These are the complete graphs $K_3$ and $K_4$, respectively. It is an immediate fact that these are the only cases were a maximally planar graph contains vertices of degree less than 3 (in case of $K_3$) or adjacent vertices of degree 3 (in case of $K_4$). Any maximally planar graph containing more than 4 vertices cannot have two adjacent vertices of degree 3 and

(1)        any degree 3 vertex belongs to a subgraph isomorphic to $K_4$.

If $Cut(i,j)$ is already an edge of $G$ such an edge exchange move would only produce duplicate edges which we are not going to consider. So, there is no legitimate move $EdgeCut(i,j)$ where both edges, the old one $ij$ and the new one $pq = Cut(i,j)$ belong to a subgraph of $G$ isomorphic to a $K_4$. Now we can restrict the definition of a *legitimate move* EdgeCut(i,j) to only those cases where Cut(i,j) is not an edge of G. As a consequence legitimate moves cannot touch the "interior" of a $K_4$ subgraph, which gives us that

(2)      the move EdgeCut(i,j) is legitimate iff neither i nor j has degree 3 in G.

So we may assume that G has more than 4 vertices. Hence the number of legitimate moves equals the number of edges in the subgraph of G that results by removing all vertices of degree 3 and their incident edges. Observe, that

(3)      removing a vertex of degree 3 still yields a maximum planar graph.

From the previous observations we know that there is a legitimate move. Consider a vertex x of degree at least 4 in G. Then, either x has only neighbours of degree at least 4 in G or each neighbour w of degree 3 defines a $K_4$ in G. In both cases we have that there are at least three edges connecting vertices of degree at least 4 in G. We can conclude that G contains in fact, at least 3 legitimate moves.

An appropriate data structure in order to describe and implement a sequence of moves will be in form of a list, say EdgeCutList, that contains one row for each edge ij of G. A row consists of the three entries, edge ij, Cut(i,j), w(EdgeCut(i,j)). These entries will be called respectively the drop edge, the add edge and the move weight.

By reference to the entries of the EdgeCutList, let pq = Cut(i,j), and let vertices r, s, t, u be given by ir = Cut(p,j), is =Cut(q,j), tj = Cut(i,q), uj = Cut(i,p). A legitimate move, EdgeCut(i,j), affects precisely 5 rows of the EdgeCutList, producing updated entries whose elements are identified as follows, where every row contains the same drop edge in the new list as in the original list except the row for pq, which replaces the original row for ij (reversing ij and pq as the drop and add edges). The entries in the table below are the entries after the move EdgeCut(i,j). The entries in the EdgeCutList before the move are shown in brackets.

| | Drop Edge | Add Edge | Move Weight |
|---|---|---|---|
| For row pj: | jp (jp) | rq (ri) | $w_{jp} - w_{rq}$   $(w_{jp} - w_{ri})$ |
| For row qj: | jq (jq) | sp (si) | $w_{jq} - w_{sp}$   $(w_{jq} - w_{si})$ |
| For row iq: | iq (iq) | tp (tj) | $w_{iq} - w_{tp}$   $(w_{iq} - w_{tj})$ |
| For row ip: | ip (ip) | uq (uj) | $w_{ip} - w_{uq}$   $(w_{ip} - w_{uj})$ |
| For row pq: | pq (ij) | ij (pq) | $w_{pq} - w_{ij}$   $(w_{ij} - w_{pq})$ |

The observation that there are only 5 rows in the list affected by a move establishes that the available edge–cut moves and their move weights can be identified highly efficiently, in constant time, from one iteration to another. The most improving edge–cut move can thus be performed in $O(\log n)$ time where n is the number of vertices in G. Here we make use of the fact that the number of edges, i.e. also the number of legitimate moves, in a planar graph is of the order $O(n)$. Furthermore the EdgeCutList is organized as a heap which can be updated in $O(\log n)$ time.

Also, by result (3), we may consider the use of edge–cut moves as a useful supplement to the **extract–insert moves**, defined as follows.

Extract a degree 3 vertex by deleting its incident edges in G, and insert the vertex in any face of the resulting graph other than the face created by the extraction, adding the 3 edges to join it to the vertices of this face.

Every maximally planar graph contains at least one vertex of degree 3, 4 or 5. However, we subsequently identify an algorithm that generates maximally planar graphs that contain no vertices of degree 3 or no vertices of degree 4. The extract–insert moves cannot operate on such graphs, and hence the edge–cut moves are relevant either as a replacement or supplement to extract–insert moves in order to provide legitimate transformations of graphs of these types. This is in analogy to the traveling salesman case where the 2–opt neighbourhood and the moves with respect to this neighbourhood is elementary and connected, i.e. from any traveling salesman tour there is a sequence of moves with respect to the 2–opt neighbourhood, that leads to an optimal solution. Similarly, while extract–insert moves cannot define a neighbourhood which is connected, the edge–cut moves do. It can be shown that for any maximally planar graph there is a sequence of edge–cut moves leading to an optimal solution, i.e. a maximum planar graph of maximum weight.

Construction algorithms also can be used to provide graphs that are candidates for a maximum weight maximally planar graph. However, we will focus on transformation approaches rather than construction approaches, using construction approaches primarily to provide diverse starting structures.

We propose additional more advanced moves that may be used to supplement the edge–cut moves, and then elaborate on how the simpler and more advanced moves can be implemented effectively.

## 2.3 Advanced Moves

For a given vertex i, let Star(i) denote the list of vertices j such that ij is an edge of G. We say Star(i) is (*facially*) *ordered* if for each two consecutive vertices r and s in this list (... ,r,s ,...), the triple (i,r,s) is a face of G, where by convention we say that the last vertex of Star(i) immediately precedes the first. Denote the collection consisting of Star(i) for each vertex i of G by Star(G). We call Star(G) ordered if every Star(i) is ordered. In addition, we call Star(G) uniformly ordered if every Star(i) is ordered by assigning a common orientation (e.g. clockwise) to the sequence in which the faces are encountered by this ordering. That means Star(G) is uniformly ordered if there is a sequence of the faces of G such that the order of every Star(i) is a subsequence. In other words, all orders Star(i) are induced by Star(G). It can be shown by induction that Star(G) always exists. There is an easy way to find the uniform order of Star(G) using the dual of a planar graph. To each planar graph G, in particular to a maximally planar one, there exists a (topologically) dual planar graph $G^d$. To each face of G there exists a vertex in $G^d$ and two vertices are connected in $G^d$ if their corresponding faces in G share a common edge. Note that in a maximally planar graph all faces are triangles and therefore two adjacent faces share exactly one common edge. Edges in G and $G^d$ correspond to each other. Regard, $G^d$ is in general not maximally planar. Now, any traveling salesman tour in $G^d$ (visiting the vertices of $G^d$ exactly once) defines a uniform order of Star(G).

One edge in G uniquely determines two faces, each face in G is uniquely determined by two of its bounding edges. Two edges defining a face of G are called *partners*. Vertices r and s are called consecutive in Star(i) if one immediately precedes the other. Assume Star(i) is ordered. Then edges ir and is are partners if and only if r and s are consecutive in Star(i).

Hereafter we assume that Star(G) is always uniformly ordered. Write h = Before(i,k) and k = After(i,h) if h immediately precedes k in Star(i) (maintaining the convention that the last vertex of the star immediately precedes the first).

Let r = Before(i,s). Then if degree(i) is at least 4, the operation of deleting the partner edges ir and is yields a face (i,q,r,s,t), where q and t are identified by q = Before(i,r) and t = After(i,s).

By means of the foregoing definition and the data structure embodied in a uniformly ordered Star(G), we will show that it is possible to implement a set of advanced moves for transforming G into a new graph G* which also is maximally planar. In particular, for each such advanced move, we will identify a set of highly efficient rules (algorithms) that correspondigly transform Star(G) into Star(G*), where Star(G*) is likewise uniformly ordered. Moreover, we will show this data structure also permits these advanced moves to be coordinated with both the edge–cut moves and the extract–insert moves, with similar efficiencies. These advanced moves in fact encompass both the edge–cut and the extract–insert moves as simple special cases.

### Form of Advanced Moves:

#### (i) Partner–Rotation Move:

For a vertex i of degree at least 4, delete any two partner edges ir and is, where r = Before(i,s). Consider the resulting face (i,q,r,s,t), and add the pair of edges that join any vertex other than i to the two non–adjacent vertices of the face. Hence this adds qs and qt, or adds ri and rt, etc. Each of these 4 possible moves, given i, r and s, may be viewed as rotating the partners ir and is to yield a corresponding pair of partner edges emanating from one of the other vertices of the pentagon.

#### (ii) Mulitlevel Extract–Insert–Replacement Move:

We are considering operations which focus on those vertices in the graph which have a degree at most 5. Hence, from a complexity point of view there are three levels A, B, and C, indicating if whether the vertex under consideration has degree 3, 4 or 5, respectively, in G.

**Extraction Operation:** For a given vertex x of degree 3, 4, or 5, extract x by deleting all edges of Star(x).

Level A extraction:  No further operation.

Level B extraction:  Denote Star(x) by (q,r,s,t) (in ordered form). Add edge qs or edge rt.

Level C extraction:  Denote Star(x) by (q,r,s,t,u). Add a pair of partner edges from any of these vertices to any two nonadjacent vertices. (This yields the 5 possibilities consisting of adding qs and qt, or rt and ru, etc.)

**Insert–Replacement Operation:** Start with a graph produced by an extraction operation. Let x be the extracted vertex.

Level A insert–replacement: Select any face that has not been newly created by the extraction, and insert vertex x by connecting it by an edge to each vertex of the face.

Level B insert–replacement: Select any edge ij of the graph other than an edge newly added by the extraction operation. Replace this edge by connecting vertex x by an edge to each of the vertices i, j, p, q, where pq = Cut(i,j).

Level C insert–replacement: Select any pair of partner edges ir and is. Replace this pair of edges by connecting x to each of the vertices i, q, r, s, t, where q = Before(i,r) and t = After(i,s) (relative to Star(i)).

The following results and observations identify the properties of the preceding moves.

A partner–rotation move is legitimate if neither of its added edges duplicates an edge of G except for one of the deleted edges. (The latter duplication reintroduces the edge previously deleted, resulting in no net change for that edge.)

Each of the two ways of duplicating a deleted edge (reintroducing ir as a result of adding the pair ri and rt, or reintroducing edge is as a result of adding the pair si and sq) yields a simple edge–cut move, and

(4)      all edge–cut moves are included among the partner–rotation moves.

A level B insert–replacement may be viewed as a step that first transforms a 3 edge bounded face into a 4 edge bounded face by deleting an edge, and then inserting vertex x within this resulting 4 edge bounded face. Similarly, a level C insert–replacement may be viewed as step that first transforms a 3 edge bounded face into a 5 edge bounded face by deleting a pair of edges, and then inserting vertex x within the 5 edge bounded face. The representation of these steps as an operation of replacing an edge or a pair of partner edges in G leads to an organization that allows them to be considered without duplications.

A level B insert–replacement can allow ij to be one of the newly added edges if a level C extraction is used. Compounding an insert–replacement operation by adding an edge that does not meet vertex x only duplicates a move considered by a lower level operation.

More complex mulitlevel moves can be defined by extrapolating the ideas of the moves indicated, but their options become combinatorically more difficult to itemize. The foregoing moves are useful because the faces, edges, and pairs of partner edges are readily isolated (by reference to Star(G) and simple associated data structures) and are all on the order of the number of edges.

## 2.4 Updates and Data Structure Management for Advanced Moves

The goal of this section is to show how Star(G) can be efficiently transformed into Star(G*) for the advanced moves, while preserving the property of being uniformly ordered. We begin by considering partner–rotation moves.

Relative to the 5 edge bounded face (i,q,r,s,t) produced by deleting the partner edges ir and is in the first step of a partner rotation move, define the focal vertex of the move to be the particular vertex q, r, s, or t that is shared in common by the two added partner edges in the second step. Denote a vertex whose star is changed by the move to be a key vertex of the move, and denote the part of the star that changes to be the key substar. (The identity and order of vertices preceding and following the key substar remains intact after executing a move, where we maintain the convention that the last vertex of the star immediately precedes the first.)

The following tables identify the changes to Star(G) that yield Star(G*) under each of the four possible focal vertices of the partner rotation move.

### Change in Star(G) focal vertex q:

| Key Node | Key Substar | Key Substar Becomes |
|---|---|---|
| i | r, s | deleted |
| q | i | s, t, i |
| r | i | deleted |
| s | i | q |
| t | i | i, q |

### Change in Star(G) focal vertex r:

| Key Node | Key Substar | Key Substar Becomes |
|---|---|---|
| i | s | deleted |
| r | i | t, i |
| s | i | deleted |
| t | i | i, r |

### Change in Star(G) focal vertex s:

| Key Node | Key Substar | Key Substar Becomes |
|---|---|---|
| i | r | deleted |
| q | i | s, i |
| r | i | deleted |
| s | i | i, q |

### Change in Star(G) focal vertex t:

| Key Node | Key Substar | Key Substar Becomes |
|----------|-------------|---------------------|
| i | r, s | deleted |
| q | i | t, i |
| r | i | t |
| s | i | deleted |
| t | i | i, q, r |

The updates specified by the preceding tables require examining the stars for either 4 or 5 key vertices upon executing a move, thus producing a very small change to Star(G). Only one table is applicable to a given move. Moreover, the organization of Star(G) allows a fast evaluation of the weight change produced by candidate moves, causing a significant gain in efficiency since such evaluations are performed many times at each iteration.

The fact that edge–cut moves result for the two cases where the focal vertex is r or s, joined with the observation that in those two cases only 4 key nodes have to be examined, indicates that the use of this updating structure can be used to implement edge–cut moves with almost the same number of operations that results by using the EdgeCutList. However, the EdgeCutList additionally provides a convenient summary of the move weights for updating precisely the subset of weights that change. The data structures indicated for the advanced moves also can benefit from auxiliary tables to identify and update the move weights that change at each iteration.

We now identify the appropriate changes for executing the multilevel extract–insert–replacement moves. The extraction steps and the insert–replacement steps are treated separately.

The changes in Star(G) for the extraction steps are as follows, where all entries of Star(x) are deleted at the end of each of the following operations:

Level A extraction: For each of the three vertices i in Star(x), delete x from Star(i).

Level B extraction: Assume without loss (by the choice of naming the vertices in Star(x)) that qs is the edge added when x is extracted. Replace x by s in Star(q) and replace x by q in Star(s). Delete x from Star(r) and Star(t).

Level C extraction: Assume without loss that qs and qt are the edges added when x is extracted. Replace x by s, t in Star(q) and replace x by q in both Star(s) and Star(t). Delete x from Star(r) and Star(u).

The changes in Star(G) for the insert–replacement steps are as shown in the following tables. (Except for vertex x, the names of vertices in this step have no relation to the names of the extraction step.) Let (i,j,k) denote the 3 edge bounded face in which x is inserted. For the higher level moves the same notation as in the previous section applies.

### Level A insert–replacement

| Key Node | Key Substar | Key Substar Becomes |
|----------|-------------|---------------------|
| i        | j           | j, x                |
| j        | k           | k, x                |
| k        | i           | i, x                |
| x        | all (empty) | i, j, k             |

### Level B insert–replacement

| Key Node | Key Substar | Key Substar Becomes |
|----------|-------------|---------------------|
| i        | j           | x                   |
| p        | j           | j, x                |
| j        | i           | x                   |
| q        | i           | i, x                |
| x        | all (empty) | i, p, j, q          |

### Level C insert–replacement

| Key Node | Key Substar | Key Substar Becomes |
|----------|-------------|---------------------|
| i        | r, s        | x                   |
| q        | r           | r, x                |
| r        | i           | x                   |
| s        | i           | x                   |
| t        | i           | i, x                |
| x        | all (empty) | i, q, r, s, t       |

As in case of the partner–rotation Moves, the foregoing changes to update the data structure of Star(G) are both small in number and fast to execute. Again the data structure of Star(G), uniformly ordered, permits highly efficient evaluation of candidate moves. This evaluation can be further accelerated by maintaining auxiliary tables to update precisely the move weights that change, provided attention is restricted to a single level of extraction and of insert–replacement. Thus, a method that restricts attention in this manner for some number of iterations, then chooses another pair of levels under a similar restriction, yields a repeating process that may allow gains in efficiency. Otherwise, an intelligent candidate list approach may be used to limit the number of moves evaluated at each iteration, relying on the efficiency of full weight evaluations to yield a search algorithm that performs effectively overall.


### 2.5    Algorithms for Generating and Improving Maximally Planar Graphs

We now show how to further exploit the foregoing moves and their updates by specifying methods and supplementary data structures that precisely identify faces, edges and pairs of partner edges, using the foregoing framework. Our development yields a class of search neighbourhoods for progressively improving a starting maximally planar graph by a descent method, or for more generally improving a collection of starting maximally planar graphs by genetic algorithms, simulated annealing or tabu search. We also directly obtain a broad class of constructive algorithms. These constructive algorithms can be used in two ways: (1) to yield initial maximally planar graphs for the other methods; (2) to give a basis for applying a strategic oscillation approach, by constructing a graph until it is maximally planar, followed by progressive waves of partial deconstruction and reconstruction.

**Constructive Generation Algorithms:**

Step 1.    Start with G consisting of a triangle.

Step 2.    At each iteration, select any vertex not yet in G, and introduce it into G by a Level A, B or C insert–replacement operation (using the updating rules mentioned).

Step 2A.   (Optional) For a selected number of moves of the current iteration, seek to improve the current G by using either the simple or advanced rules of the preceding sections.

Step 3.    Return to the start of Step 2, unless no vertices remain to be added to G.


The foregoing class of constructive algorithms can be extended to a strategic oscillation approach by employing a controlled set of Level A, B and C extraction moves, e.g., making use of tabu search memory.

**Fundamental Supplementary Data Structures:**

In addition to maintaining a uniformly ordered Star(G), which is assured by the rules of the preceding sections, we maintain exactly the set of current edges as maintained in the EdgeCutList – that is, the edges which are listed as "drop edges" in that list, and which are just the edges of G. However, these edges will be recorded in an ordered form as follows. Define an OrderedEdgeList to be a listing of all edges ij of G in the form such that i < j.

Observe, the ordering that produces an OrderedEdgeList also can be applied to an EdgeCutList to produce an ordered EdgeCutList. This usefully integrates the data structures for the edge–cut moves and the more advanced moves.

To identify the edge pq = Cut(i,j) for the EdgeCutList, it suffices to access Star(i), letting p = Before(i,j) and q = After(i,j). (This is true independently of the assumption that Star(G) is uniformly ordered. We also do not assume in this case that p < q.)

So, it is straightforward that Level A extraction moves can be implemented by keeping a list of vertices whose stars have 3 elements, while Level B extraction moves can be implemented by reference to the complementary list of vertices, identifying each edge ij of G from an OrderedEdgeList and identifying the 4 edge bounded face (i,q,j,p) via the Before, After, and Cut operations.

The Level C extraction moves can be implemented by identifying each pair of partner edges ij and ik by the one–one correspondence of these pairs with each consecutive two entries j and k of Star(i). Specifically, starting with each vertex i and setting j = First(i), where First(i) names any specific vertex in Star(i), all partner edges ij, ik are generated by the following rule:

    set k = After(i,j)  (edges ij and jk are current partner edges)

    stop if k = First(i) otherwise set j = k and repeat.

This also shows that the total number of paired partner edges equals the sum of the cardinalities of the stars of G, hence twice the number of edges of G because each edge connecting consecutive vertices belongs to exactly two faces.

The key result that allows an OrderedEdgeList to be joined with the uniformly ordered Star(G) structure, to efficiently identify the faces of G (and hence to complete the remaining components of the advanced multilevel moves), is as follows.

The faces of G can be generated by the following steps.

(i)        Examine each edge ij of the OrderedEdgeList.

(ii)       Let k = After(i,j) (implicitly identifying the face (i,j,k)).

(iii)      Consider the face (i,j,k) explicitly if and only if j > k.

The set of faces explicitly considered is exactly the set of all faces of G.

The foregoing results create a full set of rules for implementing and updating an integrated set of simple and advanced moves, both for constructing G and improving it afterward.

## 2.6   Ejection Chains

Ejection chain procedures can be developed effectively by reference to the foregoing results. We consider two types of ejection chain processes: (1) successive edge–cut moves, where each operation of adding an edge to cut (and delete) a second edge leads to another operation of the same type; and (2) successive vertex ejection moves, where each operation of extracting a vertex and moving it to a new location ejects another vertex at that location, thus similarly leading to another operation of the same type.

A vertex ejection chain can either be a circular chain or a cap–and–anchor chain. The latter chain begins with a capping operation, which replaces the first extracted vertex with a set of edges to cap (fill) the resulting hole, and ends with an anchor operation which attaches the last extracted vertex in a new location without ejecting any further vertices. The capping operation corresponds to the same type of operation performed with a Level A, B or C extraction move, adding 0, 1 or 2 edges to fill the face left by a 3, 4, or 5 degree node. The anchor operation corresponds to the same type of operation performed with a Level A, B, or C insert–replacement move, deleting 0, 1 or 2 edges to allow a vertex to be inserted in their place. The beginning and ending moves of a cap–and–anchor ejection chain can be more complex, but it is useful to restriction attention to these alternatives since they can be executed efficiently by the special data structure and updating steps previously described.

We observe that the vertex ejection chains of the maximally planar graph problem are analogous to the vertex ejection chains proposed for traveling salesman problems, which similarly are subdivided into circular chains and cap–and–anchor chains. We note that ejections for the traveling salesman problem include subpath ejections as well as vertex ejections (on the basis that certain subpaths provide components that preferably should not be broken, and hence should be treated as a single unit). Similarly, ejections for the maximally planar graph problem may include cluster ejections, in which a clustered set of vertices and their interconnecting edges are treated as a single unit. These more general ejections for both traveling salesman and maximally planar graph problems may be conceived as instances of moves that eject subgraphes.

We examine the basic features of ejection chains based on successive edge–cut operations and on successive vertex ejection operations in the following two subsections.

**Edge Ejection Chains**. An edge-cut ejection chain is a rudimentray type of ejection chain. Without adequate control, it can become a concatenation of simple edge-cut moves that are idependently available in the graph G. In this case, although the moves are sequentially linked by contiguous vertices and edges, each individual edge-cut move of the sequence can be executed separately in G without affecting the existence of the other moves. Such ejection chains are decomposable, i.e., capable of being broken into smaller units. Where such decomposable chains occur, it might be more appropriate simply to consider the individual units separately rather than to "assemble" them into larger chains. However, if G is locally optimal relative to the smaller units – in this case relative to the basic edge-cut moves themselves – then it is not necessarily locally optimal relative to all chains that decompose into these units. This is apparent because the objective function value is evaluated after each possible decomposition of the complete chain into its subparts. So, similarly to tabu search and simulated annealing bad moves with respect to the basic edge-cut moves may be accepted and absorbed by better moves in the chain. In this way the final outcome probably improves and can overcome local optima in which the basic (edge-cut) moves become stucked. We may say that the decomposed units implicitly generate the larger chains.

Experience on other problems however showed that ejection chains (here in the domain of edge-cut moves) often yield substantially better results if successions of moves that are naturally dependant on each other are created, where one move does not exist except by first executing a predecessor move. This form of dependency also creates a dependence among evaluations that requires progressivly more effort to calculate updated evaluations for edge-cut moves as the length of the chain grows.

In general, to allow an efficient calculation of ejection chain alternatives, we seek chains that exhibit a first level dependence, whereby each step after the first requires some predecessor for its existence. We also seek chains that have a second level independence, where the conditions defining the structure and evaluation of a given step depend only on the first level change in G created by a single identifiable predecessor, and does not depend on cumulative changes introduced by interactions among multiple predecessors. This latter condition is important for efficient analysis.

In the case of edge-cut moves, we achieve second level independence by considering ejection chains whose components are paired edge-cut moves in a broader sense. These paired (2 step) moves, where the second relies on the first for its existence, can be generated and evaluated with particular efficiency, involving only a minute fraction of the effort required to examine all subsets of two successive edge–cuts, which include both independent as well as dependent components.

We are not creating edge–cuts in the original sense (except for some special cases), but we are creating sequences of added and dropped edges in G each of it defining only a part of a larger compound move. The chains created rely on these series of edges, hence the name edge ejection chains. We are using the fact that the number m of edges in a maximum planar graph is independent of the topological representation and for a fixed number n of vertices also fixed. The number of edges is $3 \cdot n - 6$, (Berge (1973)). Deleting any edge ij in G violates the number of edges in G respectively the maximum planarity, it leaves a hole and, obviously, a 4 edge bounded face. In order to "repair this hole" it can be "filled" if another edge, rs say, of G is deleted and "inserted" as the new edge $Cut(i,j) = pq$. Again the number of edges in G equals $m - 1$, so the newly created hole and 4 edge bounded face which results from the deletion of rs has also to be filled. temptatively Cut(rs) is introduced in order to check the objective function value. If it is increased the resulting graph creates a candidate for a new starting graph for newly generated ejection chains. In that case two edges of G are replaced by two new ones, a simple 2–opt step comparable to the traveling salesman case. Instead of immediately creating a new starting graph Cut(rs) is introduced on the cost of the deletion of another edge, etc. The ejection chain is increased like that until its maximum size is reached. The best subsequence of partial moves including a final repair step leads to the new maximum planar graph, a starting graph to generate a new ejection chain. Regard, there are two main things characterizing an ejection chain. The chain is build by successively creating a bigger move which is a compound move a simple incomplete edge exchanges. Each subpart of the chain is incomplete however a complete move is created, temptatively, by an additional step closing the currently open 4 edge bounded face. The most improving of this subsequences defines the new graph. A choice of a step, i.e the insertion of an edge and the subsequent deletion of another one is done such that the increase (deterioration) of the objective function is as large (small) is possible.

We are not going to describe the data structure in detail as it resembles closely to the edge–cut moves' data structure, it is even more simple (less changes) for a single step (deletion or insertion of an edge) of the chain.

**Vertex Ejection Chains.** The class of circular and fill–and–anchor ejection chains based on ejecting vertices (or vertex clusters) has a special advantage over edge ejection chains. In particular, while the component steps of vertex ejections, and of filling and anchoring seem more complex than those of edge ejections, they can easily be organized into subcollections that exhibit the desired first level dependence and second level independence that permits highly efficient evaluation of their outcomes. In fact, it is possible to generate nondecomposable chains of considerable length. Creation and definition of vertex ejection chains, and the temptatively orgainized feasibility are completely in analogy to the edge ejection chain. Vertex ejection chains consist of a series of extract and succeeding insert–replacement operations where a vertex is extracted and the created "hole" (a 3, 4, or 5 edge bounded face) is filled by a new vertex via an insert–replacement operation of a different level to the extract operation. As a consequence parter–rotation moves are necessary to fill a created hole completely if the insert–replacement move is of a lover level than the extract move. As a consequence, vertex ejection chains lead directly to the ability to create shortest path network models to analyze the consequences of alternative moves, and to construct ejection chains that exhibit a combinatorial leverage effect.

Combinatorial leverage occurs where the investment of small amount of effort yields a very high payoff, as where a low order polynomial degree of effort yields solutions that dominate a high order polynomial or even an exponential number of alternative solutions.

The foregoing results are counterparts of the theorems for combinatorial leverage for traveling salesman problems. Their form discloses that ejection chains of this type yield combinatorial leverage outcomes for many different classes of graph theory problems.

## 2.7   Computational Results and Comments

The described algorithms were tested on a randomly generated set of 60 test instances and problem sizes of 20, 40, 80, and 100 vertices. We followed the line of Leung (1992) for the generation of the edge weights. The edge weights were taken from a normal distribution with mean 100 and standard deviation $\sigma = 10$, 20, and 30. For each pair of values n and $\sigma$, 5 instances were generated. The computational results in Table 6 are the average over all five instances in each test set. An upper bound is computed based on the fact that a maximally planar graph has $3 \cdot n - 6$ edges. Hence, the total weight of the $3 \cdot n - 6$ edges of maximum weight provides a simple upper bound. The described algorithms were implemented as greedy local search methods except for the initial solution generation. Starting solutions were obtained using steps 1 and 2 of the construction algorithm (without Step 2A), where each step of the algorithms is performed randomly. That means a next vertex is chosen randomly, one of the three insert–replacement operations are chosen randomly, and the actual place of insertion, i.e. face, edge, or edge pair, in the current partial solution is chosen at random. All improvement heuristics have been run greedily such that always the best possible move is performed. The resulting algorithms are performed until a local optimum is reached. So we ran an edge–cut local search (Edge–Cut), and partner–rotation local search (Part–Rot), and a multilevel extract–insert–replacement local search (MultLevExIns) heuristic. Furthermore three types of circular ejection chains are performed, an edge–cut (Edge–Cut–EC), an edge based (Edge–EC), and a vertex based (Vertex–EC) ejection chain heuristic. The vertex based ejection chain however only considers ejection chains were a vertex is replaced by another one, this again by another one, etc. A vertex that is ejected is completely replaced, i.e. the new vertex is connected to all previous neighbours of the former vertex. The results in the table always show the average percentage over 5 runs (f) below the upper bound. The times (sec) are seconds on a PC486, 33 Mhz. We can see that the local search approaches become really bad for the larger problems instances. Might be that for the larger instances a worse upper bound is obtained, so that we can conclude that the edge based ejection chain algorithm is very robust and gave the best results. The vertex based ejection chain is no alternative except for its simplicity and its running time. Although partner rotation and multilevel operations perform quite well we would not give them preference because of the implementation complexity. So ejection chains based on edge cuts or edges are the winners from the implementation point of view and their results.

| Problem | | Edge-Cut | | Part-Rot | | MultLevExIns | | Edge-Cut-EC | | Edge-EC | | Vertex-EC | | Leung (1992) | |
|---|---|---|---|---|---|---|---|---|---|---|---|---|---|---|---|
| n | σ | f | sec | f | sec | f | sec | f | sec | f | sec | f | sec | f | sec |
| 20 | 10 | 97.2 | 4.2 | 96.2 | 9.3 | 98.4 | 47.0 | 90.6 | 131.2 | 99.1 | 135.6 | 95.2 | 86.0 | 97.5 | 0.9 |
| 20 | 20 | 97.4 | 5.3 | 94.4 | 7.4 | 97.6 | 49.7 | 98.5 | 134.1 | 98.5 | 138.9 | 94.1 | 84.7 | 96.0 | 0.9 |
| 30 | 30 | 96.6 | 5.0 | 96.7 | 14.9 | 98.9 | 51.1 | 98.9 | 134.8 | 98.9 | 136.0 | 93.7 | 75.5 | 93.8 | 0.8 |
| 40 | 10 | 95.1 | 17.0 | 95.8 | 21.4 | 96.4 | 77.3 | 96.5 | 154.6 | 96.6 | 174.1 | 95.3 | 120.4 | 96.8 | 37.3 |
| 40 | 20 | 91.2 | 19.2 | 93.7 | 24.9 | 96.7 | 81.5 | 96.9 | 149.8 | 97.4 | 153.1 | 92.5 | 101.9 | 94.1 | 38.1 |
| 40 | 30 | 88.4 | 19.2 | 90.7 | 27.0 | 86.4 | 74.9 | 89.1 | 153.0 | 91.2 | 182.7 | 90.4 | 115.5 | 92.4 | 36.7 |
| 80 | 10 | 94.7 | 61.2 | 94.3 | 77.4 | 93.0 | 249.4 | 95.1 | 543.0 | 97.4 | 653.5 | 97.4 | 400.1 | | |
| 80 | 20 | 86.3 | 65.7 | 82.0 | 93.4 | 92.8 | 225.5 | 92.9 | 533.0 | 94.0 | 701.0 | 94.0 | 523.6 | | |
| 80 | 30 | 84.3 | 69.0 | 85.4 | 84.5 | 88.0 | 237.6 | 91.8 | 556.5 | 95.2 | 770.0 | 95.2 | 495.0 | | |
| 100 | 10 | 96.0 | 89.0 | 92.1 | 114.0 | 95.3 | 309.0 | 97.5 | 664.6 | 99.1 | 841.6 | 96.3 | 570.4 | | |
| 100 | 20 | 84.9 | 93.0 | 84.9 | 121.4 | 86.9 | 277.5 | 84.6 | 601.5 | 93.6 | 888.9 | 93.6 | 610.2 | | |
| 100 | 30 | 83.0 | 84.4 | 85.9 | 117.5 | 85.9 | 173.5 | 85.9 | 590.2 | 89.5 | 634.2 | 86.1 | 510.6 | | |

Table 6. Results of the layout heuristics.

# 3.  Workload Balancing

## 3.1  Introduction and Background

Most of the graph based approaches for layout planning rely on a 0–1 decision of adjacency. Non–adjacent machines or non–adjacent transistors my be placed arbitrarily far, i.e. this process tends to ignore some of the adjacencies of the original graph. In particular for chip planning this is not a desirable effect as certain electronic devices have to be placed such that a maximum distance is not exceeded, although such devices need not be adjacent. The physical distance of any two transistors should be inversely proportional to the transfer rate of signals. Items with a high communication rate to several others should be placed towards the centre of a layout. Such a constructive approach is provided in this section. An item, machine, or manufacturing cell with large aggregate relationship is chosen first to be located centrally. Next the 'strongest competitor' of the first choice is located next to the already located electronic device, machine or manufacturing cell. Again the next one is chosen, etc. The problem can be modelled by means of graph theory. Vertices of a network represent transistors, machines, or manufacturing cells and a pair of vertices is adjacent in case of a communication. The edge weight may generally be understood as a distance which, for instance, might be inversely proportional to the transfer rate of signals. Hence, the problem arises as to find in a certain sense best locations on the network (under competition). The forementioned hierarchical approach to choose the always best central location leads to a network location problem with two competitors, which we are going to consider in this final section [298].

Network location problems occur when new facilities are to be located on a network. The network of interest may be for instance a road network or a railway network. For a given network location problem the new facilities are idealized as points and may be located anywhere on the network. Usually some objective function is to be minimized.

Problems, concerning the location of an emergency facility, and whose solution involves minimizing the greatest (weighted) distance from any vertex to the facility are known as *minimax location problems*. A solution is called a center. Problems with the aim to minimize the total sum of the distances from the

---

[298]     see also Bauer et al. (1993)

vertices of a network to a central facility are called *minisum problems*. A solution to such a problem which is closely related to the problem of locating a depot in a road network, where the vertices represent customers to be supplied by the depot, is said to be a median.[299]    The study of competitive location models dates back to the early work of Hotelling (1929). In the last years, competitive aspects of facility location and the relation to voting theory have increasingly been investigated [300].

Suppose that voters located on a network must reach a decision upon locating a single facility. Each voter being a customer of the facility wishes to have the facility as close as possible to his location because the benefit enjoyed by the user is a decreasing function of the distance travelled. It seems to be impossible to meet the wishes of all users simultaneously . There are several kinds of compromises to be found [301].    A solution associated with the voting process is any point  c  such that no strict majority of users prefers another point to  c . Such a point is called a Condorcet point (cf. above mentioned literature). It does not exist in general as the triangle network with a user at each vertex and equal edge lengths illustrates.

We will mainly consider a generalization of the Condorcet concept and define an optimal point as a location such that there exists no competitor with higher expected value. It is assumed that the customer visits a nearer facility more frequently than a remote one (see subsection 3.3).

Obviously, network location problems under competition can also be considered as clustering or partitioning problems of the vertices with respect to their weights (respectively the vertices representing items) of the network (hence in line with the first section in this chapter). Let us illustrate that in case of two competitors. The Condorcet model immediately divides the set of weighted vertices into two subsets not necessarily of equal size. The generalization in form of a probabilistic model (i.e. searching for optimal points) divides the set of weighted vertices into two non–intersecting clusters which are not uniquely

---

299    The literature on network location problems of this kind is quite extensive (see Domschke / Drexl (1985)). For the main results see Hansen / Thisse / Labbé (1987).

300    see Hansen / Thisse (1981), Wendell / McKelvey (1981), Hansen / Thisse / Wendell (1986, 1990), Hakimi (1983, 1986, 1990), Labbé (1985), Bandelt (1985), Bandelt / Labbé (1986), Hansen / Labbé (1988), Hansen / Thisse / Labbé (1987), Dählmann (1989), De Palma et al. (1989), Labbé / Hakimi (1991). A survey on competitive location models is provided in Eiselt / Laporte (1989), Eiselt et al. (1993). For a general approach concerning 0–1 programming problem formulation and the minimization of travel and setup costs see Dobson / Karmarkar (1987).

301    cf. Hakimi (1983, 1986, 1989)

determined. Consider two competing locations (determining the creation of clusters), one is the optimal point in the network and the other is the strongest competitor. Then a vertex can randomly be assigned to one of the two locations with a probability inversely proportional to the distance to these locations. Thus we can generate different clusters of vertices all of which having an almost equal sum of vertex weight (about the expected value). This is in particular useful in order to successively assign program parts to the different processors on a transputer network. The number of interacting processors often is a power of 2 which in particular suggests to use the two competitor model in order to divide the processor set under the objective of workload balancing.

Consider the location of a central tool magazine in a flexible manufacturing environment. A machine corresponds to a vertex in the network model and the machine workload indicates the vertex weight. Clearly, each machine should have the tool magazine as close as possible to its location. The outcome is a voting process in which the desire to vote for a magazine location at a certain place is inversely proportional to the machine location distance. Hence a best location would maximize the expected value of votes such that there exists no rival point preferred to the considered location.

The section is organized as follows. In the next subsection main results on competitive location theory provided in the literature and concerning the existence of best locations on vertices are worked up. It also includes a description of the Condorcet model, while the model of our interest is introduced in subsection 3.3. Subsection 3.4 contains some minor results on optimal locations. They are mainly given to make algorithms more efficient. In subsection 3.5 we prove that optimal locations only occur at vertices and a maximal rejection of some point is achieved by a vertex. The results of these theorems are used in subsection 3.6 to provide efficient algorithms for finding optimal locations or suboptimal locations if optimal ones do not exist. Examples and counterexamples concerning future research are contained in subsection 3.7 while the remarks in our conclusions describe other interesting research directions.

## 3.2   The Condorcet Model

There is a finite number of users located at the vertices of the network N [302]. At each vertex there may be several users or none at all. The demand is described by a weight function $\pi$ from V to the set of non-negative real numbers where $\pi$ is different from the zero function. For a subnetwork N' of N we denote by $\pi(N')$ or $\pi(V')$ the sum $\sum_{u \in V'} \pi(u)$ where V' is the vertex set of N' . The weighted distance sum of a point x in the network N with weight function $\pi$ is given by

$$D(x) := \sum_{u \in V} d(u,x) \; \pi(u) \; .$$

A point $x \in N$ is called a median of N if $D(x) \leq D(y)$ for all $y \in N$ .

Location of a facility is allowed at any point of the network. To find a median it is enough to consider vertices as Hakimi (1964) showed.

---

[302]   The model is described using the notation of Bandelt (1985). A network $N = (V,E,\lambda)$ consists of a finite set V , a finite family E of two–element subsets of V and a mapping $\lambda : E \longrightarrow \mathbb{R}^+$. The pair (V,E) is a graph in the usual sense (cf. Swamy / Thulasiraman (1981)). The value $\lambda(e)$ is the length of edge e . The points x of N ($x \in N$) are the elements of the edges (including all vertices). Two points x and y on an edge e ($x,y \in e$) determine a subedge xy of e , the length of which is denoted by $\lambda(xy)$ . A path P(x,y) joining two points $x \in uv_1$ and $y \in v_n w$ is either a subedge or a sequence of edges and (at most two) subedges passing at most once through each point where P(x,y) contains x and y but no proper connected subset of P(x,y) does. The points x and y are the end points of P(x,y) . The length of P(x,y) is equal to the sum of the lengths of the edges and subedges. If the length of P(x,y) is minimum among all paths connecting x and y , then P(x,y) is a shortest path; its length is the distance d(x,y) between x and y . A cycle consists of an edge e joining vertices u and v and some path $P(u,v) \neq e$ connecting u to v . A network is connected if for any two points x and y there exists a path joining x and y . A connected network without cycles is a tree network.
Let V' be a subset of the vertex set of N . The network $N' = (V',E',\lambda')$ is the subnetwork of N on the vertex set V' if E' is a subset of E such that each edge of E joining u and v belongs to E' if and only if u and v are in V' and $d(u,v) = \lambda(uv)$ . The mapping $\lambda'$ is the restriction of $\lambda$ to E' .
We assume that all networks considered are connected. Furthermore there are no loops (= edges uu) at the vertices and without loss of generality each edge length is at least 1 . For a network $N = (N,E,\lambda)$ and a vertex u of N the vertex–deleted subnetwork N–u consists of vertex set V – {u} (for short : V–u) and all edges of E which are not incident with vertex u . A vertex u of N is a cut–vertex of N if the vertex–deleted subnetwork N–u consists of a greater number of components (= maximal connected subnetworks) than N .

**Theorem 1.**   Let   $N = (V,E,\lambda)$   be a network with vertex weight function $\pi : V \longrightarrow \mathbb{R}_0^+$ . Then there exists a vertex which is a median of  N . If there is no subset   $V' \subset V$   such that   $\pi(V') = \frac{\pi(N)}{2}$   then every median of  N  is a vertex.    □

We derive a similar result for optimal locations in a network, see also Hakimi (1986 or 1990).

Before we present the formal definition of a Condorcet point we provide the following notation. The vertex set  V  of a network  $N = (V,E,\lambda)$  with vertex weight function  $\pi$  is partitioned into three sets with respect to any pair  x , y  of points :

$$[x \succ y] := \{u \in V \,|\, d(u,x) < d(u,y)\} \,,$$
$$[x \sim y] := \{u \in V \,|\, d(u,x) = d(u,y)\} \,.$$

Then   $\pi([x \succ y])$   is the sum of the weights of all those vertices which prefer  x to  y . A point  x  is said to be a *β–Condorcet* point, where   $0 \leq \beta \leq 1$ , if $\pi([y \succ x]) + \beta \, \pi([y \sim x]) \leq \frac{\pi(N)}{2}$   for all points  $y \in N$ . Different choices of  $\beta$ result in different voting equilibria. A 1–Condorcet point is called a Condorcet point. Since Condorcet points need not exist, one can consider a substitute, see Bandelt / Labbé (1986). A  point    x    for which  the  quantity $\tilde{\rho}(x) := \frac{1}{\pi(N)} \max_{y \in N} \{\pi([y \succ x] + \beta \, \pi([y \sim x])\}$  is minimal, is called a *β–Simpson* point. It means that the maximal relative rejection  $\tilde{\rho}(x)$  of point  x  is minimal. The next theorem may be derived from Hansen / Thisse (1981).

**Theorem 2.**    Let   $N = (V,E,\lambda)$   be a network with vertex weight function $\pi : V \longrightarrow \mathbb{R}_0^+$ . If there is no subset  V'  of  V  such that   $\pi(V') = \frac{\pi(N)}{2}$   then every β–Condorcet point is a vertex for  $0 \leq \beta \leq 1$ .    □

The proof of this theorem is similar to the proof of Theorem 1. Hansen / Thisse / Wendell (1986) show :

**Theorem 3.**    Let   $N = (V,E,\lambda)$   be a network with vertex weight function $\pi : V \longrightarrow \mathbb{R}_0^+$ . Let  u  be a cut–vertex of  N  and  $N_1 ,..., N_k$  be the components of the vertex–deleted subnetwork  N–u . If  $\pi(N_j) \leq \frac{\pi(N)}{2}$  for all  $1 \leq j \leq k$ then  u  is a median and a β–Condorcet point of  N  for all  $0 \leq \beta \leq 1$ .    □

Finally, let $N$ be a tree network where $C$ denotes the set of all Condorcet points and $M$ denotes the set of all medians. Let $A := \{x \in N \mid \pi(N_j) \le \frac{\pi(N)}{2}$, $1 \le j \le k\}$ where $N_j$ are the components which result from $N$ after deletion of point $x$. If $x$ is not a vertex then the deletion of $x$ means the deletion of the complete edge (without the edge defining vertices) from $N$.

**Theorem 4.**    In a tree network $N = (V,E,\lambda)$ with vertex weight function $\pi : V \longrightarrow \mathbb{R}_0^+$ one has $C = M = A$.    □

The proof of $C = M$ can be found in Bandelt (1985), Hansen / Thisse (1981). The other equation may be derived from Goldman / Witzgall (1970) or in a more general setting from Bandelt (1990b).

Hence the existence of a $\beta$–Condorcet point depends upon the structure of the network and the distribution of the users. The distribution of the users however is irrelevant if the network contains no cycles. Wendell and McKelvey (1981) proved that, on a tree network, medians and $\frac{1}{2}$ – Condorcet points, also known as plurality points, coincide. Bandelt (1985) took these results, obtained for tree networks, a stage further. He characterized these networks, which always have a Condorcet point or a plurality point for any user distribution [303]. He also answered the question as to when Condorcet points and medians coincide for any distribution of users [304]. Other results on Condorcet points and polynomial algorithms to determine the, possibly infinite, set of Condorcet points in a network, may be found in Labbé (1985), and Hansen / Labbé (1988).

In the next two sections we take the concept of $\beta$–Condorcet point a little further and show how far the results of theorems 1 to 4 are transferable.


## 3.3    The Model

In the previous section the definition of Condorcet or plurality points in a network assumed that the location of the facility results from a voting procedure in which each voter (= customer) prefers to have the facility as close as possible to him. A solution associated with the voting process is any point where no strict majority of customers prefers another point. In other words each customer

---

[303]    see also Dählmann (1989)

[304]    see also Bandelt (1990a)

located at a vertex of the network visits only that facility closest to him.

This model is a very restrictive one and we are going to investigate some generalizations. We consider the situation where there are a given number of customers at fixed locations (vertices) in a network, and where each customer will purchase (without loss of generality) one unit of a commodity from the facility closer to his location more frequently than from a remote one [305]. Again, let all customers at a vertex behave in the same way and facility locations may be on any point of the network. As in the Condorcet model we consider only two facilities in competition. The best location in the network of a new facility competing with an existing one is to determine such that some objective function is optimized.

A customer located at vertex $u$ of the network $N$ visits a facility choosing among two competitive ones $x$ and $y$. We assume that the probability visiting the facility (at point) $x$ is proportional to $\dfrac{1}{d(u,x)^k}$, $k \in \mathbb{N} \cup \{\infty\}$, if $u \neq x$. The probability is $1$ if $u = x$. Let $u_1, ..., u_n$ be the vertices of the network $N$. Then, for $x \neq y$,

$$p_i^k(x,y) := \frac{d(u_i,y)^k}{d(u_i,x)^k + d(u_i,y)^k} , \ k \in \mathbb{N} \cup \{\infty\} .$$

is the probability that the customers located at $u_i$ purchase at $x$. For very large $k$ the decisions of the customers are alike to the deterministic decision in the Condorcet case. Hence, $k$ may be seen as a parameter for customer loyality.

$$E^k(x,y) := \sum_{i=1}^{n} \pi(u_i) \frac{d(u_i,y)^k}{d(u_i,x)^k + d(u_i,y)^k} , \ k \in \mathbb{N} \cup \{\infty\} , \ x \neq y ,$$

is the expected number of customers visiting facility $x$. We drop $k$ if it is equal $1$.

Each firm wants to choose a facility location that is best possible for itself. A location at a point $x$ which will guarantee the firm at least as many customers as its competitor, i.e. where $E^k(x,y) \geq \dfrac{\pi(N)}{2}$ for all $y \in N$, $y \neq x$, regardless of where its competitor locates, will be called a k-optimal location, the point (vertex) is called a k-optimal point (vertex). We say that the location $x$ dominates location $y$ or $y$ is rejected by $x$ if $E^k(x,y) > \dfrac{\pi(N)}{2}$. The

---

[305]   cf. Hakimi (1983), Pesch (1988), De Palma et al. (1989)

maximal relative k-rejection of a point $x \in N$ is given by the quantity
$\rho^k(x) := \sup_{x \neq y \in N} \frac{E^k(y,x)}{\pi(N)}$ . If $k = 1$ we denote the maximal relative rejection
by $\rho(x)$. As we shall see later k-optimal points need not exist generally, hence
we introduce a definition analogous to the Simpson point. A point $x$ is called
k-suboptimal if its maximal relative rejection is minimal, i.e. if $\rho^k(x) = \min_{y \in N} \rho^k(y)$ .[306]

## 3.4 Basic Results

The reader may have noticed that a k-optimal point is a generalized plurality
point. Indeed, if $k$ is large enough both models are very similar. For $k \to \infty$
k-optimal and plurality points coincide as $\lim_{k \to \infty} E^k(x,y) = \pi([x \succ y]) + \frac{1}{2}$
$\pi([x \sim y])$ if $x \neq y$ [307]. This is easily seen because

$$\lim_{k \to \infty} p_i^k(x,y) = \lim_{k \to \infty} \frac{d(u_i,y)^k}{d(u_i,x)^k + d(u_i,y)^k} = \begin{cases} 1 & \text{if } d(u_i,x) < d(u_i,y) \\ \frac{1}{2} & \text{if } d(u_i,x) = d(u_i,y) \\ 0 & \text{if } d(u_i,x) > d(u_i,y) \end{cases}$$

This close relationship between k-optimal points and plurality points arises the
question whether k-optimal points exist at all and how to find them. Trivial
examples are easily derived : for instance every vertex $u$ of a network with
$\pi(u) \geq \frac{1}{2} \pi(N)$ is a k-optimal point. Non-trivial examples are provided later (see
subsection 3.7). The vertices $y_2$, $y_4$, and $y_6$ in Figure 1 of De Palma et al.
(1989) are optimal while there is no equilibrium according to their approach.
However, k-optimal vertices need not exist. Consider a triangle network and
assume that all edge lengths and all vertex weights are equal $1$. Then this
cycle contains no optimal points if $k \geq 2$. However for $k = 1$ all vertices are
optimal.

---

[306] From a marketing point of view the problem receives attention as that of optimally
positioning products under price competition, see Choi et al. (1990) and Lederer /
Thisse (1990). The existence of equilibria on trees for two or three competing facilities
has been treated by Eiselt / Laporte (1991, 1993). Multi-criteria aspects are
introduced in Karkazis (1989).

[307] cf. Pesch (1988)

In the definition of a Simpson point we know that the maximum always exists. It would be convenient to know for $\rho^k(x)$ that the supremum is a maximum, too. But this is easily achieved. As $N$ is a finite network we only need some continuity assertions on $E^k(x,y)$ and complete $E^k(x,y)$ for $x = y$ continuously, where $E^k(x,y)$ is considered as a function of only the first (or the second) argument and $\pi(z) := 0$ for all points $z$ of $N$ which are not vertices.

**Proposition 1.**    Let $N$ be a (finite) network. Then

$$\lim_{x \to y} E^k(x,y) = \frac{\pi(N) - \pi(y)}{2} \; , \; y \in N \; , \; \text{and}$$

$$\lim_{y \to x} E^k(x,y) = \frac{\pi(N) + \pi(x)}{2} \; , \; x \in N \; , \; \text{for all} \; k \in \mathbb{N} \; .$$

**Proof.**    By definition we have

$$E^k(x,y) = \sum_{u_i \in V} \frac{d(u_i,y)^k \; \pi(u_i)}{d(u_i,y)^k + d(u_i,x)^k} = \sum_{\substack{u_i \in V \\ u_i \neq y}} \frac{\pi(u_i)}{1 + \left[\frac{d(u_i,x)}{d(u_i,y)}\right]^k}$$

Hence we immediately get the first assertion. For the second assertion let $\epsilon := d(x,y)$ such that $\epsilon < \min_{\substack{u_i \in V \\ u_i \neq x}} d(x,u_i)$. Application of the triangle inequality

yields :

$$E^k(x,y) \leq \sum_{\substack{u_i \in V \\ u_i \neq x}} \frac{\pi(u_i)}{1 + \left[\frac{d(u_i,x)}{d(u_i,x) + \epsilon}\right]^k} + \pi(x) \quad \text{and}$$

$$E^k(x,y) \geq \sum_{\substack{u_i \in V \\ u_i \neq x}} \frac{\pi(u_i)}{1 + \left[\frac{d(u_i,x)}{d(u_i,x) - \epsilon}\right]^k} + \pi(x)$$

Since for $\epsilon \to 0$ the right parts of the above inequalities are equal we get :

$$\lim_{y \to x} E^k(x,y) = \frac{1}{2} \sum_{\substack{u_i \in V \\ u_i \neq x}} \pi(u_i) + \pi(x) = \frac{\pi(N) + \pi(x)}{2}$$

This proves our second assertion.   □

Now we extend the definition of $E^k(x,y)$ where $E^k(y,y) := \frac{\pi(N) - \pi(y)}{2}$ for all $y \in N$ . This also has a reasonable interpretation in practice. If a new facility $x$ is established at the same site, for instance in a shopping center, where another one $y$ already exists then the probabilities to visit any of the facilities are equal. Only customers located at shopping center $y$ , for instance employees, would prefer $y$ .

**Proposition 2.**   Let $N$ be a (finite) network. The expected values $E^k(x,.)$ and $E^k(x_0,y)$ , $x_0 \notin V$ , are continuous functions on $N$ each in one variable.

**Proof.**   Since the proof is similar to that of Proposition 1, especially for the first assertion, we shall only show that $E^k(.,y)$ is continuous. Let $x_0$ be a point in $N$ . If we define $\epsilon := d(y,y_0)$ , $y_0 \in N$ , where $\epsilon < \min_{\substack{u_i \in V \\ u_i \neq y_0}} d(y_0,u_i)$ then we get

$$E^k(x_0,y_0) = \lim_{\substack{\epsilon \to 0 \\ u_i \neq y_0}} \sum_{u_i \in V} \frac{\pi(u_i)}{1 + \left[\frac{d(u_i,x_0)}{d(u_i,y_0) - \epsilon}\right]^k} \leq \lim_{y \to y_0} E^k(x_0,y) \leq$$

$$\leq \lim_{\substack{\epsilon \to 0 \\ u_i \neq y_0}} \sum_{u_i \in V} \frac{\pi(u_i)}{1 + \left[\frac{d(u_i,x_0)}{d(u_i,y_0) + \epsilon}\right]^k} = E^k(x_0,y_0) .   □$$

The above propositions permit to redefine the maximal relative k–rejection of a point $x \in N$ as

$$\rho^k(x) := \max_{y \in N} \frac{E^k(y,x)}{\pi(N)}$$

because the point set of N is compact.

We may assume that there are no multiple edges in the network N. For instance an edge connecting two vertices u and v which is not a shortest path from u to v either contains no suboptimal points or (in the trivial case) there is also a suboptimal point at the u and v connecting shortest path. The next result is basic to look by hand where suboptimal points are located.

A subnetwork M of a network N is *gated* if there is a map g from N to the vertex set of M with the following property [308]: for each $x \in N$ and for each $y \in M$ the gate to x in M, namely the vertex g(x) of M, satisfies

$$d(x,y) = d(x,g(x)) + d(g(x),y)$$

By N–M we denote the subnetwork of N which results from N when we successively delete all vertices u of M from N. Now we can formulate

**Proposition 3.**     Let   M   be a gated subnetwork of a network   N   with $\pi(N) > 0$. Let   x   be a point which does not belong to   M   and where $\pi(g(x)) \geq \pi(N-M) \geq 0$ then $E(g(x),z) > E(x,z)$ for all $z \in M$, $z \neq g(x)$.

**Proof.**     Assume that   $\pi(N-M) > 0$. We have

$$E(g(x),z) = \pi(g(x)) + \sum_{\substack{u_i \in M \\ u_i \neq g(x)}} \frac{d(u_i,z)\ \pi(u_i)}{d(u_i,z) + d(u_i,g(x))} +$$

$$+ \sum_{u_i \in N-M} \frac{d(u_i,z)\ \pi(u_i)}{d(u_i,z) + d(u_i,g(x))} >$$

$$> \pi(g(x)) \frac{d(g(x),z)}{d(g(x),z) + d(g(x),x)} + \sum_{\substack{u_i \in M \\ u_i \neq g(x)}} \frac{d(u_i,z)\ \pi(u_i)}{d(u_i,z) + d(u_i,x)} +$$

$$+ \sum_{u_i \in N-M} \frac{d(u_i,z)\ \pi(u_i)}{d(u_i,z) + d(u_i,x)} = E(x,z) \ \text{ if } \ \pi(g(x)) +$$

$$+ \sum_{u_i \in N-M} \frac{d(u_i,z)\ \pi(u_i)}{d(u_i,z) + d(u_i,g(x))} >$$

---

[308]    cf. Goldman / Witzgall (1970), Bandelt (1990a)

$$> \pi(g(x)) \frac{d(g(x),z)}{d(g(x),z) + d(g(x),x)} + \sum_{u_i \in N-M} \frac{d(u_i,z) \ \pi(u_i)}{d(u_i,z) + d(u_i,x)} \ .$$

The latter is equivalent to $\dfrac{\pi(g(x)) \ d(g(x),x)}{d(x,z)} = \dfrac{\pi(g(x)) \ d(g(x),x)}{d(g(x),z) + d(g(x),x)} >$

$$\sum_{u_i \in N-M} \frac{d(u_i,z) \ \pi(u_i)[d(u_i,g(x)) - d(u_i,x)]}{[d(u_i,z) + d(u_i,g(x))][d(u_i,z) + d(u_i,x)]} \ .$$

Application of the triangle inequality and division by $d(u_i,z)$ yields the following upper bound for the right hand side of the last inequality :

$$\sum_{u_i \in N-M} \frac{\pi(u_i) \ d(g(x),x)}{d(u_i,z) + d(u_i,g(x)) + d(u_i,x)}$$

As $d(u_i,z) + d(u_i,g(x)) + d(u_i,x) > d(x,z)$ , for $u_i \in N-M$ , the last sum is bounded by

$$\frac{d(g(x),x)}{d(x,z)} \sum_{u_i \in N-M} \pi(u_i) \leq \frac{\pi(g(x)) \ d(g(x),x)}{d(x,z)}$$

If $\pi(N-M) = 0$ then the sums over $u_i \in N-M$ disappear in the first inequality above. As $d(u_i,g(x)) < d(u_i,x)$ , for all $u_i \in M$ , and $\dfrac{d(g(x),z)}{d(g(x),z) + d(g(x),x)} < 1$ the assertion is also true for all $z \in N$ , $x \neq z$ . This proves the proposition. □

The last proposition is not true for $k \geq 2$ . Consider Figure 6 where each vertex weight and each edge length is equal 1 . Let $x$ be a point at distance $\frac{1}{4}$ from $g(x)$ .

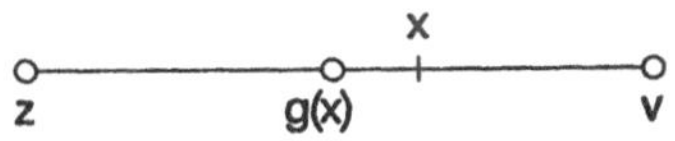

Figure 6

Then $\quad E^k(g(x),z) = 1 + \dfrac{2^k}{2^k + 1}$ , $\quad E^k(x,z) = \dfrac{4^k}{4^k + 1} + \dfrac{8^k}{8^k + 3^k}$ . Hence

$E^k(x,z) > > E^k(g(x),z)$ if and only if $24^k + 16^k + 3^k + 2 \cdot 6^k < 32^k$ which is met if $k \geq 2$. However, $E^k(g(x),x) > \frac{\pi(N)}{2}$ for all $k \in \mathbb{N}$ and all $x \in N{-}M$. Moreover, if $g(x)$ is a unique gate for all points in $N{-}M$ then $x$ cannot be suboptimal. A cut–vertex $u$ of a network $N$ is a k–optimal point for all $k \in \mathbb{N}$ if the sum of the vertex weights of each component $N_j$ of $N$, which results from $N$ after deletion of $u$, is at most $\pi(u)$. This is in line with Theorem 3.

### 3.5   Main Results

As mentioned earlier k–optimal points need not exist. In this section we shall see that they always coincide with vertices whenever they exist. These vertices are among the end vertices of closed intervals. The set of all points $x$ on shortest paths between two vertices $u$ and $v$ of a network $N$ is called the interval $I(u,v)$ between $u$ and $v$, that is : $I(u,v) = \{x \,|\, d(u,v) = d(u,x) + d(x,v)\}$ [309]. The interval $I(u,v)$ is closed if for any two points $x \in I(u,v)$ and $y \notin I(u,v)$ every shortest path from $x$ to $y$ passes through $u$ or $v$.

Now, we show that k–optimal points are always vertices. An optimal point can be found in polynomial time because the maximal relative rejection of a point is reached at a vertex. Hence comparisons with points which are not vertices are superfluous if $k = 1$. This is the message of the following theorem the proof of which may take use of convexity properties [310].

**Theorem 5.**   Let $N$ be a network with weight function $\pi : V \longrightarrow \mathbb{R}_0^+$ and at least two positive vertex weights. Let $I(u,v)$ be a closed interval of $N$ where $\pi(I(u,v)) = \pi(\{u,v\})$ and let $y$ be a point of this interval where $u \neq y \neq v$. Then $E(u,x) > E(y,x)$ or $E(v,x) > E(y,x)$ for all $x \in N$.

**Proof.**   Suppose $E(u,x) \leq E(y,x)$ and $E(v,x) \leq E(y,x)$ for some $x \in N$. We may assume that $x \neq u$. The case that $x \neq v$ is similar. Define the vertex sets $\mathcal{U}$ and $\mathcal{V}$, where

$$\mathcal{U} := \{u_i \notin I(u,v) \,|\, d(u_i,u) + d(u,y) \leq d(u_i,v) + d(v,y)\} \cup \{u\} \text{ and}$$
$$\mathcal{V} := \{u_i \notin I(u,v) \,|\, d(u_i,v) + d(v,y) < d(u_i,u) + d(u,y)\} \cup \{v\}$$

---

[309]   cf Mulder (1980)

[310]   see also Hakimi (1986)

Regard, $u \notin \mathcal{V}$ and $v \notin \mathcal{U}$. Now $E(y,x) \geq E(u,x)$ is equivalent to

$$\sum_{u_i \in \mathcal{V}} \frac{d(u_i,x) \ \pi(u_i)}{d(u_i,x) + d(u_i,y)} - \sum_{u_i \in \mathcal{V}} \frac{d(u_i,x) \ \pi(u_i)}{d(u_i,x) + d(u_i,u)} \geq$$

$$\geq \sum_{u_i \in \mathcal{U}} \frac{d(u_i,x) \ \pi(u_i)}{d(u_i,x) + d(u_i,u)} - \sum_{u_i \in \mathcal{U}} \frac{d(u_i,x) \ \pi(u_i)}{d(u_i,x) + d(u_i,y)}$$

which is equivalent to

$$\sum_{u_i \in \mathcal{V}} d(u_i,x) \ \pi(u_i) \left[ \frac{1}{d(u_i,x) + d(u_i,v) + d(v,y)} - \frac{1}{d(u_i,x) + d(u_i,u)} \right] \geq$$

$$\geq \sum_{u_i \in \mathcal{U}} d(u_i,x) \ \pi(u_i) \left[ \frac{1}{d(u_i,x) + d(u_i,u)} - \frac{1}{d(u_i,x) + d(u_i,u) + d(u,y)} \right].$$

Considering $d(u_i,u) \leq d(u_i,v) + d(v,u)$ we get :

$$\sum_{u_i \in \mathcal{V}} d(u_i,x) \ \pi(u_i) \left[ \frac{1}{d(u_i,x) + d(u_i,v) + d(v,y)} - \right.$$
$$\left. \frac{1}{d(u_i,x) + d(u_i,v) + d(v,u)} \right] \geq$$

$$\geq \sum_{u_i \in \mathcal{U}} d(u_i,x) \ \pi(u_i) \left[ \frac{1}{d(u_i,x) + d(u_i,u)} - \frac{1}{d(u_i,x) + d(u_i,u) + d(u,y)} \right].$$

which is equivalent to :

$$\sum_{u_i \in \mathcal{V}} d(u_i,x) \ \pi(u_i) \left[ \frac{1}{[d(u_i,x) + d(u_i,v) + d(v,y)][d(u_i,x) + d(u_i,v) + d(v,u)]} \right] \geq$$

$$\geq \sum_{u_i \in \mathcal{U}} d(u_i,x) \ \pi(u_i) \left[ \frac{1}{[d(u_i,x) + d(u_i,u)][d(u_i,x) + d(u_i,u) + d(u,y)]} \right].$$

In analogy we derive from $E(y,x) \geq E(v,x)$ the inequality :

$$\sum_{u_i \in \mathcal{U}} d(u_i,x) \ \pi(u_i) \left[ \frac{1}{[d(u_i,x) + d(u_i,u) + d(u,y)][d(u_i,x) + d(u_i,u) + d(u,v)]} \right] \geq$$

$$\geq \sum_{\substack{u_i \in \mathcal{V} \\ u_i \neq v}} d(u_i,x)\, \pi(u_i) \left[ \frac{1}{[d(u_i,x) + d(u_i,v)][d(u_i,x) + d(u_i,v) + d(v,y)]} \right] +$$

$$\frac{\pi(v)}{d(v,x) + d(v,y)}$$

Furthermore :

$$\sum_{\substack{u_i \in \mathcal{V} \\ u_i \neq v}} d(u_i,x)\, \pi(u_i) \left[ \frac{1}{[d(u_i,x) + d(u_i,v)][d(u_i,x) + d(u_i,v) + d(v,y)]} \right] +$$

$$\frac{\pi(v)}{d(v,x) + d(v,y)} \geq$$

$$\geq \sum_{u_i \in \mathcal{V}} d(u_i,x)\, \pi(u_i) \left[ \frac{1}{[d(u_i,x) + d(u_i,v) + d(v,y)][d(u_i,x) + d(u_i,v) + d(u,v)]} \right]$$

and

$$\sum_{u_i \in \mathcal{U}} d(u_i,x)\, \pi(u_i) \left[ \frac{1}{[d(u_i,x) + d(u_i,u)][d(u_i,x) + d(u_i,u) + d(u,y)]} \right] \geq$$

$$\geq \sum_{u_i \in \mathcal{U}} d(u_i,x)$$

$$\pi(u_i) \left[ \frac{1}{[d(u_i,x) + d(u_i,u) + d(u,y)][d(u_i,x) + d(u_i,u) + d(u,v)]} \right] .$$

Since N has at least two positive vertex weights at least one of the last two inequalities is strict. Hence the last four inequalities cannot be satisfied simultaneously, a contradiction. This proves the theorem.  □

**Theorem 6.** Let N be a network with weight function $\pi : V \longrightarrow \mathbb{R}_0^+$ and $\pi(N) > 0$. Let $I(u,v)$ be a closed interval of N where $\pi(I(u,v)) = \pi(\{u,v\})$ and let y be a point of this interval where $u \neq y \neq v$. Then, y is not a k–optimal point for each $k \in \mathbb{N}$.

**Proof.** According to Theorem 5 we may assume $k \geq 2$. In analogy to the preceding theorem suppose y is k–optimal for some $k \geq 2$. That means $E^k(y,u) \geq \frac{\pi(N)}{2}$ and $E^k(y,v) \geq \frac{\pi(N)}{2}$. We may assume that $d(y,u) \leq d(y,v)$. Let z be a point in the interval $I(u,v)$ where z is at the distance $2\,d(y,u)$ to u. Now, we devide the vertex set of N into two distinct subsets $\mathcal{U}$ and $\mathcal{V}$ where

$$\mathcal{U} := \{u_i \in V \mid \pi(u_i) \neq 0 \text{ and } d(u_i,u) \leq d(u_i,z)\} \text{ and}$$
$$\mathcal{V} := \{u_i \in V \mid \pi(u_i) \neq 0 \text{ and } d(u_i,z) < d(u_i,u)\} .$$

Obviously, $v \notin \mathcal{U}$ and $u \notin \mathcal{V}$. We have

$$E^k(y,z) = \sum_{u_i \in \mathcal{U}} \frac{d(u_i,z)^k \, \pi(u_i)}{d(u_i,z)^k + d(u_i,y)^k} + \sum_{u_i \in \mathcal{V}} \frac{d(u_i,z)^k \, \pi(u_i)}{d(u_i,z)^k + d(u_i,y)^k} =$$

$$= \sum_{u_i \in \mathcal{U}} \frac{d(u_i,z)^k \, \pi(u_i)}{d(u_i,z)^k + [d(u_i,u) + d(u,y)]^k}$$

$$+ \sum_{u_i \in \mathcal{V}} \frac{d(u_i,z)^k \, \pi(u_i)}{d(u_i,z)^k + [d(u_i,z) + d(z,y)]^k} \leq$$

$$\leq \sum_{u_i \in \mathcal{U}} \frac{[d(u_i,u) + d(u,z)]^k \, \pi(u_i)}{[d(u_i,u) + d(u,z)]^k + [d(u_i,u) + d(u,y)]^k} +$$

$$+ \sum_{u_i \in \mathcal{V}} \frac{d(u_i,z)^k \, \pi(u_i)}{d(u_i,z)^k + [d(u_i,z) + d(z,y)]^k} .$$

In analogy we derive

$$E^k(y,u) \leq \sum_{u_i \in \mathcal{U}} \frac{d(u_i,u)^k \, \pi(u_i)}{d(u_i,u)^k + [d(u_i,u) + d(u,y)]^k} +$$

$$+ \sum_{u_i \in \mathcal{V}} \frac{[d(u_i,z) + d(z,u)]^k \, \pi(u_i)}{[d(u_i,z) + d(z,u)]^k + [d(u_i,z) + d(z,y)]^k} .$$

The assumption $E^k(y,z) \geq \frac{\pi(N)}{2}$ and $E^k(y,u) \geq \frac{\pi(N)}{2}$ yields

$$\sum_{\substack{u_i \in \mathcal{V} \\ u_i \neq z}} \frac{\pi(u_i)}{1 + \left[1 + \frac{d(z,y)}{d(u_i,z)}\right]^k} \geq \frac{\pi(N)}{2} - \pi(\mathcal{U}) + \pi(\mathcal{U}) -$$

$$\sum_{u_i \in \mathcal{U}} \frac{[d(u_i,u) + d(u,z)]^k \, \pi(u_i)}{[d(u_i,u) + d(u,z)]^k + [d(u_i,u) + d(u,y)]^k} =$$

$$= \frac{\pi(N)}{2} - \pi(\mathscr{U}) + \sum_{u_i \in \mathscr{U}} \frac{[d(u_i,u) + d(u,y)]^k \, \pi(u_i)}{[d(u_i,u) + d(u,z)]^k + [d(u_i,u) + d(u,y)]^k}$$

which is equivalent to

$$\pi(\mathscr{Y}) \geq \sum_{\substack{u_i \in \mathscr{Y} \\ u_i \neq z}} \frac{\pi(u_i)}{1 + \left[1 + \frac{d(u,z)}{2\,d(u_i,z)}\right]^k} + \frac{\pi(N)}{2} - \sum_{u_i \in \mathscr{U}} \frac{\pi(u_i)}{1 + \left[1 + \frac{d(u,z)}{2\,[d(u_i,u) + \frac{1}{2}\,d(u,z)]}\right]^k}$$

Similarly, we have :

$$\sum_{\substack{u_i \in \mathscr{U} \\ u_i \neq u}} \frac{\pi(u_i)}{1 + \left[1 + \frac{d(u,z)}{2\,d(u_i,u)}\right]^k} \geq \frac{\pi(N)}{2} - \pi(\mathscr{Y}) + \sum_{u_i \in \mathscr{Y}} \frac{\pi(u_i)}{1 + \left[1 + \frac{d(u,z)}{2\,[d(u_i,z) + \frac{1}{2}\,d(u,z)]}\right]^k} .$$

We also get :

$$\sum_{u_i \in \mathscr{U}} \frac{\pi(u_i)}{1 + \left[1 + \frac{d(u,z)}{2\,[d(u_i,u) + \frac{1}{2}\,d(u,z)]}\right]^k} \geq \sum_{\substack{u_i \in \mathscr{U} \\ u_i \neq u}} \frac{\pi(u_i)}{1 + \left[1 + \frac{d(u,z)}{2\,d(u_i,u)}\right]^k}$$

and

$$\sum_{u_i \in \mathscr{Y}} \frac{\pi(u_i)}{1 + \left[1 + \frac{d(u,z)}{2\,[d(u_i,z) + \frac{1}{2}\,d(u,z)]}\right]^k} \geq \sum_{\substack{u_i \in \mathscr{Y} \\ u_i \neq z}} \frac{\pi(u_i)}{1 + \left[1 + \frac{d(u,z)}{2\,d(u_i,z)}\right]^k} .$$

Since $\pi(\mathscr{U}) > 0$ or $\pi(\mathscr{Y}) > 0$ at least one of the last two inequalities is strict.

Hence, we have a contradiction.   □

Our theorem especially says that no optimal point is on an edge if it is not one
of the edge defining end vertices and this is just in line with Theorems 1 to 4.
If $k = 1$ the maximal relative rejection $\rho(x)$ of a point $x$ is attained at a
vertex or equals $\frac{1}{\pi(N)} \frac{1}{2} [\pi(N) - \pi(x)]$. However, for $k \geq 2$ it does not suffice
only to consider vertices to compute $\rho(x)$ as Figure 7 illustrates.

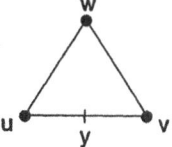

Figure 7

We assume that all vertex and edge weights are equal 1 and $y$ is the
midpoint on the edge $uv$. Then $E^k(w,u) = E^k(w,v) = \frac{3}{2}$; but
$E^k(y,w) = \frac{2}{1 + (\frac{1}{2})^k} \geq \frac{2}{1 + \frac{1}{4}} = \frac{8}{5} > \frac{3}{2}$ if $k \geq 2$. Hence, there exists no
k-optimal point, for $k \geq 2$, but all vertices are optimal. Now, let the vertex
weight of $w$ be $1 + \epsilon$ where $\epsilon < \frac{1}{5}$. Then, for $k \geq 2$ we see, if both end
vertices of an edge are rejected by some point, the edge connecting these end
vertices is not necessarily rejected. This is not possible for $k = 1$.

### 3.6  Polynomial Algorithms

With our main theorem in hand we are able to describe polynomial algorithms
which determine optimal or suboptimal points. As input data we need an
n–vector $\pi$ of vertex weights and an $n \times n$ distance matrix $(d_{ij})$ where $d_{ij}$
describes the length of a shortest path connecting vertex $u_i$ and vertex $u_j$.
Computation of the distance matrix can be achieved by the algorithm of
Floyd–Warshall or the Dijkstra algorithm in $\mathcal{O}(n^3)$ time [311]. With the distance
matrix and vector $\pi$ in hand we can easily compute the expected values in

---

[311]   see Dijkstra (1959), Syslo / Deo / Kowalik (1983)

linear time. All algorithms are described using PASCAL keywords.

Now, we can describe the first algorithm which determines all optimal points of a network. The vertex $u_i$ chosen in the initialization is recommended to be a median or a vertex such that $\pi(u_i)$ is maximum. Now, successively all vertices of $\mathcal{V} = V$ are tested and thereafter deleted. Vertex set $V_p$ contains all vertices which are already recognized as optimal points. Hence, when the algorithm terminates $V_p$ contains all optimal points. Vertex sets $V_1$ and $V_0$ contain vertices which do not belong to $V_p$, i. e. which are not already recognized as optimal. The vertices of $V_0$ are compared with vertex $u_i$ which is tested whether it is optimal. Whenever a vertex of $V_0$ is dominated by $u_i$ this vertex is deleted from $V_0$. Otherwise, if a vertex $u_j$ of $V_0$ dominates $u_i$ or is as good as $u_i$ then $u_j$ is memorized; $j_0$ indicates the index of that vertex $u_j \in V_0$ which rejects $u_i$ at most among all considered vertices of $V_0$. Variable $M$ gets the maximum rejection of $u_i$ among all considered vertices.

## ALGORITHM 1

```
begin
    {Initialization}
    choose any vertex u_i in N ;
    V_p := ∅ ; V_1 := V ; V_0 := V_1 ; 𝒱 := V ;
    {Iteration}
    while V_1 ≠ ∅ do
    begin
        M := 0 ; j_0 := 0 ;
        for all u_j∈V_1 , j ≠ i  do
        begin  {check if u_i dominates u_j }
            EW := E(u_j,u_i) ;
            if EW < π(N)/2  then  V_0 := V_0 - u_j
            else
                if EW > M  then begin  M := EW ; j_0 := j end
        end;
        if M ≤ π(N)/2  then
        begin
            j := 1 ;
```

$$\textbf{while } j \leq n \textbf{ and } M \leq \frac{\pi(N)}{2} \textbf{ do}$$
$$\textbf{begin}$$
$$\qquad \textbf{if } u_j \in \mathcal{V} - V_1 \textbf{ then } M := \max \{M, E(u_j,u_i)\} ;$$
$$\qquad j := j + 1$$
$$\textbf{end}$$
$$\textbf{end};$$
$$\textbf{if } M \leq \frac{\pi(N)}{2} \textbf{ then } V_p := V_p \cup \{u_i\} ;$$
$$V_0 := V_0 - u_i ; V_1 := V_0 ; \mathcal{V} := \mathcal{V} - u_i ; i := j_0$$
$$\textbf{end}$$
$$\textbf{end.}$$

Leaving the for ... do iteration $M$ is the value of the maximal rejection of $u_i$ by a vertex $u_{j_0}$ in $V_1$ if some vertex of $V_1$ dominates $u_i$. Otherwise $M$ is zero. If no vertex of $V_1$ dominates $u_i$ then the vertices not occuring in $V_1$ must be considered because one of these vertices could dominate $u_i$. Only vertices which do not belong to $\mathcal{V}$ are done. When the main while ... do iteration is at its end a new vertex $u_i$ from $V_1$ is chosen for the next iteration. The new vertex $u_i$ rejected the old one at most.
The algorithm has time complexity $\mathcal{O}(n^3)$ where $n$ is the number of vertices in $N$.

From Theorem 5 we infer that for any vertex $w \in N$ and any point $x$ in the interval $I(u,v)$ where $\pi(I(u,v)) = \pi(\{u, v\})$ and $u \neq x \neq v$ holds :

$$E(w,u) < E(w,x) \text{ or } E(w,v) < E(w,x) .$$

Especially, for any two points $y$ and $y'$ on an edge every point $x$ between $y$ and $y'$ is rejected by $w$ more than $y$ or more than $y'$. This is just the definition of strong quasiconcavity if $E(w,x)$ is considered as a function depending on one variable $x$, where $x$ belongs to an edge $uv$. Hence, to determine suboptimal points we need to find all points $x$ of any edge $uv$ which are rejected by vertex $w_k$ at most $\rho_k$, where $\rho_k$ is the smallest relative rejection of some point already known in iteration $k$. In the first iteration $\rho_0$ takes the smallest relative rejection of some vertex. Points rejected more than $\rho_k$ cannot be suboptimal. Moreover, since $E(.,x)$ is strongly quasiconcave these points are in one open interval of $uv$. At iteration $k$ the interval with endpoints $x_k$, $y_k$ which possibly contains suboptimal points is reduced. Notice, that $\rho_k$ will always be smaller or equal to min $\{\rho(x_k)\,,\,\rho(y_k)\}$ . Thus, at iteration $k$ we need to find at most one point $\alpha_k$ of uv, $x_k \neq \alpha_k \neq y_k$ where $E(w_k,\alpha_k) = \rho_k$ and $E(w_k,x_k) \geq \rho_k$ or $E(w_k,y_k) \geq \rho_k$ . Finding this point is achieved by the Fibonacci Search Method [312]. The method is based on the Fibonacci sequence $\{F_\nu\}$ defined as :

$$F_{\nu+1} = F_\nu + F_{\nu-1}\,,\quad \nu = 1\,,\,2\,,... \text{ and } F_0 = F_1 = 1$$

At iteration $k$ suppose that the interval of the edge uv which contains all points $x$ where $E(w_k,x) \leq \rho_k$ is defined by the points $x_k$ , $y_k$ . For convenience we may assume that $x_k$ and $y_k$ are defined by the distance to u . Consider the two points $\lambda_k$ and $\mu_k$ given below, and defined by the distance to u , where p is the total number of functional evaluations planned.

$$d(u,\lambda_k) = d(u,x_k) + \frac{F_{p-k-1}}{F_{p-k+1}}\, d(x_k,y_k)\,,\quad k = 1\,,...,\,p-1$$

$$d(u,\mu_k) = d(u,x_k) + \frac{F_{p-k}}{F_{p-k+1}}\, d(x_k,y_k)\,,\quad k = 1\,,...,\,p-1$$

The new interval defined by the points $x_{k+1}$ , $y_{k+1}$ where $E(w_k,x) \leq \rho_k$ only if $d(u,x_{k+1}) \leq d(u,x) \leq d(u,y_{k+1})$ is given by the points

$$\begin{array}{ll} \lambda_k\,,\,y_k & \text{if } E(w_k,\lambda_k) \geq \rho_k > E(w_k,\mu_k) \\ x_k\,,\,\lambda_k & \text{if } E(w_k,\lambda_k) \geq \rho_k > E(w_k,x_k) \\ x_k\,,\,\mu_k & \text{if } E(w_k,\lambda_k) < \rho_k \leq E(w_k,\mu_k) \\ \mu_k\,,\,y_k & \text{if } E(w_k,y_k) < \rho_k \leq E(w_k,\mu_k) \end{array}$$

If $E(w_k,\lambda_k) < \rho_k > E(w_k,\mu_k)$ we split the whole interval $[x_k,y_k]$ into two parts $[x_k,\lambda_k]$ and $[\lambda_k,y_k]$ .

Hence we get the following procedure, where P always is a set of possibly suboptimal points on the edge that contains the subedge $x_ky_k$ . For convenience indices refering to distinct edges are neglected. Notice, that in view of Theorem 5 there are at most $n + 1$ suboptimal points on an edge. Each vertex of N rejects at most one subedge connecting two suboptimal points.

**begin**

     P := ∅ for all edges of N ; $\rho_0 := \min\limits_{u \in V} \rho(u)$ ;

     $E_p := E$ ; $\{E_p$ contains all subedges which may have suboptimal points$\}$

     **while** $E_p \neq \emptyset$ **do**

     **begin**

---

[312]    cf. Bazaraa / Shetty (1979)

**repeat**
>   choose any subedge $x_1y_1$ from $E_p$ ; $E_p := E_p - x_1y_1$

**until** $\rho(x_1) \leq \rho_0$ **or** $\rho(y_1) \leq \rho_0$ ;

$k := 1$;

**while** $|d(u,x_k) - d(u,y_k)| > \epsilon$ **and** $|P| < n + 2$ **do**

**begin**
>   let $w_k$ be a vertex such that $E(w_k,x_k) = \rho(x_k)$ ;
>   define the points $\lambda_k$ and $\mu_k$ by

$$d(u,\lambda_k) := d(u,x_k) + \frac{F_{p-k-1}}{F_{p-k+1}} \, d(x_k,y_k) \text{ and}$$

$$d(u,\mu_k) := d(u,x_k) + \frac{F_{p-k}}{F_{p-k+1}} \, d(x_k,y_k) ;$$

$\rho_k := \min \{\rho(x_k) , \rho_{k-1}\}$ ;

**if** $\rho(x_k) < \rho_{k-1}$ **then** $P := \{x_k\}$

**else**
>   **if** $\rho(x_k) = \rho_{k-1}$ **then** $P := P \cup \{x_k\}$ ;

**if** $E(w_k,\lambda_k) < \rho_k > E(w_k,\mu_k)$ **then**
>   **begin** $E_p := E_p \cup \{\lambda_k y_k\}$ ; $x_{k+1} := x_k$ ; $y_{k+1} := \lambda_k$ **end**

**else**
>   **if** $E(w_k,\lambda_k) \geq \rho_k > E(w_k,\mu_k)$ **then begin** $x_{k+1} := \lambda_k$ ; $y_{k+-}$ :
>   **else**
>   >   **if** $E(w_k,\mu_k) \geq \rho_k > E(w_k,y_k)$ **then**
>   >   >   **begin** $x_{k+1} := \mu_k$ ; $y_{k+1} := y_k$ **end**;

**if** $|d(u,x_{k+1}) - d(u,y_{k+1})| > \epsilon$ **then**
>   **begin** exchange $x_{k+1}$ and $y_{k+1}$ ; $k := k + 1$ **end**

**else**

**begin**

$\rho_0 := \rho(\alpha_k)$ where $d(u,\alpha_k) = \frac{1}{2} (d(u,x_k) + d(u,y_k))$ ;

**if** $\rho_0 < \rho_k$ **then** $P := \{\alpha_k\}$

**else**
>   **if** $\rho_0 = \rho_k$ **then** $P := P \cup \{\alpha_k\}$

**end**

**end;**

**if** $|d(u,x_k) - d(u,y_k)| > \epsilon$ **then begin** $E_p := E_p \cup x_k y_k$ ; $\rho_0 := \rho_k$ en

**if** $|P| \geq n + 2$ **then** $P := \emptyset$ ;

    end
end.

In the above procedure we assumed that the value $\rho$ is already known. Therefore, we need some procedure to compute it efficiently.

The next algorithm determines vertices with minimum rejection in a network. Vertex set $V_p$ contains all vertices with smallest rejection among all considered vertices. Vertex $u_{i_1}$ has the smallest current rejection. Vertex $u_{j_0}$ rejects $u_{i_1}$ most.

## A L G O R I T H M   2

begin
    {Initialization}
    choose any vertex $u_i$ in $N$ ;
    $V_p := \varnothing$ ; $V_1 := V$ ; $\mathcal{V} := V$ ;
    $\rho(u_i) := 0$ ;
    for $j := 1$ to $n$ , $j \neq i$ , do
    begin
        $\rho(u_j) := E(u_i,u_j) / \pi(N)$ ;
        if $\rho(u_i) < 1 - \rho(u_j)$ then begin $\rho(u_i) := 1 - \rho(u_j)$ ; $j_0 := j$ end
    end;
    $\mathcal{V} := \mathcal{V} - u_i$ ; $V_p := \{u_i\}$ , $V_1 := V_1 - u_i$ ;
    $i_1 := i$ ; $i := j_0$;
    {Iteration}
    while $V_1 \neq \varnothing$ do
    begin
        for all $u_j \epsilon V_1$ , $j \neq i$ do
        begin
            EW $:= E(u_i,u_j) / \pi(N)$ ;
            if EW $> \rho(u_j)$ then $\rho(u_j) := $ EW ;
            if $\rho(u_j) > \rho(u_{i_1})$ then $V_1 := V_1 - u_j$ ;
            if $1 - $ EW $> \rho(u_i)$ then $\rho(u_i) := 1 - $ EW
        end;
        $j := 1$ ;
        while $\rho(u_i) \leq \rho(u_{i_1})$ and $j \leq n$ do

```
        if u_j ∈ 𝒱 - V_1  then
        begin
            EW := E(u_j,u_j) / π(N) ;
            if  1 - EW > ρ(u_j)  then  ρ(u_j) := 1 - EW ;
            j := j + 1
        end;
    if ρ(u_i) ≤ ρ(u_{i_1})  then
    begin
        i_1 := i ;
        if  ρ(u_i) = ρ(u_{i_1})  then  V_p := V_p ∪ {u_i}  else  V_p := {u_i}
    end ;
    𝒱 := 𝒱 - u_i ; V_1 := V_1 - u_i ;
    ρ := 1 ;
    for all  u_j ∈ V_1  do
    begin
        if  ρ(u_j) < ρ  then begin  ρ := ρ(u_j) ; i_0 := j  end;
        if  ρ(u_j) > ρ(u_{i_1})  then  V_1 := V_1 - u_j
    end;
    i := i_0
end;
```

$$\text{if } \frac{\pi(N) - \pi(u_{i_1})}{2\ \pi(N)} > \rho(u_{i_1}) \text{ then } \rho(u_{i_1}) := \frac{\pi(N) - \pi(u_{i_1})}{2\ \pi(N)}$$

```
end.
```

If there exists an optimal vertex then the above algorithm finds it. Therefore the last instruction is necessary. The time complexity of Algorithm 2 is $\mathcal{O}(n^3)$ . Hence finding suboptimal points is at most of complexity $\mathcal{O}(n^5)$ . However, finding optimal points is more quickly achieved with the first algorithm. The reader may notice that the algorithms may be improved using the results of subsection 3.4 to reduce the considered network first.

## 3.7  Examples and Counterexamples

Throughout this paper the reader may have had several ideas worthy to be examined but until yet not proved. In this subsection we provide examples and counterexamples supporting or destroying such ideas.

Consider the network with given vertex weights and edge lengths as in Figure 8.

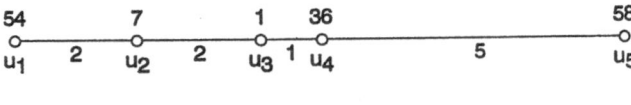

Figure 8

Then we get :

$$E(u_5,u_4) = 54 \cdot \frac{1}{3} + 7 \cdot \frac{3}{11} + \frac{1}{7} + 58 = 78 + \frac{4}{77} > 78 = \frac{\pi(N)}{2} \ ,$$

$$E(u_1,u_5) = 54 + 7 \cdot \frac{4}{5} + \frac{3}{5} + 36 \cdot \frac{1}{2} = 78.2 > 78 \ ,$$

$$E(u_4,u_1) = 7 \cdot \frac{2}{5} + \frac{4}{5} + 36 + 58 \cdot \frac{2}{3} = 78 + \frac{4}{15} > 78 \ ,$$

$$E(u_1,u_2) = 54 + \frac{1}{3} + 36 \cdot \frac{3}{8} + 58 \cdot \frac{4}{9} = 93 + \frac{11}{18} > 78 \ ,$$

$$E(u_4,u_3) = 54 \cdot \frac{4}{9} + 7 \cdot \frac{2}{5} + 36 + 58 \cdot \frac{6}{11} = 94 + \frac{24}{55} > 78 \ .$$

Hence Theorem 6 yields that the network has no optimal point. Thus tree networks, moreover paths, need not have optimal points. Value $\rho = 78 + \frac{4}{77}$ equals the rejection of vertex $u_4$ by vertex $u_5$ . Networks which are not trees and without optimal points are easy to find. For instance consider the network in Figure 9.

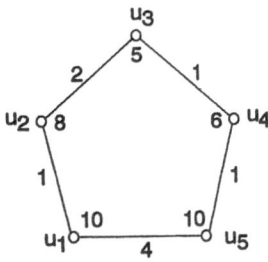

Figure 9

In Figure 9 vertex $u_1$ is dominated by vertex $u_4$, $u_4$ is dominated by $u_5$, $u_2$ rejects $u_5$, and $u_1$ rejects $u_2$. Vertex $u_1$ also dominates vertex $u_3$. In Figure 8 and Figure 9 vertex $u_4$ and vertex $u_2$, respectively, are the only suboptimal points.

Next, we show that not only for $k = 1$ but for each $k \in \mathbb{N}$ there exists a path which has no k–optimal point. For $k = 1$ Figure 8 proves the assertion. If $k \geq 2$ we consider the network in Figure 10. Figure 10 gives a path where each edge length equals $1$. The vertex weights are $\pi(u_1) = 4$, $\pi(u_2) = 2$, $\pi(u_3) = 1$, and $\pi(u_4) = 5$. Let $x$ be a point at the edge $u_3 u_4$ where $d(x, u_3) = \frac{1}{3}$.

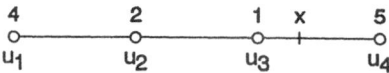

Figure 10

Then for each $k \geq 2$ we have $\dfrac{\pi(N)}{2} = 6$ and $E^k(u_2, u_1) = 2 + \dfrac{2^k}{2^k + 1} + 5$

$$\frac{3^k}{3^k + 2^k} = = 2 + \frac{1}{1 + (\frac{1}{2})^k} + \frac{5}{1 + (\frac{2}{3})^k} \geq 2 + \frac{4}{5} + 5 \cdot \frac{9}{13} > 6 \,,$$

$$E^k(u_2, u_4) = 4 \frac{3^k}{3^k + 1} + 2 + \frac{1}{2} = \frac{4}{1 + (\frac{1}{3})^k} + 2.5 > 6 \,,$$

$$E^k(u_2, u_3) = 4 \frac{2^k}{2^k + 1} + 2 + 5 \frac{1}{2^k + 1} > 6 \,,$$

$$E^k(x, u_2) = \frac{4}{(\frac{7}{3})^k + 1} + \frac{1}{(\frac{1}{3})^k + 1} + 5 \frac{2^k}{(\frac{2}{3})^k + 2^k} = \frac{4}{(\frac{7}{3})^k + 1} + \frac{6}{(\frac{1}{3})^k + 1} > 6$$

for all $k \geq 2$.

Hence, for each $k \geq 2$ there is a path without an optimal vertex. Again we see that the maximum rejection of some point may be achieved by a point which is not a vertex. The above example says something more. Similar results as achieved for Condorcet points and medians in tree networks in subsection 3.3 do not hold for optimal points. All points of the edge $u_2 u_3$ are $\beta$–Condorcet points, for $0 \leq \beta \leq 1$, and medians but even if $k = 1$ only vertex $u_4$ is optimal.

If the set of $\beta$-Condorcet points consists of only one vertex this need not be k-optimal even if k is very large. Consider Figure 11 where $d(u_1,u_2) = 2$, $d(u_2,u_3) = 1$, $\pi(u_1) = 5 = = \pi(u_3)$, $\pi(u_2) = 1$.

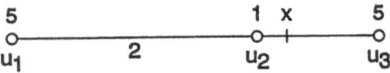

Figure 11

Let $x := x_k$ be a point at the edge $u_2u_3$ at distance $\epsilon_k$ to $u_2$. Vertex $u_2$ is the only $\beta$-Condorcet point for $0 \leq \beta \leq 1$. But

$$E^k(x_k,u_2) = 5\frac{2^k}{2^k + (2+\epsilon_k)^k} + 5\frac{1}{1 + (1-\epsilon_k)^k} =$$

$$= 5 + 5\frac{1 - (1-\epsilon_k)^k (1+\frac{1}{2}\epsilon_k)^k}{(1 + (1+\frac{1}{2}\epsilon_k)^k)(1 + (1-\epsilon_k)^k)} > \frac{11}{2}$$

if and only if $\dfrac{1 - (1-\epsilon_k)^k (1+\frac{1}{2}\epsilon_k)^k}{(1 + (1+\frac{1}{2}\epsilon_k)^k)(1 + (1-\epsilon_k)^k)} > \dfrac{1}{10}$. If $\epsilon_k := \dfrac{2}{k}$ then

$$\lim_{k \to \infty} \frac{1 - (1-\frac{2}{k})^k (1+\frac{1}{k})^k}{(1 + (1+\frac{1}{k})^k)(1 + (1-\frac{2}{k})^k)} = \frac{1 - e^{-2} e^1}{(1 + e^1)(1 + e^{-2})} \approx 0.15 > 0,1.$$

Hence there is an integer $k_0$ such that for each $k \geq k_0$ we have $E^k(x_k,u_2) > \frac{11}{2}$.

Thus, $u_2$ is not a k-optimal point for $k \geq k_0$. Even for very large k plurality points and optimal points do not coincide. Moreover any vertex u (esp. $\beta$-Condorcet point $u_2$) dominates points at a distance less than $\epsilon_k$ to u as Proposition 1 and 2 say. Thus for each k (esp. for each $k \geq k_0$) there exists an $\epsilon_k$ such that u dominates all points x at a distance less than $\epsilon_k$ to u. We have just seen that a vertex u with positive weight dominates all points x close enough to u. If x is a more remote point to u then there exists a k sufficiently large such that a $\beta$–Condorcet point, $0 \leq \beta \leq \frac{1}{2}$, (also with zero vertex weight) dominates point x in the network. This is a consequence of the next proposition.

**Proposition 4.** Let $N = (V,E,\lambda)$ be a network and let u be a vertex of N If u is a k–optimal point for infinitely many k then u is a $\beta$–Condorcet point with $0 \leq \beta \leq \frac{1}{2}$.

**Proof.** First we prove for any point $x \in N$ : Let $y$, $y \neq x$, be a point in N such that $\beta \, \pi([y \sim x]) + \pi([x \succ y]) > \frac{\pi(N)}{2}$ with $0 \leq \beta \leq \frac{1}{2}$. Then there is an integer $k_y$ such that $E^k(x,y) > \frac{\pi(N)}{2}$ for all $k \geq k_y$.

$$E^k(x,y) \geq \frac{1}{2} \cdot \pi([x \sim y]) + \sum_{u_i \in [x \succ y]} \pi(u_i) \frac{d(u_i,y)^k}{d(u_i,x)^k + d(u_i,y)^k} =$$

$$= \frac{1}{2} \cdot \pi([x \sim y]) + \sum_{u_i \in [x \succ y]} \pi(u_i) \frac{1}{1 + \dfrac{d(u_i,x)^k}{d(u_i,y)^k}} \xrightarrow{k \to \infty}$$

$$\frac{1}{2} \cdot \pi([x \sim y]) + \pi([x \succ y]) \geq \beta \, \pi([y \sim x]) + \pi([x \succ y]) > \frac{\pi(N)}{2}.$$

If y is no $\beta$–Condorcet point, $0 \leq \beta \leq \frac{1}{2}$, then there is a point x in N , $x \neq y$, such that $\beta \, \pi([y \sim x]) + \pi([x \succ y]) > \frac{\pi(N)}{2}$. Thus there is an integer $k_y$ such that x dominates y for all $k \geq k_y$. Hence we know :

> If y is no $\beta$–Condorcet point with $0 \leq \beta \leq \frac{1}{2}$ then there is an integer $k_y$ such that y is no k–optimal point for each $k \geq k_y$, $k \in \mathbb{N}$.

We know that k–optimal points are always vertices. If a vertex v is no $\beta$–Condorcet point, $0 \leq \beta \leq \frac{1}{2}$, then there is an integer $k_v$ such that v is not a k–optimal point for each $k \geq k_v$. Since the network N is finite there exists only a finite number of vertices. Let u be k–optimal where $k \geq \max_{v \in V} \{k_v\}$, then u is a $\beta$–Condorcet point with $0 \leq \beta \leq \frac{1}{2}$. $\quad \square$

We describe another tree network in which the only $\beta$–Condorcet point (vertex) u dominates each other vertex if k is large enough. However, this does not suffice that the $\beta$–Condorcet point u is also k–optimal for k large enough. For each point $y \in N$ there exists an integer $k_y$ which depends on y such that $E^k(u,y) \geq \frac{\pi(N)}{2}$ for all $k \geq k_y$. We cannot find an integer k sufficiently large and independent on the points y of N where $E^k(u,y) \geq \frac{\pi(N)}{2}$.

Consider the graph in Figure 12 where $d(u_3,u_2) = d(u_3,u_4) = 2$, $d(u_1,u_3) = 1$, $\pi(u_1) = \frac{3}{2}$, $\pi(u_2) = \pi(u_4) = 1$, $\pi(u_3) = 0$. Let $\epsilon_k := d(x,u_3)$ where $x \in u_1 u_3$.

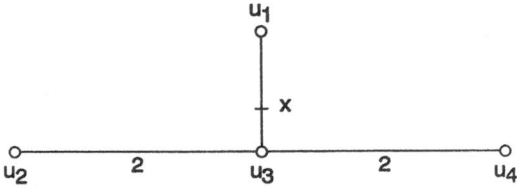

Figure 12

Vertex $u_3$ is the only $\beta$–Condorcet point for $0 \leq \beta \leq 1$. $E^k(u_3,u_j) > \frac{7}{4}$ for each $j \in \{1,2,4\}$ and k sufficiently large.

$$E^k(x,u_3) = 2\, \frac{2^k}{2^k + (2+\epsilon_k)^k} + \frac{3}{2}\, \frac{1}{1 + (1-\epsilon_k)^k} =$$

$$= \frac{3}{2}\, \frac{1 + (1+\frac{1}{2}\epsilon_k)^k + 1 + (1-\epsilon_k)^k}{(1 + (1+\frac{1}{2}\epsilon_k)^k)\,(1 + (1-\epsilon_k)^k)} + \frac{1}{2}\, \frac{1}{1 + (1+\frac{1}{2}\epsilon_k)^k} =$$

$$= \frac{3}{2} + \frac{3}{2}\, \frac{1 - (1+\frac{1}{2}\epsilon_k)^k\,(1-\epsilon_k)^k}{(1 + (1+\frac{1}{2}\epsilon_k)^k)\,(1 + (1-\epsilon_k)^k)} + \frac{1}{4} + \frac{1}{4}\, \frac{1 - (1+\frac{1}{2}\epsilon_k)^k}{1 + (1+\frac{1}{2}\epsilon_k)^k}$$

Hence $E^k(x,u_3) > \frac{7}{4}$ is equivalent to $\dfrac{1 - (1-\frac{1}{2}\epsilon_k - \frac{1}{2}\epsilon_k^2)^k}{((1+\frac{1}{2}\epsilon_k)^k - 1)\,(1 + (1-\epsilon_k)^k)} > \dfrac{1}{6}$

Using the rule of l'Hospital we get :

$$\lim_{\epsilon \to 0} \frac{1 - (1-\frac{\epsilon}{2} - \frac{\epsilon^2}{2})^k}{((1+\frac{\epsilon}{2})^k - 1)(1 + (1-\epsilon)^k)} = \frac{1}{2} \lim_{\epsilon \to 0} \frac{1 - (1-\frac{\epsilon}{2} - \frac{\epsilon^2}{2})^k}{(1+\frac{\epsilon}{2})^k - 1} =$$

$$= \frac{1}{2} \lim_{\epsilon \to 0} \frac{-k (1-\frac{\epsilon}{2} - \frac{\epsilon^2}{2})^{k-1} (-(\frac{1}{2} + \epsilon))}{\frac{1}{2} k (1+\frac{\epsilon}{2})^{k-1}} = \frac{1}{2}$$

Thus, for each $k \in \mathbb{N}$ there exists a point $x_k \in u_1 u_3$ at distance $\epsilon_k > 0$ to $u_3$ such that $E^k(x_k, u_3) > \frac{7}{4}$. Thus, for each $k \in \mathbb{N}$, the Condorcet point $u_3$ is not a k-optimal point.

## 3.8   Conclusions

Throughout this paper we have seen that several questions are still open. If $k = 1$ the maximal rejection of some point is always achieved by a vertex. However, this is not true for $k \geq 2$. Is there still a finite set of points which has to be examined in order to determine the maximal rejection? Is there always a k-suboptimal point, $k \geq 1$, which is a vertex? Is there an upper bound near $\frac{1}{2}$ for the maximal rejection rate of a suboptimal point? These and other research questions arise in case of two competing facilities.

A consideration of m competitive facilities will result in a high complicated model description. Therefore a useful generalization is to consider two groups X and Y of companies to be located, where X consists of facilities $x_1, ..., x_p$ and Y of $y_1, ..., y_q$. Assume that each customer chooses in each group the facility next to him. Hence in the considered model only the distance function changes to $d(u_i, X) := \min_{1 \leq j \leq p} \{d(u_i, x_j)\}$ and $d(u_i, Y) := \min_{1 \leq j \leq q} \{d(u_i, y_j)\}$. If there is a unique median and the customer tastes are sufficiently dispersed then the median is an optimal location as De Palma et al. (1989) showed.

Also the question arises as to: What is the smallest integer $q$ such that the location of $q$ new facilities $y_1, ..., y_q$ in competition to $x_1, ..., x_p$ is optimal? A completely different approach to the above problems will be achieved by an integer programming formulation as Dobson / Karmarkar (1987) did.

# EPILOGUE

We have considered several kinds of local search based machine learning with applications to automated manufacturing. This thesis is one step in order to pave the way for future research dedicated to machine learning in automated manufacturing. What are the characteristics that make difficult problems so difficult? Is it possible to generate rules the application of which would yield near optimal solutions for machine scheduling problems such as job shop scheduling or problems from flexible manufacturing? If there is a possibly infinite set of rules (probably encoded in strings over a finite alphabet) are we able to give these rules a reasonable interpretation which can lead to completely new models and more efficient heuristics in problem solving [213]? The answer of these questions could provide a much deeper insight into complex problems arising in production and operations management. However, up to now these problems are far from being understood and have first to be answered for the more "simpler" types of planning and scheduling problems which then can serve as building blocks for complex systems.

---

[213]  see Glover (1978), 1990b, 1993) and Hilliard et al. (1990), Park et al. (1992), Smith (1983), Holland (1980), Bottou / Vapnik (1992)

# REFERENCES

E.H.L. Aarts and J.H.M. Korst (1989a): Simulated Annealing and Boltzmann Machines. Wiley, Chichester.

E.H.L. Aarts and J.H.M. Korst (1989b): Boltzmann machines for travelling salesman problems. European Journal of Operational Research 39, 79–95.

E.H.L. Aarts, J.H.M. Korst, and P.J.M. van Laarhoven (1988): A quantitative analysis of the simulated annealing algorithm: a case study for the traveling salesman problem. Journal of Statistical Physics 50, 187–206.

E.H.L. Aarts and P.J.M. van Laarhoven (1985): A new polynomial–time cooling schedule. Proc. IEEE International Conf. Computer–Aided Design, Santa Clara, 206–208.

E.H.L. Aarts and P.J.M. van Laarhoven (1987): Simulated annealing: a pedestrain review of the theory and some applications. In: (P.A. Devijver and J. Kittler, eds.) Pattern Recognition and Applications, Springer, Berlin.

E.H.L. Aarts and P.J.M. van Laarhoven (1989): Simulated annealing: an introduction. Statistica Neerlandica 43, 31–52.

E.H.L. Aarts, P.J.M. van Laarhoven, and N.L.J. Ulder (1991): Local–search–based algorithms for job shop scheduling. Working paper, University of Eindhoven.

P. Ablay (1987): Optimieren mit Evolutionsstrategien. Spektrum der Wissenschaft 7, 162–173.

D.H. Ackley (1987): A Connectionist Machine for Genetic Hillclimbing. Kluwer, Dordrecht.

N. Adam and J. Surkis (1980): Priority update intervals and anomalies in dynamic ratio type job shop scheduling rules. Management Science 26, 1227–1237.

J. Adams, E. Balas, and D. Zawack (1988): The shifting bottleneck procedure for job shop scheduling. Management Science 34, 391–401.

R.H. Ahmadi and C.S. Tang (1991): An operation partitioning problem for automated assembly system design. Operations Research 39, 824–835.

A.V. Aho, J.E. Hopcroft, and J.D. Ullman (1974): The Design and Analysis of Computer Algorithms. Addison Wesley, Reading, Mass..

S.B. Akers (1956): A graphical approach to production scheduling problems. Operations Research 4, 244–245.

J.R.A. Allwright and D.B. Carpenter (1989): A distributed implementation of simulated annealing for the travelling salesman problem. Parallel Computing 10, 335–338.

S.G. de Amorim, J.-P. Barthélemy, and C.C. Ribeiro (1992): Clustering and clique partitioning: simulated annealing and tabu search approaches. Journal of Classification 9, 17–41.

M.R. Anderberg (1973): Cluster Analysis for Applications. Academic Press, New York.

A. Anily and A. Federgruen (1987a): Simulated annealing methods with general acceptance probabilities. Journal of Applied Probability, 657–667.

S. Anily and A. Federgruen (1987b): Ergodicity in parametric nonstationary Markov chains: an application to simulated annealing methods. Operations Research 35, 867–874.

S. Anily and A. Federgruen (1991): Structured partitioning problems. Operations Research

39, 130–149.

D. Applegate and W. Cook (1991): A computational study of the job–shop scheduling problem. ORSA Journal on Computing 3, 149–156.

P. Arabie and L.J. Hubert (1990): The bond energy algorithm revisited. IEEE Transactions on Systems, Man, and Cybernetics 20, 268–274.

G.C. Armour and E.S. Buffa (1963): A heuristic algorithm and simulation approach to the relative location of facilities. Management Science 9, 294–309.

V. Arumugam and S. Ramani (1980): Evaluation of value time sequencing rules in a real world job shop. Journal Operational Research Society 31, 895–904.

S. Ashour (1972): Sequencing Theory. Springer, Berlin.

S. Ashour and S.R. Hiremath (1973): A branch–and–bound approach to the job–shop scheduling problem. International Journal of Production Research 11, 47–58.

S. Ashour, T.E. Moore, and K.–Y. Chiu (1974): An implicit enumeration algorithm for the nonpreemptive shop scheduling problem. AIIE Transactions 6, 62–72.

R.G. Askin, S.H. Cresswell, J.B. Goldberg, and A.J. Vakharia (1991): A Hamiltonian path approach to reordering the part–machine matrix for cellular manufacturing. International Journal of Production Research 29, 1081–1100.

R.G. Askin and S.P. Subramanian (1987): A cost–based heuristic for group technology configuration. International Journal Production Research 25, 101–113.

J.G. Augustson and ‚J. Minker (1970): An analysis of some graph theoretical cluster techniques. Journal of the ACM 17, 571–588.

A. Bachem (1980): Komplexitätstheorie im Operations Research. Zeitschrift für Betriebswirtschaft 50, 812–844.

T. Bäck and F. Hoffmeister (1992): Extended selection mechanisms in genetic algorithms. Working paper, University of Dortmund.

T. Bäck, F. Hoffmeister, F. Kursawe, G. Rudolph, and H.–P. Schwefel (1990): Four Contributions to the Development of Evolution Strategies and Genetic Algorithms. Report 368, University of Dortmund.

T. Bäck, F. Hoffmeister, and H.–P. Schwefel (1991): A survey of evolution strategies. Proc. 4th. International Conf. on Genetic Algorithms (R.K. Belew and L.B. Booker, eds.), Morgan Kaufmann, 2–9.

S. Bagchi, S. Uckun, Y. Miyabe K. Kawamura (1991): Exploring problem–specific recombination operators for job shop scheduling. Proc. 4th. International Conf. on Genetic Algorithms (R.K. Belew and L.B. Booker, eds.), Morgan Kaufmann, 10–17.

J.E. Baker (1987): Reducing bias and inefficiency in the selection algorithm. Proc. 2nd International Conf. on Genetic Algorithms and Their Applications (J.J. Grefenstette, ed.). Lawrence Erlbaum Ass., 14–21.

K.R. Baker (1974): Introduction to Sequencing and Scheduling. Wiley, New York.

K.R. Baker (1975): A comparative study of flow shop algorithms. Operations Research 23, 62–73.

K.R. Baker (1984): Sequencing rules and due–date assignments in a job shop. Management Science 30, 1093–1104.

K.R. Baker, E.L. Lawler, J.K. Lenstra, and A.H.G. Rinnooy Kan (1983): Preemptive scheduling of a single machine to minimize maximum cost subject to release dates and precedence constraints. Operations Research 31, 381–386.

M.S. Bakshi and S.R. Arora (1969): The sequencing problem. Management Science 16,

B247–B263.

E. Balas (1969): Machine sequencing via disjunctive graphs: an implicit enumeration algorithm. Operations Research 17, 941–957.

E. Balas (1985): On the facial structure of scheduling polyhedra. Mathematical Programming Study 24, 179–218.

E. Balas and N. Christofides (1981): A restricted Lagrangian approach to the traveling salesman problem. Mathematical Programming 21, 19–46.

E. Balas, J.K. Lenstra, and A. Vazacopoulos (1992): One machine scheduling with delayed precedence constraints. Management Science Research Report MSRR–589, Carnegie Mellon University, Pittsburgh, PA.

K.N. Balasubramanian and R. Panneerselvam (1993): Covering technique–based algorithm for machine grouping to form manufacturing cells. International Journal of Production Research 31, 1479–1504.

H.–J. Bandelt (1985): Networks with Condorcet solutions. European Journal of Operational Research 20, 314–326.

H.–J. Bandelt (1990a): Single facility location on median networks. Mathematics of Operations Research, to appear.

H.–J. Bandelt (1990b): Centroids and medians of finite metric spaces. Working Paper, University of Hamburg.

H.–J. Bandelt and M. Labbé (1986): How bad can a voting location be? Social Choice and Welfare 3, 125–145.

F. Barahona, M. Grötschel, M. Jünger, and G. Reinelt (1988): An application of combinatorial optimization to statistical physics and circuit layout design. Operations Research 36, 493–513.

J.R. Barker and G.B. McMahon (1985): Scheduling the general job–shop. Management Science 31, 594–598.

E.R. Barnes (1982): An algorithm for partitioning the nodes of a graph. SIAM Journal on Algebraic and Discrete Methods 3, 541–550.

E.R. Barnes, A. Vanelli, and J. Walker (1988): A new heuristic for partitioning the nodes of a graph. SIAM Journal on Discrete Mathematics 1, 299–305.

J.W. Barnes, M. Laguna, and F. Glover (1992): An overview of tabu search approaches to production scheduling problems. Proc. Symposium Intelligent Scheduling Systems, San Francisco, 30–50.

J.P. Barthélemy and B. Monjardet (1981): The median procedure in cluster analysis and social choice theory. Mathematical Social Sciences 1, 235–267.

J.J. Bartholdi and L.K. Platzman (1988): Heuristics based on spacefilling curves for combinatorial problems in Euclidean space. Management Science 34, 291–305.

D.L. Battle and M. Vose (1993): Isomorphisms of genetic algorithms. Artificial Intelligence 60, 155–165.

A. Bauer, W. Domschke, and E. Pesch (1993): Competitive location on a network. European Journal of Operational Research 66, 372–391.

N.S. Bazaraa and C.N. Shetty (1979): Nonlinear Programming. Wiley, New York.

J.C. Bean (1992): Genetics and random keys for sequencing and optimization. Working paper 92–43, University of Michigan, Ann Arbor.

J.C. Bean and A.B. Hadj–Alouane (1992): A dual genetic algorithm for bounded integer programs. Working paper 92–53, University of Michigan, Ann Arbor.

P. van Beek (1992): Reasoning about qualitative temporal information. Artificial Intelligence 58, 297–326.

C.J.P. Bélisle, H.E. Romeijn, and R.L. Smith (1990): Hide–and–seek: a simulated annealing algorithm for global optimization. Working paper, Erasmus University, Rotterdam.

M. Bellmore and J.C. Mallone (1971): Pathology of traveling–salesman subtour–elimination algorithms. Operations Research 19, 278–307.

M. Bellmore and G.L. Nemhauser (1968): The traveling salesman problem: a survey. Operations Research 16, 538–558.

J.F. Benders (1962): Partitioning procedures for solving mixed variables programming problems. Numerische Mathematik 4, 238–252.

E. Bensana, G. Bel, and D. Dubois (1988): OPAL: a multi–knowledge–based system for industrial job–shop scheduling. International Journal of Production Research 28, 795–819.

J.J. Bentley (1992): Fast algorithms for geometric traveling salesman problems. ORSA Journal on Computing 4, 387–411.

C. Berge (1973): Graphs and Hypergraphs. North Holland, Amsterdam.

E.K. Berkley and A.S. Kiran (1991): A simulation study of sequencing rules in a Kanban–controlled flow–shop. Decision Sciences 22, 559–582.

P. Bertier and B. Roy (1965): Trois exemples numeriques d'application de la procedure SEP. Note de travail No. 32 de la Direction Scientifique de la SEMA.

A. Bertoni and M. Dorigo (1993): Implicit parallelism in genetic algorithms. Working paper, No. 93–001, University Berkeley, CA

A.D. Bethke (1980): Genetic Algorithms as Function Optimizers. Dissertation, University of Michigan, Ann Arbor.

H.–G. Beyer (1992): Some aspects of the 'evolution strategy' for solving TSP–like optimization problems appearing at the design studies of a 0.5TeV $e^+e^-$–linear collider. In: (R. Männer and B. Manderick, eds.) Parallel Problem Solving from Nature 2, Elsevier Publishers, 361–369.

J.N. Bhuyan, V.V. Raghavan, and V.K. Elayavalli (1991): Genetic algorithms for clustering with an ordered representation. Proc. 4th International Conf. on Genetic Algorithms (R.K. Belew and L.B. Booker, eds.), 408–415.

W. Bibel (1987): Automated Theorem Proving. Vieweg, Braunschweig, 2nd ed.

J.R. Bitner and E.M. Reingold (1975): Backtrack programming techniques. Communications ACM 18, 651–656.

J.H. Blackstone, D.T Phillips, and G.L. Hogg (1982): A state of the art survey of dispatching rules for manufacturing job shop operations. International Journal Production Research 20, 27–45.

R.G. Bland and D.F. Shallcross (1989): Large traveling salesman problems arising from experiments in X–ray crystallography: a preliminary report on computation. Operations Research Letters 8, 125–128.

J. Blazewicz (1987): Selected topics in scheduling theory. Annals of Discrete Mathematics 31, 1–60.

J. Blazewicz, R.E. Burkard, G. Finke. and G.L. Woeginger (1992): Vehicle scheduling in two–cycle flexible manufacturing systems. Working paper, University of Grenoble.

J. Blazewicz, W. Domschke, and E. Pesch (1993): The job shop scheduling problem: conventional and new solution techniques. Working paper, University of Limburg,

Maastricht.

J. Blazewicz, M. Dror, and J. Weglarz (1991): Mathematical programming formulations for machine scheduling: a survey. European Journal of Operational Research 51, 283–300.

J. Blazewicz, K. Ecker, G. Schmidt, and J. Weglarz (1993): Scheduling in Computer and Manufacturing Systems. Springer, Heidelberg.

L. Bodin and L. Levy (1991): The arc partitioning problem. European Journal of Operational Research 53, 393–401.

W.J. Boe and C.H. Cheng (1991): A close neighbour algorithm for designing cellular manufacturing systems. International Journal of Production Research 29, 2097–2116.

E. Bonomi and J.–I. Lutton (1984): The n–city traveling salesman problem: statistical mechanics and the Metropolis algorithm. SIAM Review 26, 551–568.

L.B. Booker, D.E. Goldberg, and J.H. Holland (1989): Classifier systems and genetic algorithms. Artificial Intelligence 40, 235–282.

A.H. Borning (1981): The programming language aspects of ThingLab. A constraint–oriented simulation laboratory. ACM TOPLAS 3, 353–397.

L. Bottou and V. Vapnik (1992): Local learning algorithms. Neural Computation 4, 888–900.

D.G. Bounds (1987): New optimization methods from physics and biology. Nature 329, 215–219.

E.K. Bowden (1969): Priority assignment in a network of computers. IEEE Transactions on Computers C–18, 1021–1026.

E.H. Bowman (1959): The scheduling sequencing problem. Operations Research 7, 621.

S.C. Boyd and W.H. Cunningham (1991): Small traveling salesman polytopes. Mathematics of Operations Research 16, 259–271.

S.C. Boyd and W.R. Pulleyblank (1990): Optimizing over the subtour polytope of the travelling salesman problem. Mathematical Programming, Ser. A 49, 163–187.

R.M. Brady (1985): Optimization strategies gleaned from biological evolution. Nature 317, 804–806.

S. Brah, J. Hunsucker, and J. Shah (1991): Mathematical modeling of scheduling problems. Journal of Information & Optimization Sciences 12, 113–137.

H. Bräsel (1990): Lateinische Rechtecke und Maschinenbelegung. Dissertation B, Technical University of Magdeburg.

H. Bräsel and F. Werner (1989): The job–shop problem – modelling by latin rectangles exact and heuristic solution. Working paper Math 8/89, Technical University of Magdeburg.

R. Brause (1991): Neuronale Netze. Teubner, Stuttgart.

A. Brindle (1981): Genetic Algorithms for Function Optimization. Dissertation, University of Alberta, Edmonton.

G.H. Brooks and C.R. White (1965): An algorithm for finding optimal or near–optimal solutions to the production scheduling problem. Journal of Industrial Engineering 16, 34–40.

P. Brucker (1981): Scheduling. Akademische Verlagsgesellschaft, Wiesbaden.

P. Brucker (1988): An efficient algorithm for the job–shop problem with two jobs. Computing 40, 353–359.

P. Brucker, J. Hurink, and F. Werner (1993): Improving neighbourhoods for local search heuristics. Working paper, University of Osnabrück.

P. Brucker and B. Jurisch (1993): A new lower bound for the job–shop scheduling problem.

European Journal of Operational Research 64, 156–167.

P. Brucker, B. Jurisch, and B. Sievers (1991): A branch and bound algorithm for the job–shop scheduling problem. Discrete Applied Mathematics, to appear.

P. Brucker, B. Jurisch, and A. Krämer (1992): The job–shop problem and immediate selection. Working paper, University of Osnabrück.

P. Brucker, B. Jurisch, and B. Sievers (1992): Job–shop (C codes). European Journal of Operational Research 57, 132–133.

B. Bruno, A. Elia, and P. Laface (1986): A rule–based system to schedule production. IEEE Computer 7, 32–40.

R. Bruns and H.–J. Appelrath (1991): Ein universelles Modell für Ablaufplanungsprobleme. Wirtschaftsinformatik 33, 516–525.

M. Bruynooghe (1981): Solving combinatorial search problems by intelligent backtracking. Information Processing Letters 12, 36–39.

M. Bruynooghe and L.M. Pereira (1984): Deduction revision by intelligent backtracking. In: (J.A. Campbell, ed.) Implementations of PROLOG, Ellis Horwood, Chichester, 194–215.

T. Bui, S. Chaudhuri, F.T. Leighton, and M. Sipser (1987): Graph bisection algorithms with good average case behaviour. Combinatorica 7, 171–191.

E.S. Buffa, G.C. Armour, and T.E. Vollman (1964): Allocating facilities with CRAFT. Harvard Business Review 42, 136–159.

Bull (1990a): Artificial Intelligence, Charme V1, Users Guide. Bull S.A., Cedoc–Dilog.

Bull (1990b): Artificial Intelligence, Charme V1, Reference Manual. Bull S.A., Cedoc–Dilog.

J.L. Burbidge (1963): Production flow analysis. The Production Engineer 42, 742–752.

J.L. Burbidge (1975): The Introduction of Group Technology. Wiley, New York.

R.E. Burkard (1979): Traveling salesman and assignment problems: a survey. Annals of Discrete Mathematics 4, 193–215.

R. Burkard and F. Rendl (1984): A thermodynamically motivated simulation procedure for combinatorial optimization problems. European Journal of Operational Research 17, 169–174.

L.I. Burke and P. Damany (1992): The guilty net for the traveling salesman problem. Computers & Operations Research 19, 255–265.

F. Burns and J. Rooker (1978): Three–stage flow–shops with recessive second stage. Operations Research 26, 207–208.

J.A. Buzacott and D.D. Yao (1986): Flexible manufacturing systems: a review of analytical models. Management Science 32, 890–905.

H.G. Campbell, R.A. Dudek, and M.L. Smith (1970): A heuristic algorithm for the n job m machine sequencing problem. Management Science 16, 630–637.

J. Carlier (1982): The one machine sequencing problem. European Journal of Operational Research 11, 42–47.

J. Carlier (1987): Scheduling jobs with release dates and tails on identical machines to minimize the makespan. European Journal of Operational Research 29, 298–306.

J. Carlier and E. Pinson (1989): An algorithm for solving the job–shop problem. Management Science 35, 164–176.

J. Carlier and E. Pinson (1990): A practical use of Jackson's preemtive schedule for solving the job shop problem. Annals of Operations Research 26, 269–287.

G. Carpaneto, M. Fischetti, and P. Toth (1989): New lower bounds for the symmetric travelling salesman problem. Mathematical Programming 45, 233–254.

G. Carpaneto and P. Toth (1980): Some new branching and bounding criteria for the asymmetric travelling salesman problem. Management Science 26, 736–743.

A.S. Carrie and J.M. Moore, M. Rocsniak, and J.J. Seppänen (1978): Graph theory and computer aided facility design. OMEGA 6, 353–361.

F. Catthoor, H. de Man, and J. Vandrewalle (1988): SAMURAI: a general and efficient simulated–annealing schedule with fully adaptive annealing parameters. Integration, the VLSI Journal 6, 147–178.

V. Cerny (1985): Thermodynamical approach to the traveling salesman problem; an efficient simulation algorithm. Journal Optimization Theory and Applications 45, 41–51.

R.D. Chamberlain, M. Edelman, M. Franklin, and E. Witte (1988): Simulated annealing on a multiprocessor. Proc. IEEE International Conf. on Computer Design, Rye Town Hill NY, 540–544.

D. Chan and D. Mercier (1989): IC insertion: an application of the travelling salesman problem. International Journal of Production Research 27, 1837–1841.

H.M. Chan and D.A. Milner (1982): Direct clustering algorithm for group formation in cellular manufacture. Journal of Manufacturing Systems 1, 65–74.

M.P. Chandrasekharan and R. Rajagopalan (1986): An ideal seed non–hierachical clustering algorithm for cellular manufacturing. International Journal of Production Research 28, 451–464.

E. Charniak and D. McDermott (1985): Introduction to Artificial Intelligence. Addison–Wesley, Reading.

J.M. Charlton and C.C. Death (1970): A generalized machine scheduling algoritnm. Operations Research Quarterly 21, 127–134.

K.M. Cheh, J.B. Goldberg, and R.G. Askin (1991): A note on the effect of neighborhood structure in simulated annealing. Computers & Operations Research 18, 537–547.

Y. Chen (1991): Improving Han and Lee's path–consistency algorithm. Proc. of the 3rd. International Conf. on Tools for AI, 346–350.

T.–S. Chiang and Y. Chow (1988): On the convergence rate of annealing processes. SIAM Journal Control and Optimization 26, 1455–1470.

N. Chiba and T. Nishizeki (1989): The Hamiltonian cycle problem is linear–time solvable for 4–connected planar graphs. Journal of Algorithms 10, 187–211.

Y. Cho and S. Sahni (1981): Preemptive scheduling of independent jobs with release and due times on open, flow and job shops. Operations Research 29, 511–522.

S.C. Choi, W.S. DeSarbo, and P.T. Harker (1990): Product positioning under price competition. Management Science 36, 175–199.

S. Chopra and M.R. Rao (1991): On the multiway cut polyhedron. Networks 21, 51–89.

S. Chopra and M.R. Rao (1993): The partition problem. Mathematical Programming 59, 87–116.

T. Christof, M. Jünger, and G. Reinelt (1991): A complete description of the traveling salesman polytope on 8 nodes. Operations Research Letters 10, 497–500.

N. Christofides (1970): The shortest Hamiltonian chain of a graph. SIAM Journal on Applied Mathematics 19, 689–696

N. Christofides (1976): Worst–case analysis of a new heuristic for the travelling salesman problem. Report 388, Graduate School of Industrial Administration, Carnegie–Mellon

University, Pittsburgh, PA.

N. Christofides (1979): The travelling salesman problem. In: (N. Christofides, A. Mingozzi, P. Toth, and C. Sandi, eds.) Combinatorial Optimisation, Wiley, New York, 131–149.

N. Christofides and P. Brooker (1976): The optimal partitioning of graphs. SIAM Journal on Applied Mathematics 30, 55–69.

N. Christofides and S. Eilon (1972): Algorithms for large–scale travelling salesman problems. Operational Research Quarterly 23, 511–518.

N. Christofides and M. Hassan (1991): Lagrangean relaxation for the maximal planar graph. Working paper, University of London.

C.H. Chu (1989): Clustering analysis in manufacturing cellular formation. Omega 17, 289–295.

C. Chu, M.C. Portmann, and J.M. Proth (1992): A splitting–up approach to simplify job–shop scheduling problems. International Journal of Production Research 30, 859–870.

V. Chvàtal (1973): Edmonds polytopes and weakly hamiltonian graphs. Mathematical Programming 5, 29–40.

W. Clark (1922): The Gantt Chart: A Working Tool of Management. The Ronald Press, 3rd. ed., Pittman, New York.

A. Claus (1984): A new formulation for the traveling salesman problem. SIAM Journal of Algebraic Discrete Methods 5, 21–25.

W.F. Clocksin and C.S. Mellish (1984): Programming in Prolog. Springer, Berlin.

E.G. Coffman, ed. (1976): Computer and Job–Shop Scheduling Theory. Wiley, New York.

J. Cohen (1990): Constraint logic programming languages. Communications ACM 33, 52–68.

J.P. Cohoon, W.N. Martin, and D.S. Richards (1991): A multi–population genetic algorithm for solving the k–partition problem on hyper–cubes. Proc. 4th. International Conf. on Genetic Algorithms (R.K. Belew and L.B. Booker, eds.), Morgan Kaufmann, 244–248.

N.E. Collins, R.W. Eglese, and B.L. Golden (1988): Simulated annealing: an annotated bibliography. American Journal of Mathematical and Management Sciences 8, 205–307.

A. Colmerauer (1987): Opening the PROLOG III universe. Byte 12, 177–182.

A. Colmerauer (1990): An introduction to PROLOG III. Communications ACM, 69–90.

A. Colorni, M. Dorigo, and V. Maniezzo (1992): An investigation of some properties of an "ant algorithm". In: (R. Männer and B. Manderick, eds.) Parallel Problem Solving from Nature 2, Elsevier Publishers, 509–520.

R.N. Conway, W.L. Maxwell, and L.W. Miller (1967): Theory of Scheduling. Addison–Wesley, Reading.

M.C. Cooper (1989/90): An optimal k–consistency algorithm. Artificial Intelligence 41, 89–95.

P.R. Cooper and M.J. Swain (1992): Arc consistency: parallelism and domain dependence. Artificial Intelligence 58, 207–235.

D.G. Corneil and M.E. Woodward (1978): A comparison and evaluation of graph theoretical clustering techniques. INFOR 16, 74–89.

G. Cornuéjols, J. Fonlupt, and D. Naddef (1985): The traveling salesman problem on a graph and some related integer polyhedra. Mathematical Programming 33, 1–27.

G. Cornuéjols, D. Naddef, and W. Pulleyblank (1985): The traveling salesman problem on graphs with 3–edge cutsets. Journal ACM 32, 383–410.

P.T. Cox (1984): Finding backtrack points for intelligent backtracking. In: (J. A. Campbell, ed.) Impementation of Prolog. Ellis Horwood, Chichester, 216–233.

Y. Crama, A.W.J. Kolen, A.G. Oerlemans, and F.C.R. Spieksma (1991): Minimizing the number of tool switches on a flexible machine. Working paper, University of Limburg, Maastricht.

Y. Crama, A. Kolen, and E. Pesch (1993): Local search in combinatorial optimization. In: (P. Braspenning, F. Thuijsman, A.J.M.M. Weijters, eds.) Artificial Neural Networks, University of Maastricht.

Y. Crama and A.G. Oerlemans (1991a): The job grouping problem for flexible manufacturing systems. Working paper, University of Limburg, Maastricht.

Y. Crama and A.G. Oerlemans (1991b): A column generation approach to job grouping for flexible manufacturing systems. Working paper, University of Limburg, Maastricht.

Y. Crama and A.G. Oerlemans (1992): A local search approach to job grouping. Working paper, University of Limburg, Maastricht.

Y. Crama and M. Oosten (1992): Models for machine–part grouping in cellular manufacturing. Working paper, University of Limburg, Maastricht.

G.A. Croes (1958): A method for solving traveling salesman problems. Operations Research 6, 791–812.

H. Crowder and M. Padberg (1980): Solving large–scale symmetric traveling salesman problems to optimality. Management Science 26, 495–509.

W.B. Crowston, F. Glover, G.L. Thompson, and J.D. Trawick (1963): Probabilistic and parametric learning combinations of local job shop scheduling rules. ONR Research Memorandum No. 117, GSIA, Carnegie–Mellon University, Pittsburg, PA.

A. Dählmann (1989): Pluralitätslösungen auf Graphen und injektiven metrischen Räumen. Dissertation, University of Oldenburg.

F. Dammeyer, P. Forst, and S. Voß (1991): On the cancellation sequence method of tabu search. ORSA Journal on Computing 3, 262–265.

F. Dammeyer and S. Voß (1993): Dynamic tabu list management using the reverse elimination method. Annals of Operations Research 41, 31–46.

D.G. Dannenbring (1977): An evaluation of flow shop scheduling heuristics. Management Science 23, 1174–1182.

G. Dantzig, R. Fulkerson, and S. Johnson (1954): Solution of a large–scale traveling salesman problem. Operations Research 2, 393–410.

G. Dantzig, R. Fulkerson, and S. Johnson (1959): On a linear programming combinatorial approach to the traveling salesman problem. Operations Research 7, 58–66.

S.K. Das (1993): A facility layout method for flexible manufacturing systems. International Journal of Production Research 31, 279–297.

S. Dausere–Peres and J.–B. Lasserre (1993): A modified shifting bottleneck procedure for job–shop scheduling. International Journal of Production Research 31, 923–932.

E. Davis (1987): Constraint propagation with interval labels. Artificial Intelligence 32, 281–331.

L. Davis (1985): Job shop scheduling with genetic algorithms. Proc. an International Conf. on Genetic Algorithms and Their Applications (J.J. Grefenstette, ed.). Lawrence Erlbaum Ass., 136–140.

L. Davis, ed. (1987): Genetic Algorithms and Simulated Annealing. Morgan Kaufmann, Los Altos, CA.

L. Davis, ed. (1991): Handbook of Genetic Algorithms. Van Nostrand Reinhold, New York.

T.E. Davis and J.C. Principe (1991): A simulated annealing like convergence theory for the simple genetic algorithm. Proc. 4th. International Conf. on Genetic Algorithms (R.K. Belew and L.B. Booker, eds.), Morgan Kaufmann, 174–181.

J.E. Day and M.P. Hottenstein (1970): Review of sequencing research. Naval Research Logistics Quarterly 17, 11–39.

W.H.E. Day (1977): Validity of clusters formed by graph theoretic cluster methods. Mathematical Biosciences 36, 199–317.

W.H.E. Day (1981): The complexity of computing metric distances between partitions. Mathematical Social Sciences 1, 269–287.

R. Dechter (1986): Learning while searching in constraint–satisfaction problems, Proc. AAAI–86, 178–183.

R. Dechter (1989/90): Enhancement schemes for constraint processing : backjumping, learning, and cutset decomposition. Artificial Intelligence 41, 273–312.

R. Dechter (1992): From local to global consistency. Artificial Intelligence 55, 87–107.

R. Dechter, A. Dechter, and J. Pearl (1990): Optimization in constraint networks. In: (R.M. Oliver and J.Q. Smith, eds.) Influence Diagrams, Belief Nets and Decision Analysis. Wiley, New York, 411–425.

R. Dechter and J. Pearl (1988): Network–based heuristics for constraint satisfaction problems. Artificial Intelligence 34, 1–38.

R. Dechter and J. Pearl (1989): Tree clustering for constraint networks. Artificial Intelligence, 353–366.

K.A. De Jong (1975): Analysis of the Behavior of a Class of Genetic Adaptive Systems. Dissertation, University of Michigan, Ann Arbor.

K.A. De Jong (1980): Adaptive system design: a genetic approach. IEEE Transactions on System, Man, and Cybernetics 10, 566–574.

K.A. De Jong (1990): Genetic–algorithm–based learning. In: (Y. Kodratoff and R.S. Michalski) Machine Learning, An Artificial Intelligence Approach, Vol. III, 'Morgan Kaufmann, San Mateo, 611–638.

K.A. De Jong and W.M. Spears (1989): Using genetic algorithms to solve NP–complete problems. Proc. 3rd International Conf. on Genetic Algorithms (J.D. Schaffer, ed.). Morgan Kaufmann Publ., 124–132.

K.A. De Jong and W.M Spears (1992): A formal analysis of the role of multi–point crossover in genetic algorithms. Annals of Mathematics and Artificial Intelligence 5, 1–26.

A. Dekker and E. Aarts (1991): Global optimisation and simulated annealing. Mathematical Programming 50, 367–394.

M. Dell'Amico (1993): Shop problems with two machines and time lags. Working paper, Politecnico di Milano.

M. Dell'Amico and M. Trubian (1993): Applying tabu–search to the job shop scheduling problem. Annals of Operations Research 41, 231–252.

M. Desrochers and G. Laporte (1991): Improvements and extensions to the Miller–Tucker–Zemlin subtour eliminiation constraints. Operations Research Letters 10, 27–36.

M. Desa, M. Grötschel, and M. Laurent (1992): Clique–web facets for multicut polytopes. Mathematics of Operations Research 17, 981–1000.

E.W. Dijkstra (1959): A note on two problems in connection with graphs. Numer. Math. 1,

269–271.

M. Dincbas, P. van Hentenryck, H. Simonis, A. Aggoun, T. Graf and F. Berthier (1988): The constraint logic programming language CHIP. Proc. International Conf. on Fifth Generation Computer Systems, Tokyo, Japan.

M. Dincbas, H. Simonis und P. van Hentenryck (1990): Solving large combinatorial problems in logic programming. Journal Logic Programming 8, 75–93.

G. Dobson and U. K. Karmarkar (1987): Competitive location on a network, Operations Research 35, 565–574.

W. Domschke (1989): Logistik: Rundreisen und Touren. Oldenbourg, München, 3rd ed.

W. Domschke and A. Drexl (1985): An international bibliography on location– and layout–planning, Springer, Berlin.

W. Domschke and A. Drexl (1990): Logistik: Standorte. Oldenbourg, München, 3rd ed.

W. Domschke and A. Drexl (1991): Einführung in Operations Research. Springer, Berlin, 2nd ed.

W. Domschke, P. Forst, and S. Voß (1992): Tabu search techniques for the quadratic semi–assignment problem. In: (G. Fandel, T. Gulledge, and A. Jones, eds.) New Directions for Operations Research in Manufacturing, Springer, Berlin, 389–405.

W. Domschke, A. Scholl, and S. Voß (1993): Produktionsplanung. Springer, Berlin.

W.E. Donath, and A.J. Hoffman (1973): Lower bounds for the partitioning of graphs. IBM Journal of Research and Development 17, 420–425.

U. Dorndorf and E. Pesch (1992a): Evolution based learning in a job shop scheduling environment. Computers & Operations Research, to appear.

U. Dorndorf and E. Pesch (1992b): Fast clustering algorithm. ORSA Journal on Computing, to appear.

U. Dorndorf and E. Pesch (1993a): Combining genetic and local search for solving the job shop scheduling problem. Symposium on Applied Mathematical Programming and Modeling APMOD93 (I. Maros, ed.), Akaprint, Budapest, 142–149.

U. Dorndorf and E. Pesch (1993b): Variable depth search based learning by reasoning from embedded schedule neighbourhoods. Working paper, University of Limburg, Maastricht.

U. Dorndorf and E. Pesch (1993c): Genetic algorithms for job shop scheduling. In: (K.–W. Hansmann et al., eds.) Operations Research Proc. 1992, Springer, Berlin, 243–250.

J. Doyle (1979): A truth maintenance system. Artificial Intelligence 12, 231–272.

A. Drexl (1988): A simulated annealing approach to the multiconstraint zero–one knapsack problem. Computing 40, 1–8.

A. Drexl and A. Sprecher (1993): Resource– and time window–constraint production scheduling with alternative process plans: an artificial intelligence approach. In: (G. Fandel et al., eds.) Operations Research in Production Planning and Control, Springer, Berlin, 307–320.

R. van Driessche, and R. Piessens (1992): Load balancing with genetic algorithms. In: (R. Männer and B. Manderick, eds.) Parallel Problem Solving from Nature 2, Elsevier Publishers, 341–350.

R. Dubes and A.K. Jain (1980): Clustering methodologies in exploratory data analysis. Advances in Computers 19, 113–228.

R.A. Dudek, S.S. Panwalkar, and M.L. Smith (1992): The lessons of flowshop scheduling research. Operations Research 40, 7–13.

G. Dueck and T. Scheurer (1990): Threshold accepting: a general purpose optimization algorithm. Journal of Computational Physics 90, 161–175.

A.E. Dunlop and B.W. Kernighan (1985): A procedure for placement of standard–cell VLSI circuits. IEEE Transactions on Computer Aided Design 4, 92–98.

R. Durbin, R. Szeliski, and A. Yuille (1989): An analysis of the elastic net approach to the traveling salesman problem. Neural Computation 1, 348–358.

R. Durbin and D. Willshaw (1987): An analogue approach to the travelling salesman problem using an elastic net method. Nature 326, 689–691.

M.E. Dyer and A.M. Frieze (1985): On the complexity of partitioning graphs into connected subgraphs. Discrete Applied Mathematics 10, 139–153.

K. Ecker (1977): Organisation von parallelen Prozessen. BI–Wissenschaftsverlag, Mannheim.

J. Edmonds (1965): Maximum matching and a polyhedron with 0, 1–vertices. Journal of Research of the National Bureau of Standards, Section B 69, 125–130.

R.W. Eglese (1990): Simulated annealing: a tool for operational research. European Journal of Operational Research 46, 271–281.

A.E. Eiben, E.H.L. Aarts, and K.H. van Hee (1991): Global convergence of genetic algorithms: a Markov Chain analysis. Lecture Notes in Computer Science 496, 4–9.

H.A. Eiselt and G. Laporte (1989): Competitive spatial models. European Journal of Operational Research 39, 231–242.

H.A. Eiselt and G. Laporte (1991): Locational equilibrium of two facilites on a tree. RAIRO 25, 5–18.

H.A. Eiselt and G. Laporte (1993): The existence of equilibria in the 3–facility Hotelling model in a tree. Transportation Science 27, 39–43.

H.A. Eiselt, G. Laporte, and J.–F. Thisse (1993): Competitive location models: a framework and bibliography. Transportation Science 27, 44–54.

S.E. Elmaghraby (1968): The machine sequencing problem – review and extensions. Naval Research Logistics Quarterly 15, 205–232.

S.E. Elmaghraby and A.N. Elshafei (1976): Branch–and–bound revised: a survey of basic concepts and their applications in scheduling. In: (W.H. Marlow, ed.) Modern Trends in Logistics Research, M.I.T. Press, Cambridge, Mass.

T.E. El–Rayah and R.H. Hollier (1970): A review of plant design techniques. International Journal of Production Research 8, 263–278.

D.A. Elvers and L.R. Taube (1983): Deterministic / stochastic assumptions in job shops. European Journal of Operational Research 14, 89–94.

H.B. Eom and S.M. Lee (1990): Decision support systems applications research: a bibliography (1971–1988). European Journal of Operational Research 46, 333–342.

J.R. Evans (1987): Structural analysis of local search heuristics in combinatorial optimization. Computers & Operations Research 14, 465–477.

U. Faigle and W. Kern (1991): Note on the convergence of simulated annealing algorithms. SIAM Journal of Control and Optimization 29, 153–159.

U. Faigle and W. Kern (1992): Some convergence results for probabilistic tabu search. ORSA Journal on Computing 4, 32–37.

U. Faigle and R. Schrader (1988): On the convergence of stationary distributions in simulated annealing algorithms. Information Processing Letters 27, 189–194.

U. Faigle, R. Schrader, and R. Suletzki (1987): A cutting–plane algorithm for optimal graph partitioning. Methods of OR 57, 109–116.

T.A. Feo, O. Goldschmidt, and M. Khellaf (1992): One–half approximation algorithms for the k–partition problem. Operations Research 40, S170–S173.

T.A. Feo and M. Khellaf (1990): A class of bounded approximation algorithms for graph partitioning. Networks 20, 181–195.

J.F.F. Ribeiro and B. Pradin (1993): A methodology for cellular manufacturing design. International Journal of Production Research 31, 235–250.

C.M. Fiduccia and R.M. Mattheyses (1982): A linear–time heuristic for improving network partitions. Proc. of the 19th Design Automation Conf., Las Vegas N.M., 175–181.

P.N. Finlay (1990): Decision support systems and expert systems: a comparison of their components and design methodologies. Computers & Operations Research 17, 535–543.

M.L. Fisher (1973): Optimal solution of scheduling problems using Lagrange multipliers: part I. Operations Research 21, 1114–1127.

M.L. Fisher, B.J. Lageweg, J.K. Lenstra, and A.H.G. Rinnooy Kan (1983): Surrogate duality relaxation for job shop scheduling. Discrete Applied Mathematics 5, 65–75.

H. Fisher and G.L. Thompson (1963): Probabilistic learning combinations of local job–shop scheduling rules. In: (J.F. Muth and G.L. Thompson, eds.) Industrial Scheduling, Prentice Hall, Englewood Cliffs.

B. Fleischmann (1985): A cutting plane procedure for the travelling salesman problem on road networks. European Journal of Operational Research 21, 307–317.

B. Fleischmann (1988): A new class of cutting planes for the symmetric traveling salesman problem. Mathematical Programming 40, 225–246.

M. Florian, P. Trépant, and G. McMahon (1971): An implicit enumeration algorithm for the machine sequencing problem. Management Science 17, B782–B792.

J.–P. Follonier (1992): On grouping parts and tools and balancing the workload in a flexible manufacturing system. Working paper, University of Lausanne.

J. Fonlupt and D. Naddef (1992): The traveling salesman problem in graphs with some excluded minors. Mathematical Programming 53, 147–172.

L.R. Ford and D.R. Fulkerson (1974): Flows in Networks, 6th ed. Princeton University Press, Princeton.

L.R. Foulds (1983): Techniques for facility layout: deciding which pairs of facilities should be adjacent. Management Science 29, 1414–1426.

L.R. Foulds, P.B. Gibbons, and J.W. Giffin (1985): Facilities layout adjacency determination: an experimental comparison of three graph theoretic heuristics. Operations Research 33, 1091–1106.

L.R. Foulds and D.F. Robinson (1976): A strategy for solving the plant layout problem. Operations Research Quarterly 27, 845–855.

L.R. Foulds and D.F. Robinson (1978): Graph theoretic heuristics for the plant layout problem. International Journal of Production Research 16, 27–37.

L.R. Foulds and D.F. Robinson (1979): Construction properties of combinatorial celtahedra. Discrete Applied Mathematics 1, 75–87.

B.L. Fox (1993a): Integrating and accelerating tabu search, simulated annealing, and genetic algorithms. Annals of Operations Research 41, 47–67.

B.L. Fox (1993b): Faster simulated annealing. Working paper, University of Colorado, Denver.

B.L. Fox (1993c): Simulated annealing with overrides. Working paper, University of Colorado, Denver.

K. Fox, B. Gavish, and S. Graves (1980): An n–constraint formulation of the (time–dependent) traveling salesman problem. Operations Research 28, 1018–1021.

M.S. Fox (1987): Constraint–Directed Search: A Case Study of Job Shop Scheduling. Pitman, London.

M.S. Fox and S.F. Smith (1984): ISIS – a knowledge based system for factory scheduling. Expert Systems 1, 25–49.

J.A. Freeman and D.M. Skapura (1992): Neural Networks Algorithms, Applications, and Programming Techniques. Addison–Wesley, Reading.

B. Freisleben and M. Härtfelder (1993): Optimization of genetic algorithms by genetic algorithms. Proc. of the '93 International Conf. on Neural Nets and Genetic Algorithms, Springer, 243–248.

S. French (1982): Sequencing and Scheduling: An Introduction to the Mathematics of the Job–Shop. Wiley, New York.

E.C. Freuder (1982): A sufficient condition of backtrack–free search. Journal of the ACM 29, 24–32.

E.C. Freuder (1985): A sufficient condition for backtrack–bounded search. Journal of the ACM 32, 755–761.

E.C. Freuder und R.J. Wallace (1992): Partial constraint satisfaction. Artificial Intelligence 58, 21–70.

H. Friedrich, J. Keßler, E. Pesch, and B. Schildt (1991): Batch scheduling on parallel units in acrylic–glass production. ZOR–Zeitschrift Operations Research 35, 321–345.

H.L. Gantt (1919): Efficiency and democracy. Trans. Amer. Soc. Mech. Engin. 40, 799–808.

M.R. Garey und D.S. Johnson (1979): Computers and Intractability. A Guide to the Theory of NP–Completeness. Freeman and Company, San Francisco.

M.R. Garey, D.S. Johnson, R. Sethi (1976): The complexity of flowshop and jobshop scheduling. Mathematics of Operations Research 1, 117–129.

M.R. Garey, D.S. Johnson, and L. Stockmeyer (1976): Some simplified NP–complete problems. Theoretical Computer Science 1, 237–267.

R.S. Garfinkel (1973): On partitioning the feasible set in a branch–and–bound algorithm for the asymmetric traveling salesman problem. Operations Research 21, 340–343.

R.S. Garfinkel (1977): Minimizing wallpaper waste, part I: a class of traveling salesman problems. Operations Research 25, 741–751.

R.S. Garfinkel and G.L. Nemhauser (1970): Optimal political districting by implicit enumeration techniques. Management Science 16, B495–B508.

J. Gaschnig (1974): A constraint satisfaction method for inference making. Proc. 12th Annual Allerton Conf. Circuit System Theory, Urbana–Champaign,IL, 866–874.

J. Gaschnig (1979): Performance Measurement and Analysis of Certain Search Algorithms. Dissertation, Carnegie–Mellon University, Pittsburgh.

B. Gavish and K.N. Srikanth (1986): An optimal solution method for large–scale multiple traveling salesman problems. Operations Research 34, 698–717.

S.B. Gelfand and S.K. Mitter (1993): Metropolis–type annealing algorithms for global optimization in $\mathbb{R}^d$. SIAM Journal Control and Optimization 31, 111–131.

M. Gendreau, A. Hertz, and G. Laporte (1992): New insertion and postoptimization procedures for the traveling salesman problem. Operations Research 40, 1086–1094.

D.H. Gensch (1978): An industrial application of the travelling salesman's subtour problem. AIIE Transactions 10, 362–370.

W.S. Gere (1966): Heuristics in job–shop scheduling. Management Science 13, 167–190.

B. Gidas (1988): Nonstationary Markov chains and convergence of the annealing algorithm. Journal of Statistical Physics 39, 73–131.

J.W. Giffin, L.R. Foulds, and D.C. Cameron (1986): Drawing a block plan from a relationship chart with graph theory and microcomputer. Computers and Industrial Engineering 10, 109–116.

B. Giffler and G.L. Thompson (1960): Algorithms for solving production scheduling problems. Operations Research 8, 487–503.

B. Giffler, G.L. Thompson, and V. van Ness (1963): Numerical experience with the linear and Monte Carlo algorithms for solving production scheduling problems. In: (J.F. Muth and G.L. Thompson, eds.) Industrial Scheduling. Prentice–Hall, Englewood Cliffs, N.J.

C.A. Glass, C.N. Potts, and P. Shade (1992): Genetic algorithms and neighbourhood search for scheduling unrelated parallel machines. Working paper No. OR47, University of Southampton.

F. Glover (1969): Integer programming over a finite additive group. SIAM Journal Control 7, 213–231.

F. Glover (1972): Cut search methods in integer programming. Mathematical Programming 3, 86–100.

F. Glover (1977): Heuristic for integer programming using surrogate constraints. Decision Sciences 8, 156–160.

F. Glover (1978): Parametric branch and bound. Omega 6, 145–152.

F. Glover (1986): Future paths for integer programming and links to artificial intelligence. Computers & Operations Research 13, 533–549.

F. Glover (1989a): Candidate list strategies and tabu search. Working paper, University of Colorado, Boulder.

F. Glover (1989b): Artificial intelligence, heuristic frameworks and tabu search. Managerial and Decision Economics 11, 365–375.

F. Glover (1989c): Tabu Search–Part I. ORSA Journal on Computing 1, 190–206.

F. Glover (1990a): Tabu Search–Part II. ORSA Journal on Computing 2, 4–32.

F. Glover (1990b): Issues and methods for applying target analysis to tabu search. Working paper, University of Colorado, Boulder.

F. Glover (1990c): Tabu search: A tutorial. Interfaces 20, 74–94.

F. Glover (1991a): Multilevel tabu search and embedded search neighbourhoods for the traveling salesman problem. Working paper, University of Colorado, Boulder.

F. Glover (1991b): Tabu search for nonlinear and parametric optimisation with links to genetic algorithms. Discrete Applied Mathematics, to appear.

F. Glover (1992a): New ejection chain and alternating path methods for traveling salesman problems. In: (O. Balci, R. Sharda, and S.A. Zenios, eds.) Computer Science and Operations Research, Pergamon Press, 491–508.

F. Glover (1992b): Simple tabu thresholding in optimization. Working paper, University of Colorado, Boulder.

F. Glover (1992c): Ejection chains, reference structures and alternating path methods for traveling salesman problems. Working paper, University of Colorado, Boulder.

F. Glover (1993): Scatter search and star–paths: beyond the genetic metaphor. Working paper, University of Colorado at Boulder.

F. Glover and H.J. Greenberg (1989): New approaches for heuristic search: A bilateral linkage with artificial intelligence. European Journal of Operational Research 13, 563–573.

F. Glover, J. Kelly, and M. Laguna (1992): Genetic algorithms and tabu search: hybrids for optimization. Computers & Operations Research, to appear.

F. Glover and M. Laguna (1993): Tabu search. In: (C. Reeves, ed.) Modern Heuristic Techniques for Combinatorial Problems, to appear.

F. Glover, M. Laguna, E. Taillard and D. de Werra, eds. (1993): Tabu Search. Annals of Operations Research 41.

F. Glover and C. McMillan (1986): The general employee scheduling problem: an integration of MS and AI. Computers & Operations Research 13, 563–573.

F. Glover and E. Pesch (1993): Efficient facility layout planning. Working paper, University of Colorado, Boulder.

F. Glover, E. Taillard, and D. de Werra (1993): A user's guide to tabu search. Annals of Operations Research, to appear.

M. Goetschalckx (1992): An interactive layout heuristic based on hexagonal adjacency graphs. European Journal of Operational Research 63, 304–321.

D.E. Goldberg (1989a): Genetic Algorithms in Search, Optimization and Machine Learning. Addison–Wesley, Reading.

D.E. Goldberg (1989b): Zen and the art of genetic algorithms. Proc. 3rd International Conf. on Genetic Algorithms (J.D. Schaffer, ed.). Morgan Kaufmann, 80–85.

D.E. Goldberg and A.L. Thomas (1986): Genetic algorithms: a bibliography 1962–1986. TCGA Report No. 86001, The Clearing House of Genetic Algorithms, University of Alabama.

B.L. Golden, L. Bodin, T. Doyle, and W. Stewart (1980): Approximate travelling salesman algorithms. Operations Research 28, 694–711.

B.L. Golden and C.C. Skiscim (1986): Using simulated annealing to solve routing and location problems. Naval Research Logistics Quarterly 33, 261–280.

A.J. Goldman and C.J. Witzgall (1970): A localization theorem for optimal facility placement. Transportation Science 4, 406–409.

S.W. Golumb and L.D. Baumert (1965): Backtrack programming. Journal of the ACM 12, 516–524.

T. Gonzales and S. Sahni (1978): Flowshop and jobshop schedules: complexity and approximation. Operations Research 20, 36–52.

M. Gorges–Schleuter (1989): ASPARAGOS, a parallel genetic algorithm and population genetics. Proc. 3rd International Conf. on Genetic Algorithms (J.D. Schaffer, ed.). Morgan Kaufmann, 422–427.

J. Grabowski, E. Nowicki, and S.S, Zdrzalka (1986): A block approach for single machine scheduling with release dates and due dates. European Journal of Operational Research 26, 278–285.

R.L. Graham, E.L. Lawler, J.K. Lenstra, and A.H.G. Rinnooy Kan (1978): Optimization and approximation in deterministic sequencing and scheduling: a survey. In: (J.K. Lenstra, A.H.G. Rinnooy Kan, and P. van Emde Boas, eds.) Interfaces Between Computer Science and Operations Research, Mathematical Centre Tracts 99, Amsterdam, 169–231.

R.L. Graham, E.L. Lawler, J.K. Lenstra, and A.H.G. Rinnooy Kan (1979): Optimization and approximation in deterministic sequencing and scheduling theory: a survey. Annals of Discrete Mathematics 5, 287–326.

G.I. Green and L.B. Appel (1981): An empirical analysis of job shop dispatch rule selection. Journal Operations Management 1, 197–203.

J.W. Greene and K.J. Supowit (1986): Simulated annealing without rejected moves. IEEE Transactions on Computer–Aided Design 5, 221–228.

H.H. Greenberg (1968): A branch and bound solution to the general scheduling problem. Operations Research 16, 353–361.

J.J. Grefenstette (1984): GENESIS: A system for using genetic search procedures. Proc. of a Conf. on Intelligent Systems and Machines, Rochester, MI, 161–165.

J.J. Grefenstette (1986): Optimization of control parameters for genetic algorithms. IEEE Transactions on Systems, Man and Cybernetics 16, 122–128.

J.J. Grefenstette (1987a): A user's guide to GENESIS. Navy Center for Applied Research in Artificial Intelligence, Naval Research Laboratory, Washington, D.C.

J.J. Grefenstette (1987b): Incorporating problem specific knowledge into genetic algorithms. In: (L. Davis, ed.) Genetic Algorithms and Simulated Annealing, Pitman, 42–60.

J.J. Grefenstette and J.E. Baker (1989): How genetic algorithms work: a critical look at implicit parallelism. Proc. 3rd. International Conf. on Genetic Algorithms (J.D. Schaffer, ed.), Morgan Kaufmann, 20–27.

J.J. Grefenstette, R. Gopal, B. Rosmaita, and D. Van Gucht (1985): Genetic algorithms for the traveling salesman problem. Proc. 1st International Conf. on Genetic Algorithms and their Applications (J.J. Grefenstette, ed.). Lawrence Erlbaum Ass., 160–168.

M. Grötschel (1977): Polyedrische Charakterisierungen kombinatorischer Optimierungsprobleme. Hain, Meisenheim am Glan.

M. Grötschel (1980): On the symmetric travelling salesman problem: solution of a 120–city problem. Mathematical Programming Studies 12, 61–77.

M. Grötschel and O. Holland (1991): Solution of large–scale symmetric travelling salesman problems. Mathematical Programming 51, 141–202.

M. Grötschel, M. Jünger, and G. Reinelt (1991): Optimal control of plotting and drilling machines: a case study. ZOR – Zeitschrift Operations Research 35, 61–84.

M. Grötschel and M. Padberg (1977): Lineare Charakterisierungen von Travelling Salesman Problemen. ZOR – Zeitschrift Operations Research 21, 33–64.

M. Grötschel and M. Padberg (1979a): On the symmetric traveling salesman problem I. Inequalities. Mathematical Programming 16, 265–280.

M. Grötschel and M. Padberg (1979b): On the symmetric traveling salesman problem. II. Lifting theorems and facets. Mathematical Programming 16, 281–302.

M. Grötschel and M. Padberg (1985): Polyhedral theory. In: (E.L. Lawler et al., eds.) The travelling salesman problem, Wiley, New York, 251–305.

M. Grötschel and W.R. Pulleyblank (1986): Clique–tree inequalities and the symmetric traveling salesman problem. Mathematics of Operations Research 11, 537–569.

M. Grötschel and Y. Wakabayashi (1989): A cutting–plane algorithm for a clustering problem. Mathematical Programming 45, 59–96.

M. Grötschel and Y. Wakabayashi (1990a): Facets of the clique partitioning polytope. Mathematical Programming 47, 367–387.

M. Grötschel and Y. Wakabayashi (1990b): Composition of facets of the clique paritioning polytope. In: (R. Bodendiek and R. Henn, eds.) Topics in Combinatorics and Graph Theory. Essays in Honour of Gerhard Ringel. Physica, Heidelberg, 277–284.

A. Gunasekaran, T. Martikainen, and P. Yli–Olli (1993): Flexible manufacturing systems: an investigation for research and applications. European Journal of Operational Research 66, 1–26.

J.N.D. Gupta and S.S. Reddi (1978): Improved dominance conditions for the three–machine flowshop scheduling problem. Operations Research 26, 200–203.

H.–W. Güsgen und J. Hertsberg (1988): Some fundamental properties of local constraint propagation. Artificial Intelligence 36, 237–247.

J. Habraken, J. Ringelberg, K. van Kempen, and E. Hoenkamp (1991): Het gebruik van disjunctieve en redundante constraints voor een realistisch scheduling probleem. Working paper, LCN Nijmegen.

B. Hajek (1988): Cooling schedules for optimal annealing. Mathematics of Operations Research 13, 311–329.

S.L. Hakimi (1964): Optimum locations of switching centers and the absolute centers and medians of a graph. Operations Research 12, 450–459.

S.L. Hakimi (1983): On locating new facilities in a competitive environment. European Journal of Operational Research 12, 29–35.

S.L. Hakimi (1986): p–Median theorems for competitive locations. Annals of Operations Research 6, 77–98.

S.L. Hakimi (1990): Locations with spatial interactions: competitive locations and games. In: (P.B. Mirchandani and R.L. Francis, eds.) Discrete Location Theory, Wiley, New York, 439–478.

A. Hammouche and D.B. Webster (1985): Evaluation of an application of graph theory to the layout problem. International Journal of Production Research 23, 987–1000.

C.C. Han and C.H. Lee (1988): Comments on Mohr und Hendersons path consistency algorithm. Artificial Intelligence 36, 125–130.

D.J. Hand (1981): Discrimination and Classification. Wiley, New York.

K.H. Hansen and J. Krarup (1974): Improvements of the Held–Karp algorithm for the symmetric traveling–salesman problem. Mathematical Programming 7, 87–96.

P. Hansen and B. Jaumard (1990): Algorithms for the maximum satisfiability problem. Computing 44, 279–303.

P. Hansen and M. Labbé (1988): Algorithms for voting and competitive location on a network. Transportation Science 22, 278–288.

P. Hansen and J.–F. Thisse (1981): Outcomes of voting and planning: Condorcet, Weber and Rawls locations. Journal Public Economics 16, 1–15.

P. Hansen, J.–F. Thisse, and M. Labbé (1987): Single facility location on networks. Annals of Discrete Mathematics 31, 113–146.

P. Hansen, J.–F. Thisse, and R.E. Wendell (1986): Equivalence of solutions to network location problems. Mathematics of Operations Research 11, 672–678.

P. Hansen, J.–F. Thisse, and R.E. Wendell (1990): Equilibrium analysis for voting and competitive location problems. In: (P.B. Mirchandani and R.L. Francis, eds.) Discrete Location Theory, Wiley, New York, 479–502.

R.M. Haralick and G.L. Elliott (1980): Increasing tree search efficiency for constraint satisfaction problems. Artificial Intelligence 14, 263–313.

J.A. Hartigan (1975): Clustering Algorithms. Wiley, New York.

M.M. Hassan and G.L. Hogg (1987): A review of graph theory application to the facilities layout problem. Omega 15, 291–300.

M. Hassan and I.H. Osman (1993): Local search algorithms for the maximal planar layout problem. Working paper, University of Kent, Canterbury.

R. Haupt (1989): A survey of priority–rule based scheduling. OR Spektrum 11, 3–16.

A.C. Hax and D. Candea (1984): Production and Inventory Management. Prentice Hall, Englewood Cliffs.

N. Hefetz and I. Adiri (1982): An efficient optimal algorithm for the two–machines, unit–time, job–shop, schedule length, problem. Mathematics of Operations Research 7, 354–360.

M. Held and R. Karp (1970): The traveling salesman problem and minimum spanning trees. Operations Research 18, 1138–1162.

M. Held and R. Karp (1971): The traveling salesman problem and minimum spanning trees: part II. Mathematical Programming 1, 6–25.

R. Helming and K. Jörnsten (1991): A simulated annealing algorithm for general zero–one programming problems. Journal of Information & Optimization Sciences 12, 295–306.

P. van Hentenryck (1987): Consistency Techniques in Logic Programming. Dissertation, University Namur.

P. van Hentenryck (1989a): Constraint Satisfaction in Logic Programming. MIT Press.

P. van Hentenryck (1989b): A logic language for combinatorial optimization. Annals of Operations Research 21, 247–274.

P. van Hentenryck, Y. Deville and C.–M. Teng (1992): A generic arc–consistency algorithm and its specializations. Artificial Intelligence 57, 291–321.

P. van Hentenryck, H. Simonis und M. Dincbas (1992): Constraint satisfaction using constraint logic programming. Artificial Intelligence 58, 113–159.

S. Heragu and A. Alfa (1992): Experimental analysis of simulated annealing based algorithms for the layout problem. European Journal of Operational Research 57, 190–202.

S.S. Heragu and A. Kusiak (1988): Machine layout problem in flexible manufacturing systems. Operations Research 36, 258–268.

J.C. Hershauer and R.J. Ebert (1975): Search and simulation selection of a job shop sequencing rule. Management Science 21, 833–843.

A. Hertz and D. de Werra (1987): Using tabu search techniques for graph coloring. Computing 39, 345–351.

A. Hertz and D. de Werra (1990): The tabu search metaheuristic: how we use it. Annals of Mathematics and Artificial Intelligence 1, 111–121.

J. Hilger, G. Harhalakis, and J.M. Proth (1991): Manufacturing cells and part families: generalization of the GP method. Information and Decision Technologies 17, 51–61.

M.R. Hilliard and G.E. Liepins (1988): Machine learning applications to job shop scheduling. Proc. Workshop on Production Planning and Scheduling. AAAI–SIGMAN, St. Paul, MN.

M.R. Hilliard, G.E. Liepins, and M. Palmer (1990): Dicovering and refining algorithms through machine learning. In: (D.E. Brown and C.C. White, eds.) Operations Research and Artificial Intelligence: The Integration of Problem–Solving Strategies, Kluwer, 59–78.

J.C. Ho and Y.–L. Chang (1991): A new heuristic for the n–job, M–machine flow–shop problem. European Journal of Operational Research 52, 194–202.

F. Hoffmeister (1991): Scalable parallelism by evolutionary algorithms. Working paper, University of Dortmund.

F. Hoffmeister and T. Bäck (1991): Genetic algorithms and evolution strategies: similarities and differences. Lecture Notes in Computer Science 496, 455–470.

J.H. Holland (1973): Genetic algorithms and the optimal allocation of trials. SIAM Journal on Computing 2, 88–105.

J.H. Holland (1975): Adaptation in Natural and Artificial Systems. The University of Michigan Press, Ann Arbor.

J.H. Holland (1980): Adaptive algorithms for discovering and using general patterns in growing knowledge bases. International Journal of Policy Analysis and Information Systems 4, 245–268.

J. Hopcroft and R. Tarjan (1974): Efficient planarity testing. Journal of the ACM 21, 549–568.

J.J. Hopfield and D.W. Tank (1985): "Neural" computation of decisions in optimization problems. Biological Cybernetics 52, 141–152.

H. Hotelling (1929): Stability in competition. Econ. Journal 39, 41–57.

D.J. Houck, J.–C. Picard, M. Queyranne, and R.R. Vemuganti (1980): The travelings salesman problem as a constrained shortest path problem: theory and computational experience. Opsearch 17, 93–109.

I. Hubert (1974): Some applications of graph theory to clustering. Psychometrica 39, 283–309.

R. Hübscher and F. Glover (1992): Applying tabu search with influential diversification to multiprocessor scheduling. Working paper, University of Colorado, Boulder.

T.S. Hundal and J. Rajgopal (1988): An extension of Palmer's heuristic for the flow–shop scheduling problem. International Journal of Production Research 26, 1119–1124.

J. Hurink (1992): Polygon Scheduling. Dissertation, University of Osnabrück.

P. Husbands, F. Mill, and S. Warrington (1991): Genetic algorithms, production plan optimisation and scheduling. Lecture Notes in Computer Science 496, 80–84.

N.L. Hyer and U. Wemmerlöv (1989): Group technology in the US manufacturing industry: a survey of current practices. International Journal of Production Research 27, 1287–1304.

E. Ignall and L. Schrage (1965): Application of the branch–and bound technique to some flow–shop scheduling problem. Operations Research 13, 400–412.

J.R. Jackson (1956): An extension of Johnson's results on job lot scheduling. Naval Research Logistics Quarterly 3, 201–203.

B. Jaumard, P.S. Ow, and B. Simeone (1988): A selected artificial intelligence bibliography for operations researchers. Annals of Operations Research 12, 1–50.

R.E. Jensen (1969): A dynamic programming algorithm for cluster analysis. Operations Research 12, 1034–1057.

D.W. Jespen and C.D. Gelatt (1983): Macro placement by Monte Carlo annealing. Proc. International Conference Computer Design, Port Chester, 495–498.

P. Jog, J.Y. Suh, and D. Van Gucht (1989): The effects of population size, heuristic crossover and local improvement on a genetic algorithm for the traveling salesman problem. Proc. 3rd International Conf. Genetic Algorithms (J.D. Schaffer, ed.), Morgan Kaufmann, 110–115.

D.S. Johnson (1983): The NP–completeness column: an ongoing guide. Journal of Algorithms

4, 189–203.

D.S. Johnson (1990a): Local optimization and the traveling salesman problem. Proc. 17th Colloq. Automata, Languages, and Programming. Springer, 446–461.

D.S. Johnson (1990b): A catalog of complexity classes. In: (J. van Leeuwen, ed.) Handbook of Theoretical Computer Science Vol. A: Algorithms and Complexity, 67–161.

D.S. Johnson, C.R. Aragon, L.A. McGeoch, and C. Schevon (1989): Optimization by simulated annealing: an experimental evaluation; Part I, Graph partitioning. Operations Research 37, 865–892.

D.S. Johnson, C.R. Aragon, L.A. McGeoch, and C. Schevon (1991): Optimization by simulated annealing: an experimental evaluation; Part II, Graph coloring and number partitioning. Operations Research 39, 378–406.

D.S. Johnson, C.H. Papadimitriou, and M. Yannakakis (1988): How easy is local search? Journal Computer System Science 37, 79–100

E.L. Johnson, A. Mehrotra, and G.L. Nemhauser (1991): Min–cut clustering. Georgia Institute of Technology, Atlanta.

L.A. Johnson and D.C. Montgomery (1974): Operations Research in Production Planning, Scheduling and Inventory Control. Wiley, New York.

M.E. Johnson (1988): Simulated Annealing & Optimisation: Modern Algorithms with VLSI, Optimal Design, and Missile Defense Applications. American Science Press, Syracuse, NY.

S.M. Johnson (1954): Optimal two– and three–stage production schedules with setup times included. Naval Research Logistics Quarterly 1, 61–68.

C.H. Jones (1973): An economic evaluation of job shop dispatching rules. Management Science 20, 293–307.

R. Jonker and T. Volgenant (1984): Nonoptimal edges for the symmetric traveling salesman problem. Operations Research 32, 837–846.

T. Jost and R. Skuppin (1989): Technical diagnosis based on numerical models using PROLOG III. Conf. Proc. ESPRIT'89, Kluwer, Dordrecht, 513–527.

M. Jünger and P. Mutzel (1993): Solving the maximum weight planar subgraph problem by branch and cut. Working paper, University of Köln.

A.B. Kahng (1989): Fast hypergraph partition. Proc. of the 26th Design Automation Conf., 762–766.

Y. Kakazu, H. Sakanashi, and K. Susuki (1992): Adaptive search strategy for genetic algorithms with additional genetic algorithms. In: (R. Männer and B. Manderick, eds.) Parallel Problem Solving from Nature 2, Elsevier, 311–320.

Y. Kamidoi, S. Wakabayashi, J. Miyao, and N. Yoshida (1991): A fast heuristic algorithm for hypergraph bisection. Proc. of the IEEE International Symposium on Circuit and Systems, 1160–1163.

Y. Kamidoi, S. Wakabayashi, and N. Yoshida (1992a): Multiple–way hypergraph partitioning with GA hybrid. Working paper, Hiroshima University.

Y. Kamidoi, S. Wakabayashi, and N. Yoshida (1992b): An efficient GA hybrid for hypergraph bisection with application to VLSI placement. Working paper, Hiroshima University.

P.–C. Kanellakis and C.H. Papadimitriou (1980): Local search for the asymmetric traveling salesman problem. Operations Research 28, 1086–1099.

J.J. Kanet and V. Sridharan (1991): PROGENITOR: a genetic algorithm for production scheduling. Wirtschaftsinformatik 33, 332–336.

J. Karkazis (1989): Facility location in a competitive environment: a PROMETHEE based multiple criteria analysis. European Journal of Operational Research 42, 294–304.

R. Karp (1975): On the computational complexity of combinatorial problems. Networks 5, 45–68.

R. Karp (1986): Combinatorics, complexity, and randomness. Communications ACM, 98–117.

R. Karp and C. Papadimitriou (1980): On linear characterizations of combinatorial optimization problems. Proc. 21st Annual Symposium on the Foundations of Computer Science, IEEE Press, New York, 1–9.

T. Kawaguchi and S. Kyan (1988): Deterministic scheduling in computer systems: a survey. Journal Operational Research Society Japan 31, 190–217.

S.T. Kedar–Cabelli (1983): Bibliography of recent machine learning research. In: (R.S. Michalski, J.G. Carbonell, and T.M. Mitchell, eds.) Machine Learning, an Artificial Intelligence Approach, Vol. II, 671–705.

J. Kelly, B. Golden, and A. Assad (1990): Using simulated annealing to solve controlled rounding problems. ORSA Journal on Computing 2, 174–185.

B.W. Kernighan (1971): Optimal sequential partitions of graphs. Journal of the ACM 18, 34–40.

B.W. Kernighan and S. Lin (1970): An efficient heuristic procedure for partitioning graphs. The Bell System Technical Journal 49, 291–307.

G.A.P. Kindervater and J.K. Lenstra (1989): The parallel complexity of TSP heuristics. Journal of Algorithms 10, 249–270.

J.R. King (1980): Machine–component grouping in production flow analysis: an approach using a rank order clustering algorithm. International Journal of Production Research 18, 213–232.

J.R. King and V. Nakornchai (1982): Machine component group formation in group technology: review and extensions. International Journal of Production Research 20, 117–123.

J. Kingdon (1992): Genetic algorithms: deception, convergence and starting conditions. Working paper, University College London.

S. Kirkpatrick (1984): Optimisation by simulated annealing: quantitative studies. Journal of Statistical Physics 34, 975–986.

S. Kirkpatrick, C.D. Gelatt Jr., and M.P. Vecchi (1983): Optimization by simulated annealing. Science 220, 671–680.

S. Kirkpatrick and R.H. Swendsen (1985): Statistical mechanics and disordered systems. Communications ACM 28, 363–373.

K.–P. Kistner and M. Steven (1990): Produktionsplanung. Physica, Heidelberg.

J. de Kleer (1986): An assumption based truth maintenance system. Artificial Intelligence 28, 127–162.

J. de Kleer (1989): A comparison of ATMS and CSP techniques. Proc. IJCAI–89, 290–296.

Y. Kodratoff and R.S. Michalski (1990): Machine Learning, An Artificial Intelligence Approach. Vol. III, Morgan Kaufmann, San Mateo.

A. Koenig, N. Wehn, and M. Glesner (1991): Partitioning on Boltzman Machines. Working Paper, Technical University, Darmstadt.

A. Kolen and E. Pesch (1991): Genetic local search in combinatorial optimization. Discrete Applied Mathematics (to appear 1994).

W.L.G. Koontz, P.M. Narendra, and K. Fukunaga (1975): A branch and bound clustering algorithm. IEEE Transactions on Computers C–24, 908–915.

T.C. Koopmans and M.J. Beckmann (1957): Assignment problems and the location of economic activities. Econometrica 25, 53–76.

H. Kopfer (1989): Heuristische Suche in Operations Research und Künstlicher Intelligenz. Habilitation Thesis, FU Berlin.

P. Kouvelis and M.W. Kim (1992): Undirectional loop network layout problem in automated manufacturing systems. Operations Research 40, 533–550.

P. Kouvelis, A.A. Kurawarwala, and G.J. Gutiérres (1992): Algorithms for robust single and multiple period layout planning for manufacturing systems. European Journal of Operational Research 63, 287–303.

J. Kral (1965): To the problem of segmentation of a program. Information Processing Machines 2, 116–127.

W. Krautter and M. Steinert (1988): A knowledge representation for model–based reasoning using PROLOG III. Conf. Proc. ESPRIT'88, North–Holland, 814–825.

M. Krejcirik (1969): Computer aided plant layout. Computer Aided Design 2, 7–17

M.W. Krentel (1990): On finding and verifying locally optimal solutions. SIAM Journal on Computing 19, 742–749.

B. Krishnamurti (1984): An improved min–cut algorithm for partitioning VLSI networks. IEEE Transactions on Computers C–33, 438–446.

J.B. Kruskal (1956): On the shortest spanning subtree of a graph and the travelling salesman problem. Proc. of the AMS 7, 48–50.

H. Kuhn (1990): Einlastungsplanung von flexiblen Fertigungssystemen. Physica, Heidelberg.

H. Kuhn (1992): Heuristische Suchverfahren mit simulierter Abkühlung. WiSt 8, 387–391.

K.R. Kumar, A. Kusiak, and A. Vannelli (1986): Grouping parts and components in flexible manufacturing systems. European Journal of Operational Research 24, 387–397.

V. Kumar (1987): Depth–first search. Encyclopaedia of Artificial Intelligence 2 (S.C. Shapiro ed.), 1004–1005.

V. Kumar (1992): Algorithms for constraint–satisfaction problems: a survey. AI Magazine 13, 32–44.

S.R.T. Kumara, R.L. Kashyap, and C.L. Moodies (1988): Application of expert systems and pattern recognition methodologies to facilities layout planning. International Journal of Production Research 26, 905–930.

K. Kurbel (1989): Entwicklung und Einsatz von Expertensystemen. Springer, Berlin.

A. Kusiak (1985): The part families problem in flexible manufacturing systems. Annals of Operations Research 3, 279–300.

A. Kusiak (1986): Application of operational research models and techniques in flexible manufacturing systems. European Journal of Operational Research 24, 336–345.

A. Kusiak (1988): EXGT–S: a knowledge based system for group technology. International Journal of Production Research 26, 887–904.

A. Kusiak and M. Chen (1988): Expert systems for planning and scheduling manufacturing systems. European Journal of Operational Research 34, 113–130.

A. Kusiak and C.H. Cheng (1990): A branch–and–bound algorithm for solving the group technology problem. Annals of Operations Research 26, 415–431.

A. Kusiak and M. Cho (1992): Similarity coefficient algorithms for solving the group technology problem. International Journal of Production Research 30, 2633–2646.

A. Kusiak and W.S. Chow (1987): Efficient solving of the group technology problem. Journal Manufacturing Systems 6, 117–124.

A. Kusiak and S.S. Heragu (1987): The facility layout problem. European Journal of Operational Research 29, 229–251.

P.J.M. van Laarhoven and E.H.L. Aarts (1987): Simulated Annealing: Theory and Applications. Reidel, Dordrecht.

P.J.M. van Laarhoven, E.H.L. Aarts, and J.K. Lenstra (1992): Job shop scheduling by simulated annealing. Operations Research 40, 113–125.

M. Labbé (1985): Outcomes of voting and planning in single facility location problems. European Journal of Operational Research 20, 299–313.

M. Labbé and S.L. Hakimi (1991): Market locational equilibrium for two competitors. Operations Research 39, 749–757.

B. Lageweg, E.L. Lawler, J.K. Lenstra, and A.H.G. Rinnooy Kan (1982): Computer aided complexity classification of combinatorial problems. Communications ACM 25, 817–822.

B. Lageweg, J.K. Lenstra, and A.H.G. Rinnooy Kan (1977): Job–shop scheduling by implicit enumeration. Management Science 24, 441–450.

B. Lageweg, J.K. Lenstra, and A.H.G. Rinnooy Kan (1978): A general bounding scheme for the permutation flow–shop problem. Operations Research 26, 53–67.

M. Laguna (1992): Tabu search primer. Working paper, University of Colorado, Boulder.

A. Langevin, F. Soumis, and J. Desrosiers (1990): Classification of travelling salesman problem formulations. Operations Research Letters 9, 127–132.

G. Laporte (1992a): The traveling salesman problem: an overview of exact and approximate algorithms. European Journal of Operational Research 59, 231–247.

G. Laporte (1992b): The vehicle routing problem: an overview of exact and approximate algorithms. European Journal of Operational Research 59, 345–358.

R.E. Larson, M.I. Dessouky, and R.E. Devor (1985): A forward–backward procedure for the single machine problem to minimize maximum lateness. IIE Transactions 17, 252–260.

G. von Laszewski (1991): Intelligent structural operators for the k–way graph partitioning problem. Proc. 4th. International Conf. on Genetic Algorithms (R.K. Belew and L.B. Booker, eds.), Morgan Kaufmann, 45–52.

J. Lauriere (1978): A language and a program for stating and solving combinatorial problems. Artificial Intelligence 10, 29–127.

E.L. Lawler (1982): Recent results in the theory of machine scheduling. In: (A. Bachem, M. Grötschel, and B. Korte, eds.) Mathematical Programming: The State of the Art. Springer, Berlin.

E.L. Lawler, J.K. Lenstra, and A.H.G. Rinnooy Kan (1982): Recent developments in deterministic sequencing and scheduling: a survey. In: (M.A.H. Dempster, J.K. Lenstra, A.H.G. Rinnooy Kan, eds.) Deterministic and Stochastic Scheduling. Proc. of the NATO Advanced Study and Research Institute on Theoretical Approaches to Scheduling Problems, Reidel, Dordrecht, 35–73.

E.L. Lawler, J.K. Lenstra, A.H.G. Rinnooy Kan, and D.B. Shmoys, eds. (1985): The Traveling Salesman Problem. Wiley, Chichester.

E.L. Lawler, J.K. Lenstra, A.H.G. Rinnooy Kan, and D.B. Shmoys (1989): Sequencing and scheduling: algorithms and complexity. CWI–report BS–R8909, CWI Amsterdam.

E.L. Lawler and D.E. Wood (1966): Branch–and–bound methods: a survey. Operations

Research 14, 699–719.

P.J. Lederer and J.–F. Thisse (1990): Competitive location on networks under delivered prices. Operations Research Letters 9, 147–153.

H. Lee and A. Garcia–Dias (1993): A network flow approach to solve clustering problems in group technology. International Journal of Production Research 31, 603–612.

R.C. Lee and J.M. Moore (1967): CORELAP – computerized relationship layout planning. Journal of Industrial Engineering 18, 195–200.

W. Leler (1988): Constraint Programming Languages: Their Specification and Generation. Addison Wesley.

T. Lengauer (1990): Combinatorial Algorithms for Integrated Circuit Layout. Teubner, Stuttgart and Wiley, Chichester.

J.K. Lenstra (1977): Sequencing by Enumerative Methods. Mathematical Center Tract 69, Mathematisch Centrum, Amsterdam.

J.K. Lenstra and A.H.G. Rinnooy Kan (1975): Some simple applications of the travelling salesman problem. Operational Research Quarterly 26, 717–733.

J.K. Lenstra and A.H.G. Rinnooy Kan (1979): Computational complexity of discrete optimization problems. Annals of Discrete Mathematics 4, 121–140.

J.K. Lenstra, R.H.G. Rinnooy Kan, and P. Brucker (1977): Complexity of machine scheduling problems. Annals of Discrete Mathematics 4, 121–140.

C. Lepape (1985): SOJA: a daily workshop scheduling system. Expert System 85, 95–211.

J. Leung (1992): A new graph–theoretic heuristic for facility layout. Management Science 39, 594–605.

M. Liang and S.P. Dutta (1993): An integrated approach to the part selection and machine loading problem in a class of flexible manufacturing systems. European Journal of Operational Research 67, 387–404.

G.E. Liepins and M.R. Hilliard (1989): Genetic algorithms: foundations and applications. Annals of Operations Research 21, 31–57.

G.E. Liepins, M.R. Hilliard, M. Palmer, and M. Morrow (1987): Greedy genetics. Proc. 2nd. International Conf. on Genetic Algorithms and their Applications (J.J. Grefenstette, ed.), 90–99.

G.E. Liepins, M.R. Hilliard, J. Richardson, and M. Palmer (1990): Genetic algorithms applications to set covering and traveling salesman problems. In: (D.E. Brown and C.C. White, eds.) Operations Research and Artificial Intelligence: The Integration of Problem-Solving Strategies, Kluwer, Dordrecht, 29–57.

G.E. Liepins and M.D. Vose (1990): Representational issues in genetic optimization. Journal of Experimental and Theoretical Artificial Intelligence 2, 4–30.

G.E. Liepins and M.D. Vose (1992): Characterizing crossover in genetic algorithms. Annals of Mathematics and Artificial Intelligence 5, 27–34.

S. Lin (1965): Computer solutions of the traveling salesman problem. Bell System Computer Journal 44, 2245–2269.

S. Lin and B.W. Kernighan (1973): An effective heuristic algorithm for the traveling salesman problem. Operations Research 21, 498–516.

J.D.C. Little, K.G. Murty, D.W. Sweeney, and C. Karel (1963): An algorithm for the traveling salesman problem. Operations Research 11, 972–989.

B. Liu (1989): Reinforcement Planning for Resource Allocation and Constraint Satisfaction.

Dissertation, University of Edingburgh.

B. Liu (1992): Using constraint programming languages to solve scheduling problems. Proc. Intelligent Scheduling Systems Symposium (W.T. Scherer and D.E. Brown, eds.), 210–216.

B. Liu and Y.–W. Ku (1992a): Practical consistency algorithms for constraint satisfaction problems. Working paper, University of Singapore.

B. Liu and Y.–W. Ku (1992b): ConstraintLisp: An object–oriented constraint programming language. Working paper, University of Singapore.

A.J. van Looveren, L.F. Gelders, and L.N. van Wassenhove (1986): A review of FMS planning models. In: (A. Kusiak, ed.) Modelling and Design of Flexible Manufacturing Systems, Elsevier Science Publishers, Amsterdam, 3–31.

J.A. Lukes (1975): Combinatorial solution to the partitioning of general graphs. IBM Journal of Research and Development 19, 170–180.

M. Lundy and A. Mees (1986): Convergence of an annealing algorithm. Mathematical Programming 34, 111–124.

A.K. Mackworth (1977): Consistency in networks of relations. Artificial Intelligence 8, 99–118.

A.K. Mackworth (1987): Constraint satisfaction. In: Encyclopedia of Artificial Intelligence (S. Shapiro, ed.), Wiley, New York, 205–211.

A.K. Mackworth and E.C. Freuder (1985): The complexity of some polynomial network consistency algorithms for constraint satisfaction problems. Artificial Intelligence 25, 65–74.

M. Malek, M. Guruswamy, and M. Pandya (1989): Serial and parallel simulated annealing and tabu search algorithms for the traveling salesman problem. Annals of Operations Research 21, 59–84.

A.S. Manne (1960): On the job shop scheduling problem. Operations Research 8, 219–223.

R. Männer and B. Manderick, eds. (1992): Parallel Problem Solving from Nature 2. Proc. 2nd. International Workshop, Elsevier.

F. Marcotorchino and P. Michaud (1980a): Optimisation en analyse des données. relationelle. In: (E. Diday, et al., eds.) Data Analysis and Informatics, North–Holland, 655–670.

F. Marcotorchino and P. Michaud (1980b): Optimisation in exploratory data analysis (preference and similarity aggregation). Methods of OR 40, 169–173.

F. Marcotorchino and P. Michaud (1981): Heuristic approach of the similarity aggregation problem. Methods of OR 43, 395–404.

F. Margot (1992): Quick updates for p–opt TSP heuristics. Operations Research Letters 11, 45–46.

K. Mathias and D. Whitley (1992): Genetic operators, the fitness landscape and the traveling salesman problem. In: (R. Männer and B. Manderick, eds.) Parallel Problem Solving from Nature 2, 219–228.

H. Matsuo, C.J. Suh, and R.S. Sullivan (1988): A controlled search simulated annealing method for the general jobshop scheduling problem. Working paper 03–04–88, University of Texas at Austin.

D.W. Matula (1977): Graph theoretic techniques for cluster analysis algorithms. In: (Y. Alair and D.R. Liek, eds.) Classification and Clustering, Springer, Berlin, 96–129.

C.S. McCahon and E.S. Lee (1992): Fuzzy job sequencing for a flow shop. European Journal of Operational Research 62, 294–301.

M. McCord Nelson and W.T. Illingworth (1992): A Practical Guide to Neural Nets.

Addison–Wesley, 4th. ed., Reading.

W.T McCormick, P.J. Schweitzer, and T.W. White (1972): Problem decomposition and data reorganization by a clustering technique. Operations Research 20, 993–1009.

G.B. McMahon (1969): Optimal production schedules for flow shops. Canadian Oper. Res. Soc. Journal 7, 141–151.

G.B. McMahon and M. Florian (1975): On scheduling with ready times and due dates to minimize maximum lateness. Operations Research 23, 475–482.

P. Mellor (1966): A review of job shop scheduling. Operations Research Quarterly 17, 161–171.

P. Mertens (1988): Wissensbasierte Systeme in der Produktionsplanung und – steuerung – eine Bestandsaufnahme. Information Management 4/88, 14–22.

P. Mertens (1991): Artificial life – generative Algorithmen. Wirtschaftsinformatik 33, 156–159.

P. Meseguer (1989): Constraint satisfaction problems : an overview. AICOM 2, 3–17.

N. Metropolis, A. Rosenbluth, M. Rosenbluth, A. Teller, and E. Teller (1953): Equation of state calculations by fast computing machines. Journal Chemical Physics 21, 1087–1092.

W. Meyer (1992): Geometrische Methoden zur Lösung von Job–Shop Problemen und deren Verallgemeinerungen. Dissertation, University of Osnabrück.

Z. Michalewicz (1992): Genetic Algorithms + Data Structures = Evolution Programs. Springer Berlin.

Z. Michalewicz, G.A. Vignaux, and M. Hobbs (1991): A nonstandard genetic algorithm for the nonlinear transportation problem. ORSA Journal on Computing 3, 307–316.

D.L. Miller and J.F. Pekny (1991): Exact solution of large asymmetric traveling salesman problems. Science 251, 754–761.

C. Miller, A. Tucker, and R. Zemlin (1960): Integer programming formulations and traveling salesman problems. Journal ACM 7, 326–329.

J. Miltenburg and W. Zhang (1991): A comparative evaluation of nine well–known algorithms for solving the cell formation problem in group technology. Journal of Operations Management 10, 44–72.

M. Minoux and E. Pinson (1987): Lower bounds to the graph partitioning problem through generalized linear programming and network flows. R.A.I.R.O. Recherche operationelle / Operations Research 21, 349–364.

S. Minton, M.D. Johnston, A.B. Philips and P. Laird (1990): Solving large–scale constraint satisfaction and scheduling problems using a heuristic repair method. Proc. AAAI–90, Boston, 17–24.

S. Minton, M.D. Johnston, A.B. Philips and P. Laird (1992): Minimizing conflicts: A heuristic repair method for constraint satisfaction and scheduling problems. Artificial Intelligence 58, 161–205.

D. Mitra, F. Romeo, and A. Sangiovanni–Vincentelli (1986): Convergence and finite–time behavior of simulated annealing. Advances in Applied Probability 18, 747–771.

S.P. Mitrofanov (1959): The Scientific Principles of Group Technology, Leningrad (in Russian, translated by the National Lending Library 1966).

I. Mitterreiter and F.J. Radermacher (1991): Some notes on experiments on the running time behaviour of some algorithms solving propositional logical problems. Working paper, FAW Ulm.

R. Mohr and T.C. Henderson (1986): Arc and path consistency revisited. Artificial

Intelligence 28, 225–233.

R.H. Möhring (1990): Graph problems related to gate matrix layout and PLA folding. In: (G. Tinhofer et al., eds.) Combinatorial Graph Theory, Springer, Berlin, 17–51.

C.L. Monma and A.H.G. Rinnooy Kan (1983): A concise survey of efficiently solvable special cases of the permutation flow shop problem. RAIRO 17, 105–119.

U. Montanari (1974): Networks of constraints: fundamental properties and applications to picture processing. Information Sciences 7, 95–132.

B. Montreuil and H.D. Ratliff (1989): Utilization cut trees and design skeletons for facility layout. IIE Transactions 21, 136–143.

B. Montreuil, H.D. Ratliff, and M. Goetschalckx (1987): Matching based interactive facility layout. IIE Transactions 19, 271–279.

J.M. Moore and R.C. Wilson (1967): A review of simulation research in job shop scheduling. Production and Inventory Management 8, 1–10.

R.E. Moore (1966): Interval Analysis, Prentice Hall, Englewood Cliffs, New Jersey.

P. Moscato and J.F. Fontanari (1990): Stochastic versus deterministic update in simulated annealing. Physics Letters A146, 204–208.

H. Mühlenbein (1989): Parallel genetic algorithms, population genetics and combinatorial optimization. Proc. 3rd International Conf. Genetic Algorithms (J.D. Schaffer, ed.). Morgan Kaufmann Publ., 416–421.

H. Mühlenbein (1991): Evolution in time and space – the parallel genetic algorithm. In: (G. Rawlins, ed.) Foundations of Genetic Algorithms, Morgan Kaufmann, 316–337.

H. Mühlenbein (1992): Parallel genetic algorithms in optimization. Working paper, GMD, St. Augustin.

H. Mühlenbein, M. Gorges–Schleuter, and O. Krämer (1987): New solutions to the mapping problem of parallel systems: the evolution approach. Parallel Computing 4, 269–279.

H. Mühlenbein, M. Gorges–Schleuter, and O. Krämer (1988): Evolution algorithms in combinatorial optimization. Parallel Computing 7, 65–85.

H. Mühlenbein and J. Kindermann (1989): Dynamics of evolution and learning – towards genetic neural networks. In: (R. Pfeiffer, ed.) Connectionism in Perspective, Elsevier, Amsterdam.

H.M. Mulder (1980): The Interval Function of a Graph. Math. Centre Tracts 132, Mathematisch Centrum, Amsterdam.

H. Müller–Merbach (1961): Die Ermittlung des kürzesten Rundreiseweges mittels linearer Programmierung. Ablauf– und Planungsforschung 2, 70–83.

H. Müller–Merbach (1970): Optimale Reihenfolgen. Springer, Berlin.

H. Müller–Merbach (1974): Heuristic methods: structures, applications, computational experience. In: (R. Cottle and J. Krarup, eds.) Optimization Methods for Resource Allocation, The English Universities Press, London, 401–416.

H. Müller–Merbach (1981): Heuristics and their design. European Journal of Operational Research 8, 1–23.

J.F. Muth and G.L. Thompson, eds. (1963): Industrial Scheduling. Prentice Hall, Englewood Cliffs.

D. Naddef (1992): The binested inequalities for the symmetric traveling salesman polytope. Mathematics of Operations Research 17, 882–900.

D. Naddef and G. Rinaldi (1991): The symmetric traveling salesman polytope and its graphical relaxation: composition of valid inequalities. Mathematical Programming 51, 359–400.

D. Naddef and G. Rinaldi (1992): The crown inequalities for the symmetric traveling salesman polytope. Mathematics of Operations Research 17, 308–326.

D. Naddef and G. Rinaldi (1993): The graphical relaxation: a new framework for the symmetric traveling salesman polytope. Mathematical Programming 58, 53–88.

B. Nadel (1988): Tree search and arc consistency in constraint satisfaction algorithms. In: (L. Kanal and V. Kumar, eds.) Search in Artifical Intelligence, Springer, New York, 287–342.

S. Nahar, S. Sahni, and E. Shragowits (1989): Simulated annealing and combinatorial optimization. International Journal of Computer Aided VLSI Design 1, 1–23.

R. Nakano and T. Yamada (1991): Conventional genetic algorithm for job shop problems. Proc. 4th. International Conf. on Genetic Algorithms (R.K. Belew and L.B. Booker, eds.), Morgan Kaufmann, 474–479.

M. Nawaz, E.E. Enscore, and I. Ham (1983): A heuristic algorithm for the m–machine, n–job flow–shop sequencing problem. Omega 11, 91–95.

G.L. Nemhauser and L.A. Wolsey (1988): Integer and Combinatorial Optimization. Wiley, New York.

K. Neumann and M. Morlock (1993): Operations Research. Hanser, München.

S.M. Ng (1991): Bond energy, rectilinear distance and a worst–case bound for the group technology problem. Journal of the Operational Research Society 42, 571–578.

A.E. Nix and M.D. Vose (1992): Modeling genetic algorithms with Markov chains. Annals of Mathematics and Artificial Intelligence 5, 79.

J. Norback and R. Love (1977): Geometric approaches to solving the traveling salesman problem. Management Science 23, 1208–1223.

M. Norman and P. Moscato (1990): A competitive–cooperative approach to complex combinatorial search. Working paper, California Institute of Technology, Pasadena, CA.

E. Nowicki and C. Smutnicki (1993): A fast taboo search algorithm for the job shop problem. Working paper, University Wroclaw.

E. Nowicki and S. Zdrzalka (1986): A note on minimizing maximum lateness in a one–machine sequencing problem with release dates. European Journal of Operational Research 23, 266–267.

B. Nudel (1983): Consistent–labeling problems and their algorithms: expected complexities and theory–based heuristics. Artificial Intelligence 21, 135–178.

A.G. Oerlemans (1992a): Production Planning for Flexible Manufacturing Systems. Dissertation, University of Limburg, Maastricht.

A.G. Oerlemans (1992b): Flexible manufacturing: concepts and models. Working paper, University of Limburg, Maastricht.

J.C. Oglivie and C.L. Olson (1972): On the use of complete subgraphs in cluster analysis. Information Processing Letters 1, 76–79.

P.J. O'Grady and C. Harrison (1985): A general search sequencing rule for job shop sequencing. International Journal of Production Research 23, 951–973.

P.J. O'Grady and U. Menon (1987): Loading a flexible manufacturing system. International Journal of Production Research 25, 1053–1068.

O. Opitz and M. Schader (1984a): Analyse qualitativer Daten: Einführung und Übersicht; Teil 1. OR Spektrum 6, 67–83.

O. Opitz and M. Schader (1984b): Analyse qualitativer Daten: Einführung und Übersicht; Teil 2. OR Spektrum 6, 133–140.

I. Or (1976): Traveling Salesman–Type Combinatorial Problems and Their Relation to the Logistics of Regional Blood Banking. Dissertation, Northwestern University, Evanston, IL.

I. Osman (1991): Metastrategy Simulated Annealing and Tabu Search Algorithms for Combinatorial Optimisation Problems. Dissertation, University of London.

I.H. Osman and C.N. Potts (1989): Simulated annealing for permutation flow–shop scheduling. Omega 17, 551–557.

J.S. Ostroff (1991): Constraint logic programming for reasoning about discrete event processes. J. Logic Programming 11, 243–270.

R. Otten and L. van Ginneken (1988): Stop criteria in simulated annealing. Proc. International Conference on Computer Aided Design, Rye Twon Hill NY, 549–552.

R.H.J.M. Otten and L.P.P.P. van Ginneken (1989): The Annealing Algorithm. Kluwer, Boston.

R. Otten and L. van Ginneken (1990): The complexity of adaptive annealing. Proc. International Conference on Computer Aided Design, Rye Town Hill NY, 404–407.

P.S. Ow and S.F. Smith (1988): Viewing scheduling as an opportunistic problem–solving process. Annals of Operations Research 12, 85–108.

M. Padberg and S. Hong (1980): On the symmetric traveling salesman problem: a computational study. Mathematical Programming Studies 12, 78–107.

M. Padberg and G. Rinaldi (1987): Optimisation of a 532–city symmetric traveling salesman problem by branch and cut. Operations Research Letters 6, 1–7.

M. Padberg and G. Rinaldi (1990): Facet identification for the symmetric traveling salesman polytope. Mathematical Programming 47, 219–257.

M. Padberg and G. Rinaldi (1991): A branch–and–cut algorithm for the resolution of large–scale symmetric traveling salesman problems. SIAM Review 33, 60–100.

M. Padberg and T.–Y. Sung (1991): An analytical comparison of different formulations of the travelling salesman problem. Mathematical Programming 52, 315–357.

A. de Palma, V. Ginsburgh, M. Labbé, and J.–F. Thisse (1989): Competitive location with random utilities. Transportation Science 23, 244–252.

D.S. Palmer (1965): Sequencing jobs through a multi–stage process in the minimum total time – a quick method of obtaining a near optimum. Operational Research Quarterly 16, 101–107.

S.S. Panwalkar and W. Iskander (1977): A survey of scheduling rules. Operations Research 25, 45–61.

C.H. Papdimitriou (1977): The Euclidean traveling salesman problem is NP–complete. Theoretical Computer Science 4, 237–244.

C.H. Papadimitriou (1992): The complexity of the Lin–Kernighan heuristic for the traveling salesman problem. SIAM Journal Computing 21, 450–465.

C.H. Papadimitriou and K. Steiglitz (1977): The complexity of local search for the traveling salesman problem. SIAM Journal Computing 6, 76–83.

C.H. Papadimitriou and K. Steiglitz (1978): Some examples of difficult traveling salesman problems. Operations Research 26, 434–443.

C.H. Papadimitriou and K. Steiglitz (1982): Combinatorial Optimisation: Algorithms and Complexity. Prentice–Hall, Englewood Cliffs.

C.H. Papadimitriou and M. Yannakakis (1993): The traveling salesman problem with distances one and two. Mathematics of Operations Research 18, 1–11.

S.–C. Park, S. Piramuthu, N. Raman, and M.J. Shaw (1992): Intelligent scheduling with machine learning. Proc. Symposium on Intelligent Scheduling Systems (W.T. Scherer and D.E. Brown, eds.), 236–267.

J. Pearl (1984): Heuristics. Addison–Wesley, Reading.

M.P. Pensini, G. Mauri, and F. Gardin (1991): Flowshop and TSP. In: (J.D. Becker, I. Eisele, and F.W. Mündemann, eds.) Parallelism, Learning, Evolution, Springer, Berlin, 157–182.

E. Pesch (1988): Retracts of Graphs. Mathematical Systems in Economics 110, Athenäum, Frankfurt.

E. Pesch (1993a): Machine learning by schedule decomposition. Working paper, University of Limburg, Maastricht.

E. Pesch (1993b): Constraint propagation based scheduling of job shops. Working paper, University of Limburg, Maastricht.

E. Pesch (1993c): The traveling salesman problem: an annotated bibliography. Working paper, University of Limburg, Maastricht.

E. Pesch, A. Drexl, and A. Kolen (1993): Automatische Wissensakquisition mittels modellbasierter Inferenz in CHARME. Working paper, University of Limburg, Maastricht.

E. Pesch, A. Drexl, and F. Salewski (1993): Zur Bedeutung der Modellbildung für die Entwicklung wissensbasierter Systeme. Working paper, University of Kiel.

E. Pesch and F. Glover (1993): Reference structures, neighbourhoods, and candidate lists for TSP ejection chains. Working paper, University of Limburg, Maastricht.

J. Perttunen (1992): On the solving strategy in composite heuristics. Operations Research Letters 12, 165–172.

C. Peterson (1990): Parallel distributed approaches to combinatorial optimisation: benchmark studies on the traveling salesman problem. Neural Computation 2, 261–269.

C. Petrie (1987): Revised dependency–directed backtracking for default reasoning. Proc. AAAI–87, 167–172.

J. Piehler (1960): Ein Beitrag zum Reihenfolgeproblem. Unternehmensforschung 4, 138–142.

E. Pinson (1988): Le problème de job–shop. Dissertation, University Paris VI, Paris.

E. Pinson (1992): The job shop scheduling problem: a concise survey and some recent developments. Working paper, University Catholique de l'Ouest, Angers, France.

J. Pintér and G. Pesti (1991): Set partition by globally optimized cluster seed points. European Journal of Operational Research 51, 127–135.

J.–F. Pique (1988): Prolog II, a step on the Prolog road. AICOM 1, 4–16.

M. Pirlot (1992): General local search heuristics in combinatorial optimisation: a tutorial. JORBEL 32, 7–68.

D.B. Porter (1968): The Gantt chart as applied to production scheduling and control. Naval Research Logistics Quarterly 15, 311–317.

C.N. Potts (1980a): An adaptive branching rule for the permutation flow–shop problem. European Journal of Operational Research 5, 19–25.

C.N. Potts (1980b): Analysis of a heuristic for one machine sequencing with release dates and delivery times. Operations Research 28, 1436–1441.

C.N. Potts, D.B. Shmoys, and D.P. Williamson (1991): Permutation vs. non–permutation flow shop schedules. Operations Research Letters 10, 281–284.

W.H. Press, S.A. Teukolsky, W.T. Vetterling, and B.P. Flannery (1992): Numerical Recipes in C. Cambridge University Press, 2nd. ed.

C. Proust (1992): De l'influence des idées de S.M. Johnson dans la résolution des problèmes d'órdonnancement de type Flowshop. Working paper, University F. Rabelais, Tours, France.

M. Queyranne and Y. Wang (1993): Hamiltonian path and symmetric travelling salesman polytopes. Mathematical Programming 58, 89–110.

Y. Rabinovich and A. Wigderson (1991): An analysis of a simple genetic algorithm. Proc. 4th. International Conf. on Genetic Algorithms (R.K. Belew and L.B. Booker, eds.), Morgan Kaufmann, 215–221.

R.E. Randelman and G.S. Grest (1986): N–city traveling salesman problem: optimization by simulated annealing. Journal Statistical Physics 45, 885–890.

M.R. Rao (1971): Cluster analysis and mathematical programming. Journal of the American Statistical Association 66, 622–626.

D. Ratliff and R. Rosenthal (1983): Order–picking in a rectangular warehouse: a solvable case of the traveling salesman problem. Operations Research 31, 507–521.

G.J.E. Rawlins (1991): Foundations of Genetic Algorithms. Morgan Kaufmann, Los Altos.

V.J. Rayward–Smith and A. Clare (1986): A branch–and–cut algorithm for the resolution of large–scale symmetric traveling salesman problems. Networks 16, 283–294.

I. Rechenberg (1973): Optimierung technischer Systeme nach Prinzipien der biologischen Evolution. Problemata, Frommann–Holzboog.

S.S. Reddi and C.V. Ramamoorthy (1972): On the flow–shop sequencing problem with no wait in process. Operational Research Quarterly 23, 323–331.

S. Règnier (1965): Sur quelques aspects mathématiques des problèmes de classification automatique. I.C.C. Bulletin 4, 175–191.

G. Reinelt (1991): TSPLIB – a traveling salesman problem library. ORSA Journal on Computing 3, 376–384.

G. Reinelt (1992): Fast heuristics for large scale geometric traveling salesman problems. ORSA Journal on Computing 4, 206–217.

J.T. Richardson, M.R. Palmer, G. Liepins, and Mike Hillard (1989): Some guidelines for genetic algorithms with penalty functions. Proc. 3rd. International Conf. Genetic Algorithms (J.D. Schaffer, ed.), Morgan Kaufmann, 191–197.

A.H.G. Rinnooy Kan (1976): Machine Scheduling Problems: Classification, Complexity and Computations. Nijhoff, The Hague.

R. Rochette and R.P. Sadowski (1976): A statistical comparison of the performance of simple dispatching rules for a particular set of job shops. International Journal of Production Research 14, 63–75.

H. Röck (1984): The three–machine no–wait flow shop problem is NP–complete. Journal of the ACM 51, 336–345.

F.A. Rodammer and K.P. White (1988): A recent review of production scheduling. IEEE Transactions on Systems, Man, and Cybernetics 18, 841–852.

M.J. Rosenblatt (1979): The facilities layout problem: a multigoal approach. International Journal of Production Research 17, 323–332.

D.J. Rosenkrantz, R.E. Stearns, and P.M. Lewis (1977): An analysis of several heuristics for the traveling salesman problem. SIAM Journal Computing 6, 563–581.

Y. Rossier, M. Troyon, and T.M. Liebling (1986): Probabilistic exchange algorithms and Euclidean traveling salesman problems. OR Spektrum 8, 151–164.

G. Rote (1988): Two Solvable Cases of the Traveling Salesman Problem. Dissertation, Technical University of Graz.

G. Rote (1992): The n–line traveling salesman problem. Networks 22, 91–108.

C. Roucairol and P. Hansen (1987): Problème de la bipartition minimale d'un graphe. R.A.I.R.O. Recherche Operationelle / Operations Research 21, 325–348.

P. Roussel–Ragot and G. Dreyfuss (1990): A problem independent parallel implementation of simulated annealing: models and experiments. IEEE Transactions on Computer–Aided Design 9, 827–835.

B. Roy and B. Sussmann (1964): Les problèmes d'ordonnancement avec contraintes disjonctives. SEMA, Note D.S. No. 9., Paris.

R.A. Ruben, C.T. Mosier, and F. Mahmoodi (1993): A comprehensive analysis of group scheduling heuristics in a job shop cell. International Journal of Production Research 31, 1343–1369.

R.A. Rutenbar (1989): Simulated annealing algorithms: an overview. IEEE Circuits Devices Magazine 5, 19–26.

N. Sadeh (1991): Look–ahead techniques for micro–opportunistic job shop scheduling. Dissertation, Carnegie Mellon University, Pittsburgh, PA.

S. Sahni and T. Gonzales (1976): "P–complete approximation problems. Journal ACM 23, 555–565.

M.S. Salvador (1978): Scheduling and sequencing. In: (J.J. Moder and S.E. Elmaghraby, eds.) Handbook of Operations Research: Models and Applications, Van Nostrand Reinhold, New York.

L.A. Sanchis (1989): Multiple–way network partitioning. IEEE Transactions on Computers C–38, 62–81.

M.W.P. Savelsbergh (1992): Computer Aided Routing. CWI Tract 75, Amsterdam.

L.E. Scales (1985): Introduction to Non–Linear Optimization. Macmillan, London.

M. Schader and U. Tüshaus (1985): Ein Subgradientenverfahren zur Klassifikation qualitativer Daten. OR Spektrum 7, 1–5.

J.D. Schaffer, R.A. Caruane, L.J. Eshelman, and D. Rajarshi (1989): A study of control parameters affecting online performance of genetic algorithms for function optimization. Proc. 3rd International Conf. Genetic Algorithms (J.D. Schaffer, ed.). Morgan Kaufmann, 51–60.

J.D. Schaffer and J.J. Grefenstette (1985): Multi–objective learning via genetic algorithms. Proc. of the 9th. International Joint Conf. on Artificial Intelligence, 593–595.

A.A. Schäffer and M. Yannakakis (1990): Simple local search problems that are hard to solve. SIAM Journal Computing 20, 56–87.

P. Schefe (1986): Künstliche Intelligenz – Überblick und Grundlagen. BI–Wissenschaftsverlag, Mannheim.

C. Schneeweiß (1987): Einführung in die Produktionswirtschaft. Springer, 2nd. ed., Berlin.

C. Schneeweiß (1992): Planung 2. Springer, Berlin.

R.J. Schönberger (1982): Japanese Manufacturing Techniques. The Free Press, New York.

E. Schönburg and F. Heinzmann (1992): PERPLEX: Produktionsplanung nach dem Vorbild der Evolution. Wirtschaftsinformatik 34, 224–232.

A. Schrijver (1986): Theory of Linear and Integer Programming. Wiley, New York.

B. Schürmann (1988): Hierarchisches Top Down Chip Planning. Informatik Spektrum 11, 57–70.

H.–P. Schwefel (1977): Numerische Optimierung von Computer–Modellen mittels der Evolutionsstrategie. Birkhäuser, Basel.

H.–P. Schwefel and T. Bäck (1992): Künstliche Evolution – eine intelligente Problemlösungsstrategie? KI 2/92, 20–27.

G.D. Scudder and T.R. Hoffmann (1985): An evaluation of value based dispatching rules in a flow shop. European Journal Operational Research 22, 310–318.

T. Sen and S.K. Gupta (1984): A state–of–art survey of static scheduling research involving due dates. Omega 12, 63–76.

J.J. Seppänen and J.M. Moore (1970): Facilities planning with graph theory. Management Science 17, B242–B253.

S.M. Shafer and D.F. Rogers (1993a): Similarity and distance measures for cellular manufacturing. Part I. A survey. International Journal of Production Research 31, 1133–1142.

S.M. Shafer and D.F. Rogers (1993b): Similarity and distance measures for cellular manufacturing. Part II. An extension and comparison. International Journal of Production Research 31, 1315–1326.

M. Shanahan and R. Southwick (1989): Search, Inference and Dependencies in Artificial Intelligence. John Wiley, New York.

J.F. Shapiro (1991): Convergent duality for the traveling salesman problem. Operations Research Letters 10, 129–136.

M.J. Shaw (1987): Applying inductive learning to enhance knowledge–based expert systems. Decision Support Systems 3, 319–332.

L.C. Shih, T. Enkawa, and K. Itoh (1992): An AI–search technique–based layout planning method. International Journal of Production Research 30, 2839–2855.

M.W. Simmen (1991): Parameter sensitivity of the elastic net approach to the traveling salesman problem. Neural Computation 3, 363–374.

J. Skorin–Kapov (1990): Tabu search applied to the quadratic assignment problem. ORSA Journal on Computing 2, 33–45.

S.A. Slotnick, J.H. May, and T.E. Morton (1992): FURNEX: modeling expert scheduling on the factory floor. Proc. Symposium on Intelligent Scheduling Systems (W.T. Scherer and D.E. Brown, eds.), 277–286.

S.F. Smith (1983): Flexible learning of problem solving heuristics through adaptive search. Proc. of the 8th. International Joint Conference on Artificial Intelligence, 422–425.

S.F. Smith, M.S. Fox, and P.S. Ow (1986): Constructing and maintaining detailed production plans: investigations into the development of knowledge–based factory scheduling systems. AI Magazine, 46–61.

T.H.C Smith, T.W.S. Meyer, and G.L. Thompson (1990): Lower bounds for the symmetric travelling salesman problem from Lagrangian relaxations. Discrete Applied Mathematics 26, 209–217.

T.H.C. Smith, V. Srinivasan, and G.L. Thompson (1977): Computational performance of three subtour elimination algorithms for solving asymmetric traveling salesman problems. Annals of Discrete Mathematics 1, 495–506.

T.H.C. Smith and G.L. Thompson (1977): A LIFO implicit enumeration search algorithm for the symmetric traveling salesman problem using Held and Karp's 1–tree relaxation. Annals of Discrete Mathematics 1, 479–493.

W.E. Smith (1956): Various optimizers for single–stage production. Naval Research Logistics Quarterly 3, 59–66.

S. Song and K. Hitomi (1992): GT cell formation for minimising the intercell parts flow. International Journal of Production Research 30, 2737–2753.

Y.N. Sotskov (1991): The complexity of scheduling problems with two and three jobs. European Journal of Operational Research 53, 326–336.

H. Spāth (1977): Partitionierende Cluster–Analyse für große Objektemengen mit binären Merkmalen am Beispiel von Firmen und deren Berufsgruppenbedarf. In: (H. Spāth, ed.): Fallstudien Cluster–Analyse. Oldenburg, München, 63–80.

H. Spāth (1985): Cluster Dissection and Analysis: Theory, FORTRAN Programs, Examples. John Wiley, Chichester.

F.C.R. Spieksma (1992): Assignment and Scheduling Algorithms in Automated Manufacturing. Dissertation, University of Limburg, Maastricht.

P.F. Stadler and W. Schnabl (1992): The landscape of the traveling salesman problem. Physical Letters A161, 337–344.

E.F. Stafford (1988): On the development of a mixed–integer linear programming model for the flowshop sequencing problem. Journal Operational Research Society 39, 1163–1174.

T. Starkweather, S. McDaniel, K. Mathias, and D. Whitley (1991): A comparison of genetic sequencing operators. Proc. 4th. International Conf. on Genetic Algorithms (R.K. Belew and L.B. Booker, eds.) Morgan Kaufmann, 69–76.

T. Starkweather, D. Whitley, K. Mathias, and S. McDaniel (1992): Sequence scheduling with genetic algorithms. In: (G. Fandel, T. Gulledge, and J. Jones, eds.) New Directions for Operations Research in Manufacturing, Springer, Berlin, 129–148.

K.E. Stecke (1983): Formulation and solution of nonlinear integer production planning problems for flexible manufacturing systems. Management Science 29, 273–288.

K.E. Stecke (1985): Design, planning, scheduling, and control problems of flexible manufacturing systems. Annals of Operations Research 3, 3–12.

G.L. Steele (1980): The Implementation and Definition of a Computer Programming Language Based on Constraints. Dissertation, MIT, Cambridge, Mass.

P.A. Stefanski, J. Wnek, and J. Zhang (1990): Bibliography of recent machine learning research (1985–1989). In: (Y. Kodratoff and R.S. Michalski, eds.) Machine Learning, an Artificial Intelligence Approach, Vol. III, 685–789.

L. Sterling and E. Shapiro (1986): The Art of Prolog. MIT Press, Cambridge, Mass.

J.M. Stern (1992): Simulated annealing with a temperature dependent penalty function. ORSA Journal on Computing 4, 311–319.

W.R. Stewart (1987): Accelerated branch exchange heuristics for symmetric traveling salesman problems. Networks 17, 423–437.

S. Stöppler and C. Bierwirth (1992): The application of a parallel genetic algorithm to the $n/m/P/C_{max}$ flowshop problem. Working paper, University of Bremen.

R.H. Storer, S.D. Wu, and I. Park (1992): Genetic algorithms in problem space for sequencing problems. In: (G. Fandel, T. Gulledge, and A. Jones, eds.) New Directions for Operations Research in Manufacturing, Proc. of the 2nd Joint US/German Conf., Hagen.

R.H. Storer, S.D. Wu, and R. Vaccari (1991): Local search in problem and heuristic space for job shop scheduling genetic algorithms. In: (G. Fandel, T. Gulledge, and A. Jones, eds.)

New Directions for Operations Research in Manufacturing, Proc. of a Joint US/German Conf., Gaithersburg, Maryland.

R.H. Storer, S.D. Wu, and R. Vaccari (1992a): New search spaces for sequencing problems with application to job shop scheduling. Management Science 38, 1495–1509.

R.H. Storer, S.D. Wu, and R. Vaccari (1992b): Problem and heuristic space search strategies for job shop scheduling. Working paper, Lehigh University, Bethlehem, PA.

S. Subramanyam, R.G. Askin (1986): An expert system approach to scheduling in flexible manufacturing systems. In: (A. Kusiak, ed.) Flexible Manufacturing Systems: Methods and Studies, North–Holland, Amsterdam, 243–256.

J.Y. Suh and D. Van Gucht (1987): Incorporating heuristic information into genetic search. Proc. 2nd International Conf. Genetic Algorithms and Their Applications (J.J. Grefenstette, ed.). Lawrence Erlbaum Ass., 100–107.

K. Sundermeyer (1991): Knowledge–Based Systems. BI Wissenschaftsverlag, Mannheim.

R. Suri (1985): An overview of evaluative models for flexible manufacturing systems. Annals of Operations Research 3, 13–21.

G.J. Sussman and G.L. Steele (1980): CONSTRAINTS – a language for expressing almost hierarchical descriptions. Artificial Intelligence 14, 1–39.

J.A. Svestka and J.–C. Jiang (1992): IS – the intelligent scheduler for job shops. Proc. Symposium on Intelligent Scheduling Systems (W.T. Scherer and D.E. Brown, eds.), 302–322.

M.N.S. Swamy and K. Thulasiraman (1981): Graphs, Networks, and Algorithms. Wiley, New York.

M.M. Syslo, N. Deo, and J.S. Kowalik (1983): Discrete Optimisation Algorithms with PASCAL Programs. Prentice–Hall, Englewood Cliffs.

W. Szwarc (1971): Elemination methods in the m×n sequencing problem. Naval Research Logistics Quarterly 18, 295–305.

W. Szwarc (1973): Optimal elemination methods in the m×n sequencing problem. Operations Research 21, 1250–1259.

W. Szwarc (1978): Dominance conditions for the three–machine flow shop problem. Operations Research 26, 203–206.

E. Taillard (1989): Parallel taboo search technique for the jobshop scheduling problem. Working paper, University of Lausanne.

E. Taillard (1990): Some efficient heuristic methods for the flow shop sequencing problem. European Journal of Operational Research 47, 65–74.

E. Taillard (1993): Benchmarks for basic scheduling problems. European Journal of Operational Research 64, 278–285.

K.Y. Tam (1992): Genetic algorithms, function optimization, and facility layout design. European Journal of Operational Research 63, 322–346.

C.S. Tang and E.V. Denardo (1988a): Models arising from a flexible manufacturing machine, part I: minimization of the number of tool switches. Operations Research 36, 767–777.

C.S. Tang and E.V. Denardo (1988b): Models arising from a flexible manufacturing machine, part II: minimization of the number of switching instants. Operations Research 36, 778–784.

A. Tate (1985): A review of knowledge–based–planning techniques. Expert Systems 85, 89–111.

H. Tempelmeier and H. Kuhn (1992): OR–Modelle zur Planung flexibler Fertigungssyteme, ein Überblick. OR Spektrum 14, 177–192.

H. Tempelmeier and H. Kuhn (1993): Flexible Manufacturing Systems. Wiley, New York.

U.A.W. Tetzlaff (1990): Optimal Design of Flexible Manufacturing Systems. Physica, Heidelberg.

J. Thiel and S. Voß (1992): Solving multiconstraint zero–one knapsack problems with genetic algorithms. Working paper, Technical University, Darmstadt.

G.L. Thompson and D.J. Zawack (1985/6): A problem expanding parametric programming method for solving the job shop scheduling problem. Annals of Operations Research 4, 327–342.

C.A. Tovey (1988): Simulated simulated annealing. American Journal of Mathematical and Management Sciences 8, 389–403.

S. Turner and D. Booth (1987): Comparison of heuristics for flow shop sequencing. Omega 15, 75–78.

U. Tüshaus (1983): Aggregation binärer Relationen in der qualitativen Datenanalyse. Mathematical Systems in Economics 82, Hain, Königstein.

N.L.J. Ulder, E.H.L. Aarts, H.–J. Bandelt, P.J.M. van Laarhoven, and E. Pesch (1991): Genetic local search algorithms for the traveling salesman problem. Lecture Notes in Computer Science 496, 109–116.

R.J.M. Vaessens, E.H.L. Aarts, and J.K. Lenstra (1992): A local search template. In: (R. Männer and B. Manderick, eds.) Parallel Problem Solving from Nature 2, Elsevier Publishers, 65–74.

D. Vanderbilt and S.G. Louie (1984): A Monte Carlo simulated annealing approach to optimization over continuous variables. Journal of Computational Physics 56, 259–271.

A. Vannelli and R.G. Hall (1993): An eigenvector solution methodology for finding part–machine families. International Journal of Production Research 31, 325–349.

M.P. Vecchi and S. Kirkpatrick (1983): Global wiring by simulated annealing. IEEE Transactions on Computer–Aided Design 2, 215–222.

S. van de Velde (1991): Machine Scheduling and Lagrangian Relaxation. Dissertation, CWI Amsterdam.

V. Venugopal and T.T. Narendran (1992): Cell formation in manufacturing systems through simulated annealing: an experimental evaluation. European Journal of Operational Research 63, 409–422.

V. Venugopal and T.T. Narendran (1993): Design of cellular manufacturing systems based on asymptotic forms of a Boolean matrix. European Journal of Operational Research 67, 405–417.

M.G.A. Verhoeven, E.H.L. Aarts, E. van de Sluis, and R.J.M. Vaessens (1992): Parallel local search and the traveling salesman problem. Working paper, University of Technology, Eindhoven.

G. Vescia (1985): Descriptive classification of cetacea: whales, porpoises and dolphins. In: (J.F. Marcotorchino, J.M. Proth, and J. Jansen, eds.) Data Analysis in Real Life Environment: Ins and Outs of Problem Solving. Elsevier Publishers, North Holland, 7–13.

A. Volgenant (1990): Symmetric traveling salesman problems. European Journal of Operational Research 49, 153–154.

T. Volgenant and R. Jonker (1982): A branch and bound algorithm for the symmetric traveling salesman problem based on 1–tree relaxation. European Journal of Operational Research 9, 83–89.

T. Volgenant and R. Jonker (1983): The symmetric traveling salesman problem and edge exchanges in minimal 1–trees. European Journal of Operational Research 12, 394–403.

M.D. Vose (1991): Generalizing the notion of schema in genetic algorithms. Artificial Intelligence 50, 385–396.

S. Voß (1993): Tabu search: applications and prospects. In: (D.–Z. Du and P.M. Pardalos, eds.) Network Optimization Problems, World Scientific Publ., Singapore, 333–353.

S. Voß (1994): Intelligent Search. Habilitation thesis, University of Technology, Darmstadt.

H.M. Wagner (1959): An integer linear programming model for machine scheduling. Naval Research Logistics Quarterly 6, 131.

Y. Wakabayashi (1986): Aggregation of Binary Relations: Algorithmic and Polyhedral Investigations. Dissertation, University of Augsburg.

R. van der Wal and A. Kolen (1992): An object–oriented approach to bracktrack search with constraint propagation. Working paper, University of Limburg, Maastricht.

D. Waltz (1975): Understanding line drawings of scenes with shadows. In: (P.H. Winston, ed.) Psychology of Computer Vision, McGraw Hill, Cambridge, Mass. 19–91.

G. Wäscher and P. Chamoni (1987): MICROLAY: An interactive computer program for factory layout planning on microcomputers. European Journal of Operational Research 31, 185–193.

P. Wayner (1991): Genetic algorithms. Byte 1/91, 361–368.

M.H.J. Webb (1971): Some methods of producing approximate solutions to travelling salesman problems with hundreds and thousands of cities. Operational Research Quarterly 22, 49–66.

L.M. Wein and P.B. Chevalier (1992): A broader view of the job–shop scheduling problem. Management Science 38, 1018–1033.

S. Weiss and C. Kulikowski (1990): Computer Systems that Learn. Morgan Kaufmann, Sussex.

U. Wemmerlöv and N.L. Hyer (1987): Research issues in cellular manufacturing. International Journal of Production Research 25, 413–431.

R.E. Wendell and R.D. McKelvey (1981): New perspectives in competitive location theory. European Journal of Operational Research 6, 174–182.

D. de Werra and A. Herts (1989) Tabu search techniques: a tutorial and an application to neural networks. OR Spektrum 11, 131–141.

H. Weyl (1935): Elementare Theorie der konvexen Polyeder. Commentarii Mathematici Helvetici 7, 509–533.

K.P. White and R.V. Rogers (1990): Job–shop scheduling: limits of the binary disjunctive formulation. International Journal of Production Research 28, 2187–2200.

D. Whitley (1989): The GENITOR algorithm and selection pressure: why rank–based allocation of reproductive trials is best. Proc. 3rd. International Conf. on Genetic Algorithms, (J.D. Schaffer, ed.), Morgan Kaufmann, 116–121.

D. Whitley (1992): Deception, dominance and implicit parallelism in genetic search. Annals of Mathematics and Artficial Intelligence 5, 49–78.

D. Whitley and J. Kauth (1988): GENITOR: a different genetic algorithm. Proc. of the Rocky Mountain Conf. on Artificial Intelligence, Denver, CO, 118–130.

D. Whitley, K. Mathias, and P. Fitzhorn (1991): Delta coding: an iterative search strategy for genetic algorithms. Proc. 4th. International Conf. on Genetic Algorithms (R.K. Belew and L.B. Booker, eds.) Morgan Kaufmann, 77–84.

D. Whitley, T. Starkweather, and D. Fuquay (1989): Scheduling problems and traveling salesmen: the genetic edge recombination operator. Proc. 3rd International Conf. Genetic Algorithms (J.D. Schaffer, ed.). Morgan Kaufmann, 133–140.

M. Widmer (1991): Job shop scheduling with tooling constraints: a tabu search approach. Journal Operational Research Society 42, 75–82.

M. Widmer and A. Herts (1989): A new heuristic method for the flow shop sequencing problem. European Journal of Operational Research 41, 186–193.

G.V. Wilson and G.S. Pawley (1988): On the stability of the travelling salesman problem algorithm of Hopfield and Tank. Biological Cybernetics 58, 63–70.

P.H. Winston (1987): Künstliche Intelligenz, Addision–Wesley, Bonn.

N. Wu and G. Salvendy (1993): A modified network approach for the design of cellular manufacturing systems. International Journal of Production Research 31, 1409–1421.

T. Yamada and R. Nakano (1992): A genetic algorithm applicable to large–scale job–shop problems. In: (R. Männer and B. Manderick, eds.) Parallel Problem Solving from Nature 2, Elsevier Publishers, 281–290.

M. Yannakakis (1990): The analysis of local search problems and their heuristics. Lecture Notes in Computer Science 415, 298–311.

M. Yannakakis (1991): Expressing combinatorial optimization problems by linear programs. Journal of Computer and System Sciences 43, 441–466.

R.E. Young, A. Greef und P. O'Grady (1992): An artificial intelligence–based constraint network system for concurrent engineering. International Journal of Production Research 30, 1715–1735.

C.T. Zahn (1964): Approximating symmetric relations by equivalence relations. SIAM Journal on Applied Mathematics 12, 840–847.

S.H. Zanakis, J.R. Evans, and A.A. Vasacopoulos (1989): Heuristic methods and applications: a categorized survey. European Journal of Operational Research 43, 88–110.

G. Zäpfel (1982): Produktionswirtschaft. De Gruyter, Berlin.

M.J. Zeestraten (1989): Scheduling Flexible Manufacturing Systems. Dissertation University of Delft.

S. Zelewski (1990a): PPS–Expertensysteme für die Terminfeinplanung und –steuerung, Teil 1: Konzepte. Information Management 1/90, 56–65.

S. Zelewski (1990b): PPS–Expertensysteme für die Terminfeinplanung und –steuerung, Teil 2: Prototypen. Information Management 2/90, 68–74.

W.H.M. Zijm (1988): Flexible manufacturing systems: background, examples and models. In: (H. Schellhaas et al., eds.) Operations Research Proceedings 1988, Springer, Heidelberg, 142–161.

H.–J. Zimmermann (1990): Fuzzy Set Theory and its Applications. Kluwer, 2nd. ed., Dordrecht.

M. Zweben, E. Davis, B. Daun, E. Drascher, M. Deale, and M. Eskey (1992): Learning to improve constraint–based scheduling. Artificial Intelligence 58, 271–296.